Insects and Colors between Art and Natural History

Emergence of Natural History

VOLUME 7

The titles published in this series are listed at *brill.com/enh*

Insects and Colors between Art and Natural History

Edited by

V.E. Mandrij
Giulia Simonini

BRILL

LEIDEN | BOSTON

We acknowledge the support by the Open Access Publication Funds of Technische Universität Berlin and the University of Konstanz.

The realization of this book was supported by VolkswagenStiftung and German Research Foundation.

Cover illustration: Detail from Eleazar Albin, Plate 71 showing an unidentified species of sallow moth with golden wings, in: Eleazar Albin and William Derham, *Insectorum Angliæ Naturalis Historia*, 1731, quarto, Staatsbibliothek zu Berlin – Preußischer Kulturbesitz, Abteilung Handschriften und Historische Drucke, 4", Lt 12255<a> : R. The whole folio is included as Fig. 1.7.

The Library of Congress Cataloging-in-Publication Data is available online at https://catalog.loc.gov
LC record available at https://lccn.loc.gov/2024916698

Typeface for the Latin, Greek, and Cyrillic scripts: "Brill". See and download: brill.com/brill-typeface.

ISSN 2452-3283
ISBN 978-90-04-52511-5 (hardback)
ISBN 978-90-04-68455-3 (e-book)
DOI 10.1163/9789004684553

This book is printed on acid-free paper and produced in a sustainable manner.

Contents

Foreword

Crimson pepper pod
add two pairs of wings, and look
darting dragonfly.
MATSUO BASHŌ

∴

In his haiku, the Japanese poet Matsuo Bashō (1644–1694) was referring to the scarlet dragonfly (*Crocothemis erythraea*), which is widely distributed in Asia. With its bright red color, it hints at the approaching autumn. One can think of this as a season, but also as a stage in life, thus the scarlet dragonfly becomes a brilliant guide and escort into a chillier and darker realm about to be entered. In both Eastern and Western literature, insects are frequently employed as a metaphor for humanity's struggles and failures in life or in art, its relationship to nature, and its biological, sexual, social, political, or religious self-image. In Bashō's poem, the bright red of the dragonfly is construed anthropologically.

But what do we actually know about the importance of color in the world of insects? As it happens, most of them cannot see a bright red; except for butterflies, almost all insects are red-blind. The spectrum visible to bees, for example, extends at most from the ultraviolet zone to green. A bee therefore cannot register red, but it does see ultraviolet light, which means that a bloom that strikes us as altogether colorless can be highly interesting to a bee if it is covered with ultraviolet patterns. The dragonfly is the exception to the rule that insects have limited color perception. It sees a much more colorful world than we do – probably more so, in fact, than any other creature on the planet: Whereas humans have only three of them, dragonflies, depending on the species, have between 11 and 30 opsins with which they take in their environment!

We also know that dragonflies adapt their color to the temperature of their surroundings: In warmer regions, their colors are lighter, so as to protect them from overheating; in colder climes, darker hues allow them to absorb warmth more effectively. Moreover, female dragonflies adapt their color to their male counterparts when population density increases. Like chameleons or cuttlefish, many insects can match their color to that of their surroundings in order to camouflage themselves. The caterpillars of the peppered moth (*Biston betularia*) are eminently capable of this, though unlike the chameleon, they do not 'see' the color of what they are resting on primarily with their eyes, but rather with their skin. In nature, coloring often helps creatures to blend in with their environment, or in other words: It allows them *not* to be seen. In the visual arts, however, insects attract the attention of viewers precisely *because* of their vibrant colors. In the seventeenth century, for example, they were occasionally distributed across the picture surface like painted jewels, thereby illustrating the variety and artistry of the Creation even at the lowliest tier of the *scala naturae*.

Insects amaze us with their myriad colors. Especially fascinating is their iridescent metallic sheen. If one examines up close the blue wing scales of a tropical butterfly or the golden wing case of a ground beetle, one discovers that the metallic shimmer they exhibit when struck by light is not created by pigmentary colors embedded in the skin, but rather by microscopic skin structures producing interference effects. The color of the Hercules beetle (*Dynastes hercules*), on the other hand, changes from greenish to black depending on the humidity level: The higher the moisture of its surroundings, the darker the beetle becomes. Brought about by the intricate structure of the beetle's shell, this phenomenon is of great interest to nanotechnological research with regard to the development of intelligent materials that can be utilized as humidity sensors.

The attempt to recreate the surface of the beetle's shell is but one example of many attempts to employ the colors of insects for human purposes, which range from the material use of their bodies as pigments and dyes to complex feats of modern engineering that imitate their luminous iridescence. As stated above, marveling at the splendor of insects is nothing new. The multitude of their colors was noted as early as in antiquity, with the natural histories of Aristotle and Pliny being two cases in point. Yet it was only with the increasing attention to color in natural philosophy and natural history in early modern Europe that the colors of insects moved into the spotlight of scientific research. Using the newly invented microscope, for example, Athanasius Kircher (1602–1680) realized, as reported in his *Ars Magna Lucis et Umbrae* (1646), that the wings of butterflies were not colored by pigments, but by optical structures. In the eighteenth century, insects and colors came together in science in several new ways. One of the most important pigments derived from the Americas was extracted from insects: Having been produced in Mexico for a long time, cochineal now came to be exported to Europe in quantity and at considerable profit.

Around the same time, color began to play an increasingly important role in the field of natural history. Initially, color had been peripheral to the century's classification efforts – it was too varied even within a given species, and too volatile to be considered a valid criterion. Carl Linnaeus (1707–1778), with his strong focus on botany, had explicitly excluded it, and for obvious reasons: Plant colors were unstable, and plant specimens preserved in a herbarium faded to a uniform brown. This was not the case in other branches of natural history, however. In zoology, Europeans had become aware of the full range of colors in birds from research carried out in their far-flung colonies, but they had always known that insects exhibited a broad range of colors that remained stable even in specimen collections. Without close attention to color, the classification of butterflies, for one, turned out to be impossible. With the rise of entomology as a distinct sub-discipline of zoology in the eighteenth century, the issue of color could no longer be ignored. It is thus no coincidence that, in the second half of the eighteenth century, several entomologists, among them Ignaz Schiffermüller (1727–1806) and Moses Harris (1730–1787), created color systems of their own. The other field of natural history in which color became prominent was mineralogy, and also here special color scales were created by Abraham Gottlob Werner (1749–1817) and the less well-known Franz Uibelacker (1742–after 1800). The relations between natural history and color knowledge went in both directions: As colors found their way into natural history, our knowledge of colors was in turn greatly enriched. The impact of insect research for color systematization was particularly strong, and the history of the science of colors cannot be fully understood without including insects.

Against this backdrop, it is no surprise that most of the essays assembled in the present volume deal with the late seventeenth and the eighteenth century. V.E. Mandrji and Giulia Simonini are to be congratulated for having taken the initiative to highlight, for the first time, the intimate connection between insects and colors.

Karin Leonhard and Friedrich Steinle

Acknowledgements

Insects and Colors between Art and Natural History originated as an online workshop during the COVID-19 pandemic in winter 2021 thanks to the digital infrastructure of the Technische Universität Berlin and of the Universität Konstanz, to which the editors of this volume are affiliated. The workshop was the result of the fortunate encounter of the two editors in 2020 as well as their efforts in the identification of a common point of contact between their research projects. At that time, V.E. Mandrij was writing their Ph.D. thesis "'One could ornament a painting with the wings of a butterfly': Otto Marseus van Schrieck (1620/21–1678) and the Lepidochromy technique" within the graduate program *Changing Frames. Art History and Art Technology in Exchange* (Universität Konstanz, founded by the VolkswagenStiftung) and Giulia Simonini was the postdoctoral fellow in the project "Colour Orders, Colourants and Colour Terminology" within the research group *Dimensions of techne in the Fine Arts* (Technische Universität Berlin, founded by the German Research Foundation).

The online workshop was a real success with ten international speakers, five panel chairs, and more than fifty people attending the event from all over the world. We take this opportunity to thank again all speakers and chairs: Florike Egmond (Leiden University), Kay Etheridge (Gettysburg College), Hanneke Grootenboer (University of Amsterdam), Erma Hermens (Hamilton Kerr Institute), Dominik Hünniger (Deutsches Hafenmuseum), Stefanie Jovanovic-Kruspel (Naturhistorisches Museum Vienna), Karin Leonhard (Universität Konstanz), Brian Ogilvie (University of Massachusetts Amherst), Hossein Rajaei (Naturkundemuseum Stuttgart), Katharina Schmidt-Loske (Biohistoricum, Museum Koenig Bonn), Friedrich Steinle (Technische Universität Berlin), Beth Tobin (University of Georgia), and Grace Touzel (Natural History Museum London).

We owe our deepest gratitude to Karin Leonhard and Friedrich Steinle for their support and assistance during all phases of this project. Our thanks go likewise to the libraries of the TU Berlin and Universität Konstanz for funding the open access publication of this book. We are also thankful to Lea Stengel for her precious assistance for the index of the manuscript, to Martin Bleisteiner for his wonderful copy-editing work, and to Alessandra Giliberto for her great help. And finally, we are grateful to the editors of the series *Emergence of Natural History* and, especially to Dominik Hünniger, who invited us to submit the proceedings of the workshop to the Brill series, without whom this book would probably have never seen the light of day.

Figures

Notes on Contributors

Harald Bruckner
is a zoologist in the entomological department at the Naturhistorisches Museum in Vienna, where he serves as a research assistant in the collections of Lepidoptera, Neuroptera, and Hemiptera.

Kay Etheridge
is Emeritus Professor of Biology at Gettysburg College and a former editor of the Brill book series, *Emergence of Natural History*. Her publications in physiology and ecology include studies on tropical bats, manatees, lizards, and salamanders. Her current scholarship centers on the integration of natural history images and the history of biology. Her most recent publications examine the biological contributions of naturalist/artist Maria Sibylla Merian, e.g. the 2021 book, *The Flowering of Ecology: Maria Sibylla Merian's Caterpillar Book.*

Stefanie Jovanovic-Kruspel
is an art historian working at the Naturhistorisches Museum in Vienna. She is deputy director of the department Archive of science history and subject editor on cultural history of the museum's scientific journal "Annalen des Naturhistorischen Museum in Wien." One of her main research interests is the history of science visualization.

Karin Leonhard
is Professor of Art History at the University Konstanz since 2015. She is one of the founding members of the "Arbeitskreis Niederländische Kunst und Kulturgeschichte." Her research focuses on Dutch painting of the seventeenth century with a special interest in colors. She also started the graduate program "Changing Frames", opening a collaboration between art historians at Konstanz Universität and conservators at Staatliche Akademie für Bildenden Künste Stuttgart (2018–2023).

V.E. Mandrij
(they/them) is a writer and an art historian who completed a doctoral thesis in the graduate program "Changing Frames. Art History and Art Technology in Exchange" (Konstanz Universität/Staatliche Akademie für Bildenden Künste Stuttgart, 2018–2023). They investigate the Dutch painter Otto Marseus van Schrieck and his technique of butterfly imprints (lepidochromy) with art historical and art technological methodologies. Mandrij was awarded a fellowship grant at the Royal Netherlands Institute in Rome (2018), at the Vossius Center

in Amsterdam (2023), presented their interdisciplinary research at international conferences and published several papers, including in the *Netherlands Yearbook for History of Art* 71.

Kimberly Schenck
is retired head of paper conservation and senior conservator at the National Gallery of Art. After receiving her MS from the Winterthur/University of Delaware Conservation Program, she joined the staff of the Baltimore Museum of Art as paper conservator. Her publication topics focus on the materials and techniques of drawing and printmaking including those used by Raphael, Henri Matisse, Giorgio Vasari, and several nineteenth-century French artists. Her most recent publication discusses workshop practice in the production of an early sixteenth-century, German manuscript.

Stacey Sell
received her Ph.D. in art history from the University of Virginia in 1993 with a dissertation on Rembrandt's drawings. She is associate curator of old master drawings at the National Gallery of Art, where her exhibitions have included *Drawing in Silver and Gold: Leonardo to Jasper Johns* (2015), *The Touch of Color: Pastels at the National Gallery of Art* (2019) and *The Artist's Vision: Romantic Traditions in Britain* (2006). Her publications include articles on topics ranging from the Victorian silverpoint revival to Joris Hoefnagel's use of lepidochromy.

Giulia Simonini
is a science and art historian, and a graduate conservator. Currently, she works as a post doc in the research group "Dimensions of techne in the Fine Arts" (Technische Universität Berlin) led by Magdalena Bushart. Her research focuses on color history, color science, botanical and zoological illustrations between the seventeenth and the early nineteenth century. Recently, she authored "Calau's Punic Wax, Lambert's Farbenpyramide (1772), and Prefabricated Watercolour Cakes" (2023).

Friedrich Steinle
is Professor of History of Science at the Technische Universität in Berlin from 2009 to 2024, after positions in Lyon and Wuppertal. His research interests span from the history of electricity and magnetism, through the history and philosophy of experimentation, to the history of color research. With Professor Magdalena Bushart, he was the co-editor of the volume *Colour Histories: Science, Art, and Technology in the 17th and 18th Centuries* (De Gruyter, 2015) and with Tanja Kleinwächter and Sarah Lowengard, he co-edited the volume *Ordering Colours in 18th and early 19th Century Europe* (Springer, 2023).

Beth Fowkes Tobin

Professor Emerita of English and Women's Studies at the University of Georgia, is the author of *The Duchess's Shells: Natural History Collecting in the Age of Cook's Voyages* (Yale, 2014), a project for which she received a Scholars Award from the National Science Foundation. She is also co-editor of a multi-authored volume on color, *The Materiality of Color: The Production, Circulation, and Application of Dyes and Pigments, 1400–1800* (Ashgate, 2012). Currently, she examines the visual culture and material practices of insect collecting in the late eighteenth and early nineteenth centuries.

CHAPTER 1

Introduction

V.E. Mandrij and Giulia Simonini

> I had begun to love insects because several of them showed me colors, and mixtures of colors, of which I had previously no idea. My intention was to use them to adorn my cabinet; I did not even think that their study could provide knowledge suitable for the useful nourishment of heart and mind.[1]

With these words, the Italian entomologist Jacques Brez (1771–1798) chronicled in *La flore des insectophiles* (1791) the reasons for his embracement of *inséctologie*: the mesmerizing colors of insects and their combinations in fascinating patterns hitherto unnoticed by him. Today classified as members of the class Insecta, these pancrustacean hexapod invertebrates were a source of amazement and knowledge not only for Brez, with their coloration serving as a leitmotif in many entomological books. The study of insects, known as entomology from 1745 onward,[2] has a long history in Europe that stretches back as far as ancient Greece and Aristotle.[3] Yet studying and marveling at the colors of insects was not the exclusive purview of naturalists: in pre-modern Europe, artists, too, were highly interested in the phenomenon and created fertile ground for this branch of natural history through the imitation, documentation, and preservation of the colors of insects in visual and applied works of art.

That the history of European entomology is inextricably linked with that of art and natural history becomes apparent in textual as well as visual sources

1 "J'avais commencé d'aimer les Insectes parce que plusieurs d'entr'eux me faisaient voir des couleurs, & des mélanges de couleurs, dont je n'avais auparavant aucune idée. Mon dessein était de les faire servir à l'ornement de mon cabinet ; je ne pensais pas même que leur étude pût fournir des connaissances propres à nourrir utilement le coeur & l'esprit" (Brez 1791, vii–viii). Our translation. Unless otherwise specified, all translations are our own.

2 The term 'entomology' was coined by Brez's friend, the Swiss naturalist Charles Bonnet (1720–1793), who later rejected it in favor of 'insectology.' See Ogilvie 2008, 6.

3 Bodson 1983.

that aim to describe, *inter alia*, the color characteristics of this particular class of animals. In fact, the study of insects inspired natural historians and artists alike to develop strategies for the recording and systematization of the countless nuances of color found on their tiny bodies. Brez was not exaggerating when he stated that the wings and exoskeletons of insects were adorned with colors he could not even imagine: The known species belonging to this class are counted in the millions.[4] Of course, the coloration of insects was not just the object of wonderment for scholars and connoisseurs, but it also plays a significant role in their survival and reproduction – no surprise, then, that many species change their color(s) depending on their sex, life stage, and surroundings, or as a result of evolution.[5]

Today, it is universally acknowledged that the life of these creatures is an important gauge for the health of the planet Earth.[6] The current climate crisis has a dramatic impact on our ecosystems and has led to a precipitous decline of insect populations. A substantial body of knowledge about insects, their functions, biotopes, behaviors, physiology, appearance, taxonomy, faculties, and remarkable metamorphoses has been assembled by naturalists and historians of science over the course of centuries. Reflecting the many shifts and changes in the study of insects, these historical sources can play an important role when it comes to understanding and preserving them today, especially with regard to species that are now endangered or threatened by extinction.

The recently heightened interest in insects in several academic disciplines has produced a wealth of studies which examine the significance of the tiny invertebrates in Western history, society, and culture, and address their complex relationship with Europeans, who variously treated them as vermin, precious objects, potent symbols, or wonders of God.[7] Accordingly, the recurring symbolism of the beehive and anthill for human society, as well as the metamorphosis of lepidopterans as a metaphor for religious belief have been investigated by historians of literature.[8] Historians of science, meanwhile, have focused on issues such as historical collecting and preservation practices, the knowledge gained through empirical research in entomology, and the

4 Stork 2018.

5 See Kay Etheridge's essay in this volume. Insects also tend to change their color(s) after their death, a fact that will be of relevance in several of the contributions assembled here.

6 Southwood 1997; Leather 2015.

7 On the symbolism of insects, see Cambefort 2007 and Segal and Alen 2020, 75–84. On the cultural history of butterflies, see Witkowski 2007; Marren 2017. On beetles, see Ratcliffe 2006. For general considerations regarding insects and other crawling creatures, see Hond, Jorink, and Mulder 2022.

8 See Thomas 2003; Hollingsworth 2005; Aloi 2007a; 2007b.

various means of documentation used to record the characteristics of insects.[9] However, what is still lacking in this richly variegated scholarly landscape is an in-depth study dedicated to the historical understanding of and engagement with insect coloration at the intersection of art history and the history of natural history.

For instance, in the history of microscopic investigation, which in turn fostered the study of insects, the colors of insects played little to no role.[10] Developed during the seventeenth century, the microscope rendered visible details previously unnoticeable to the naked eye. Among the pioneers of the microscopical investigation of insects we can count the English Fellow of the Royal Society Robert Hooke (1635–1703), the Italian *linceo* Francesco Stelluti (1577–1652), as well as the Dutch natural historians Jan Swammerdam (1637–1680) and Antoni van Leeuwenhoek (1632–1723). While the microscope allowed Swammerdam, Leeuwenhoek, and his rival microscopist Nicolaas Hartsoeker (1656–1725) to identify American cochineal, used to produce the precious red pigment carmine, as an insect (*Dactylopius coccus*), Karin Leonhard has persuasively shown that coloration tends to disappear through the microscopical lens and that the microscopists' world was essentially colorless.[11]

Neither insect coloration nor insect pigments in artworks have received sufficient attention from art historians. Although the focus of Janice Neri's *The Insect and the Image: Visualizing Nature in Early Modern Europe, 1500–1700* is not on colors, its author stresses that color discrepancies between images and real specimens were a discussion topic among scholars.[12] Neri also emphasizes that artists carefully selected the species they depicted according to their

9 On the study of insects in general, see Bodenheimer 1928; Essig 1936; Geisthardt, Michael 1990; Aguilar 2006; Jorink 2007; 2010; Ogilvie 2008; 2014; 2020; Onaga and Hünniger 2024. On collecting insects, see Barnard 2018; Jarvis 2018; Hünniger 2018. The entomological enterprise of Merian, in particular, has attracted a great deal of scholarly intention, see Schmidt-Loske 2007; 2010; Wettengl 2013; Roth, Bushart, and Sonnabend 2017; Schmidt-Loske 2021; Etheridge 2021; Roemer et al. 2022. On the early modern reception of Aristotle's insect studies, see Cobb 2006; Reynolds 2019.

10 On microscopic studies of insects, see Ball 1966; Chapman 2005; Jorink 2007; 2010; Neri 2011; Doherty 2012; Jervis 2013; Lawson 2016; Jorink 2018; Pater and Jorink 2022.

11 Butler Greenfield 2005, 185–202; Leonhard 2017. Interestingly, Jorink (2018, 143) reports that Marcello Malpighi's original drawings documenting his microscopic examinations of insects were in colors. However, this information was lost in the printing process.

12 The English entomologist and painter Alexander Marshall (*c.*1620–1682) added numerous annotations to his exemplar of Thomas Moffet's *Theater of Insects*. He also completed Moffet's remarks concerning the locust by glue-pasting three drawings of this insect's various species in different colors. While one seems to have been executed after observing a real model, the purple color of another matches Moffet's description. See Neri 2011, 65–66.

colors and patterns in order to display their artistic skills and please their viewers. The artist and pioneer entomologist Maria Sibylla Merian (1647–1717), for instance, often depicted only one caterpillar per species, choosing "the stage where the insect displayed the most vibrant colors or intricate patterns."[13] An interdisciplinary team of physicists and art historians developed new technologies to paint the iridescent colors of Japanese jewel beetles (*Chrysochroa fulgidissima*) thereby retracing similar attempts made 3,000 years ago in China.[14] Many of the most expensive red dyestuffs used in Europe were extracted from scale insects (*Kermes vermilio, Coccus ilicis, Dactylopius coccus,* and *Porphyrophora polonica*).[15] Textile clippings dyed with colorants obtained from these insects were used to produce lake pigments for oil and watercolor painting. Investigations into the use of these dyes in paintings have provided a fresh perspective on their importance for the history of European painting.[16]

While the existing literature only hints at the interconnectedness of insects and colors at the nexus of art and natural history, with the present book we hope to launch an interdisciplinary discussion on this subject. Originating from a homonymous workshop supported by the Technische Universität Berlin and the Konstanz Universität held online in November 2021, this publication scrutinizes the functions, methods, and strategies of recording colors in art and natural history. The contributions assembled in this collection address a broad variety of topics such as the collaboration between artists and naturalists, the media and materials employed to name and paint insect colors (including iridescent ones), and the terminology developed to describe them.

In this introductory chapter, we will present the theme of this volume from three different angles: First, the amazement and amusement that artists and natural historians like Brez experienced in the observation of insects and their colors; second, the different media and techniques explored and tested to visually represent the colors of insects in the early modern period; and third, the strategies developed to standardize colors for taxonomical purposes. In the final section, we will provide a brief sketch of the content of the individual contributions.

13 Neri 2011, 157.

14 They also listed many countries around the globe where beetle wings were used to imitate emeralds. See Schenk and Parker 2011; Schenk, Wilts, and Stavenga 2013.

15 Cardon 2007, 607–650.

16 Kirby 1977; Kirby, Spring, and Higgitt 2005; 2007; Kirby 2015; Kirby, van Bommel, and Verhecken 2014; Roque 2021.

1 Amazement and Amusement

Since the days of antiquity, insects have elicited a complex response, which often takes the form of a mixture of disgust and fascination. Aristotle (384–322 BCE) believed that insects were spontaneously generated from rotting organic waste and dirt, and were thus inherently filthy.[17] According to their presumed mode of reproduction and the fact that they were perceived as bloodless, he also ranked them among the lowest creatures in the hierarchy of animals.[18] Nonetheless, the Greek philosopher acknowledged that "there is somewhat of the marvellous" in all natural things, including "meaner animals with a bad grace" such as insects, and emphasized the importance of their study.[19] In the eleventh book of *Naturalis Historia*, Pliny the Elder (23/24–79 CE) referred with amazement to nature's craftsmanship in creating insects. Pliny's conviction that "nature is nowhere more wholly present than in the smallest things"[20] gave rise to the dictum *maxima in minimis* so beloved by early modern physico-theologians, who believed that the smallest things in nature demonstrated the infinite power of God, and this in turn contributed to the establishment of entomology as an academic discipline.[21] In the footsteps of Pliny, early modern scholars, too, turned their inquisitive gaze on the tiny bodies of insects, their complex forms, and especially the bewildering variety of their colors. Naturalists used many metaphors, analogies, and comparisons to emphasize the admirable beauty of colorful insects, often describing them as artworks made by nature. The Italian natural historian Ulisse Aldrovandi (1522–1605) opened his *De animalibus insectis* (1602) by stressing that their colors showed so much variation that they could not even be imitated by Apelles nor by any other superlative painter.[22] Likewise, the English physician Thomas Moffet (1553–1604) stated dragonflies "do set forth Natures elegancy beyond the expression of Art."[23] He also described a specimen of a moth that the Flemish botanist Carolus Clusius (1526–1609) had sent him from Vienna as "easier to wonder at and admire, than with fit expressions to describe."[24] Although moths usually display comparatively subdued colors, Moffet

17 Jorink 2007, 149; 2010, 184.
18 Cerami 2018, 144; Hond 2022, 8.
19 Aristotle 1961, 98–99; Jorink 2010, 181.
20 "rerum natura nusquam magis quam in minimis tota sit" Plinius Secundus 2013, 20. Translation from Ogilvie 2020, 171.
21 Ogilvie 2020, 172.
22 Aldrovandi 1602, [i].
23 Moffett 1658, 942–943.
24 Moffett 1658, 959.

enthusiastically remarked that the colors of Clusius's specimen were so splendid that it looked "as if Nature in adorning of this had spent her whole painters' shop."[25] In the same vein, the English naturalist John Ray (1627–1705) maintained that butterflies existed because "to contemplate their exquisite beauty and variety is to experience the truest pleasure. To gaze enquiringly at such elegance of color and form devised by the ingenuity of nature and painted by her artist's pencil is to acknowledge and adore the imprint of the art of God."[26]

When authors wanted to praise the magnificent colors of insects, the hand of God or nature was often compared to that of a skilled painter or craftsman. Christoph Arnold's closing poem in Merian's *Der Raupen wunderbare Verwandelung* (1679) is a case in point: "Brightly colored little birds, ... lovelier than we could view them painted by a master's hand."[27] The analogy was perpetuated in the eighteenth century by many authors of entomological compilations, including Eugen Johann Christoph Esper (1742–1810), who remarked that "the splendor of their colors, and the regularity of the pattern with which the hand of the Creator painted them, appeals to our eye as much as their varied size."[28] The entomologists Pieter Cramer (1721–1776) and Caspar Stoll (1720/30–1791/95) similarly described an unidentified blue butterfly and other specimens as "masterpieces of creation, ... inimitable by the brush of the artist."[29] They also compared the harmonious color combinations created by nature on many moths to "the brush of Le Brun" and to "the graces of Boucher,"[30] but for the most part, the beauty of these little creatures was held to surpass artisanal creations. It was against this backdrop that the French naturalist Jacques-Louis-Florentin Engramelle (1734–1814) proclaimed that the

25 Moffett 1658, 959.

26 "Quis enim eximiam earum pulchritudinem & varietatem contemplans mira voluptate non afficiatur? Quis tot colorum & schematum elegantias naturae ipsius ingenio excogitatas & artifici penicillo depictas curiosis oculis intuens, divinae artis vestigia eis impressa non agnoscat & miretur?" Ray and Lister 1710, 109. English translation from Raven 1942, 407.

27 "Diese bunte Vögelein ... schöner könnte sie nicht mahlen auch die beste Meisters-Hand," Merian 1679, 1:[103–104]. English translation from Etheridge 2021, 136.

28 "Die Pracht ihrer Farben, und das regelmässige der Zeichnung, mit welchen sie die Hand des Schöpfers gemahlt, rezt unser Auge eben so sehr, als die abwechselnde Grösse derselben," Esper 1777, 1:5.

29 "Chefs d'oeuvre de la création, ... inimitable au pinceau de l'Artiste," Cramer and Stoll 1779, viii.

30 "le pinceau de Le Brun," "les graces de Boucher," Cramer and Stoll 1779, viii.

eyes of lepidopterans looked like "facet-cut diamonds, of which all the art of the lapidary could not give any idea."[31]

Insects, and especially butterflies, were also analogized to precious stones not only due to the shape of their eyes, but also because of the brilliant, shimmering colors of their bodies. Swammerdam, for instance, declared that the wings of butterflies looked as if they were studded with pearls, diamonds, sapphires, turquoises, and rubies.[32] Likewise, Ray wrote in his *Historia Insectorum* (1710) that butterflies served "to adorn the world and delight the eyes of men: To brighten the countryside like so many golden jewels."[33]

The colors of lepidopterans and other insects were considered so astounding and mesmerizing that the above-quoted Brez freely admitted to having had no idea of their existence before seeing them on these animals.[34] Early natural historians frequently likened their brilliance to the colors of the rainbow, until, from the second half of the eighteenth century onward, Isaac Newton's (1642–1727) spectrum became the preferred point of comparison. For instance, Moffet observed that the iridescent blue eyespots of the peacock butterfly (*Aglais io*) "shine curiously like stars, and do cast about them sparks of the colour of the Rainbow."[35] Swammerdam even opined that the colors on the wings of butterflies "outperform the colors of the rainbow."[36] The English artist and naturalist Benjamin Wilkes (*c.*1690–1749) introduced his book on English moths and butterflies with a poem by Henry Baker (1698–1774) in which the "splendid plumes" of a butterfly were described as "show[ing] the various Glories of the painted Bow."[37] In the wake of Newton's *Opticks* (1704), spectral colors gradually established themselves as the new epitome of chromatic brilliance. This comparison was especially beloved by Cramer and Stoll, who used it twice in their introduction to *De Uitlandsche Kapellen voorkomende in de Drie Waereld-Deelen Asia, Africa en America* (1779), enthusing that "the butterfly appears decorated with the most superb colors, whose brightness seems to equal that of the prism of Newton."[38]

31 "des diamans taillés à facettes, dont tout l'art du Lapidaire ne sçaroit donner aucune idée," Ernst and Engramelle 1779a, xxvii.

32 Swammerdam 1669, 134.

33 "ad ornatum Universi, & ut hominibus spectaculo sint: Ad rura illustranda velut tot bracteae inservientes," Ray and Lister 1710, 109. English translation from Raven 1942, 407.

34 See the epigraph and note one above.

35 Moffett 1658, 967.

36 Swammerdam 1669, 134.

37 Wilkes 1749, xiii.

38 Cramer and Stoll 1779, xviii.

Iridescent or "changing" colors were observed from the very outset of the study of insects. For example, the Dutch painter and proto-entomologist Johannes Goedaert (1617–1668) noticed that the colors of many insects oscillated between green and blue and between red and yellow, exactly like those of the peacock's tail.[39] In *Metamorphosis insectorum surinamensium* (1705), Merian compared an area on the wings of a moth species of the tribe Attacini depicted on Plate 52 to "a piece of Muscovy glass."[40] The iridescent colors produced by the refraction of light on thin transparent layers of Muscovy glass had been observed and described forty years earlier by Hooke.[41] The German miniature painter and entomologist August Johann Rösel von Rosenhof (1705–1759) observed that "the special shine and shimmer" exhibited by lepidopterans was caused by "the dust feathers with which they are covered, and which, according to their different shape and composition, refract and reflect the incident light in different ways."[42] The colors of insects were also compared to precious metals. Metallic colors were in evidence at many stages of an insect's life. Accordingly, Goedaert explained that the chrysalis was sometimes called aurelia because its chromatic appearance resembled gold.[43] Merian, too, used analogies with precious metals to describe caterpillars which looked "as if they had been clad or sprayed with gold," or had the appearance of "silver or mother of pearl."[44] The Jesuit entomologists Michael Denis (1729–1800) and Ignaz Schiffermüller (1727–1806) detected gold- and silver-like colors on the chrysalis and wings of many butterflies, while Cramer and Stoll claimed that "the gold appearing on the chrysalis of diurnal butterflies) is a veritable metal."[45]

39 Goedaert 1662a, 5 [n.p.].

40 "een plek van Moscovisch Glas," Merian 1705, 52. We are grateful to Katharina Schmidt-Loske for drawing our attention to this detail.

41 In *Micrographia* (1665), Hooke remarked: "I found that up and down in several parts of them I could plainly perceive several white specks or flaws, and others diversly coloured with all the Colours of the Rainbow." Hooke 1665, 48. See also Roos 2016, 575.

42 "der besondere Glantz und Schimmer"; "an denen Staubfederlein, womit sie bewachsen, und welche nach ihrer verschiedenen Figur und Zusammensetzung, die einfallende Lichtstralen auf verschiedene Art brechen, und zurückwerfen," Rösel von Rosenhof 1746, 1:102.

43 Goedaert 1662b, 17.

44 "Als wann sie mit Gold überzogen, oder damit gesprintzelt"; "sind anzusehen, wie Silber oder Perlemutter" Merian 1679, 1:[iii]. English translation from Etheridge 2021, 142.

45 Denis and Schiffermüller 1775, 37; 1776, 37. On the colors of lepidopterans and the work of these two Jesuit scholars; see Stefanie Jovanovic-Kruspel and Harald Bruckner's essay in this volume. "L'or qui pare les Chrysalides des Papillons Diurnes, fuit un véritable métal," Cramer and Stoll 1779, xviii.

In entomological publications, parallels were sometimes drawn between the coloration of insects and the pigments and dyestuffs employed by painters. For the most part, these comparisons are found in books authored by painter-entomologists such as Merian, Rösel von Rosenhof, Jacob L'Admiral (1700–1770), and Moses Harris (1730–1787), where insects' colors are described as similar to ultramarine, cinnabar, umber, verdigris, and Saxon blue.[46] It is only in a few instances that other authors resorted to similar terminology – Moffet, for example, referred to the coloration of the collar and body of a moth with the word "vermillion,"[47] and Engramelle described the coloration of the lower abdominal wings of the grayling (*Hipparchia semele*) as "marbled in bister color."[48]

2 Techniques and Media

Representations of insects were very often advertised as having been painted *ad vivum*.[49] The phrase portrayed these images as more accurate and reliable because they were allegedly the result of first-hand observation of real specimens.[50] Accuracy was particularly important in the context of images intended for natural-historical purposes, since they frequently replaced objects that were not or no longer accessible for study.[51] Given that many of the examples discussed in this volume are strongly rooted in the history of natural history, the term "accuracy" is particularly apposite: It implies that the appearance, features, and colors of the original specimens are precisely depicted in minute detail, resulting in images that can be described as true-to-life or true-to-nature reproductions.[52]

While a trained hand had no difficulty to render the shape, volume, chiaroscuro, proportions, and other details of insects' bodies, the same hand might

46 "Ultramaryn" Merian 1705, 53; "Zinnober-roth" Rösel von Rosenhof 1746, 1:158; "Omberverwig" L'Admiral 1740, 30; "Spaansgroen" L'Admiral 1740, 21; "Saxon-blue" Harris 1776, vii.

47 Moffett 1658, 962.

48 "Marbré de couleur de bistre," Ernst and Engramelle 1779b, 76.

49 *Naer het leven, nach dem Leben*, and *au vif* are translations – into Dutch, German, and French, respectively – of the Latin term *ad vivum*. On this concept, see the essays in Balfe et al. 2019. For instance, Merian (1705), Rösel von Rosenhof (1746), Jan Christiaan Sepp (1762), as well as Pieter Cramer and Caspar Stoll (1779) emphasized in the titles of their works that they had drawn and colored all their engravings *naer het leven/nach dem Leben*.

50 Egmond 2020, 34.

51 Kusukawa 2011, 191 and 195.

52 On truth to nature and truth to life in natural history images, see Daston and Galison 2007.

have found it challenging to translate the coloration of insects with pigments. Some of their colors were notoriously difficult to imitate with brushstrokes, but artisans and painters proved to be ingenious in developing techniques involving a broad variety of media and supports. No matter which media painters used, employing, mixing, and handling pigments in a skillful way was a major component of their education.[53] With paints, artists were able to depict, with varying accuracy, all material properties of things. It is therefore not surprising that many treatises on painting dedicated several chapters to the use of colors. In some instances, the authors of these treatises also added a dedicated section on the art of painting insects. Interestingly, in *The Art of Drawing with the Pen, and Limming in Water Colours* (1606), a book addressed to amateurs eager to learn how to paint, Henry Peacham (1576–1643) maintained that insects are "not hard to be laid in colours: Because the colours of many of them are simple, and without composition, as perfect red, black, blew, yellow, &c."[54] In *De Zichtbaere Werelt* (1678), the painter and art theorist Samuel van Hoogstraten (1627–1678) discussed the depictions of crawling creatures in the chapter dedicated to coloring, emphasizing the intertwined relationship between drawing and colors.[55] He further stated that when a painter is sketching animals, including vermin, after observing real models, "their natural coloration will greatly adorn your work."[56] Authored by the still life painter Willem Beurs (1656–1693/1713), *De groote waereld in 't kleen geschildert* (1692) listed the appropriate pigments for use in oil painting, and described how to represent flies, spiders, beetles, butterflies, worms, bees, and so on.[57] Beurs even contended that the winged insects with the most beautiful colors were to be found in his native country.[58]

Long before these authors published their opinions about how to render the colors of insects, painters had started to include the tiny creatures in their compositions. For example, miniatures of winged and crawling insects appeared on the margins of medieval manuscripts painted in gouache and watercolor. A folio from the *The Hours of Engelbert of Nassau*, approximately dated in the last quarter of the fifteenth century and attributed to the Master of Mary of Burgundy, shows a damselfly in a *trompe-l'oeil* manner, as if a living

53 Wiersma 2020, 4–5.

54 Peacham 1606, 43.

55 Hoogstraten 1678, 218. See also Leonhard 2020, 165–166.

56 "haer natuerlijk kolorijt zal uw werk geweldich versieren," Hoogstraten 1678, 230. Translation in Hoogstraten, Brusati, and Jacobs 2021, 265. See also Hermens 2022, 140.

57 Beurs 1692, 104–110.

58 Beurs 1692, 104.

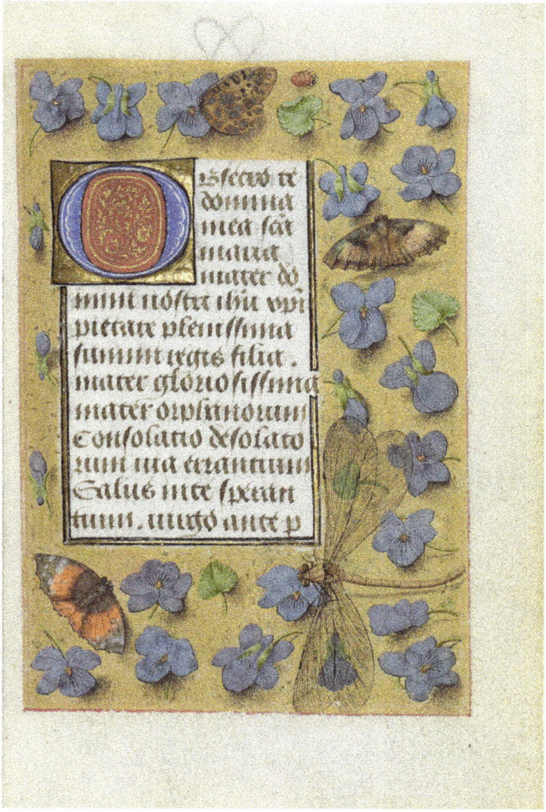

FIGURE 1.1
Vienna Master of Mary of Burgundy, Miniature with heath violets and insects in the margins in: *The Hours of Engelbert of Nassau*, c.1475–1485, body color, watercolor, and ink on parchment, 137 × 95 mm (leaf), 60 × 42 mm (ruled area), CC-BY-NC 4.0, Bodleian Libraries, Oxford, Bodleian Library, MS. Douce 219, fol. 20r HTTPS://DIGITAL.BODLEIAN.OX.AC.UK/OBJECTS/16585905-F800-43C2-AD2B-A2D828C19FA3/SURFACES/C0B1663B-CC9B-437E-9EB5-0D7896F95213/

insect was resting on the page (Fig. 1.1).[59] The technique of miniature painting required a pronounced sensibility for color, a mastery of brushwork, and meticulous attention to details – only then was it possible to depict not only the tiniest anatomical details of insects, but also their rich variety of colors.[60] While insects remained confined to the margins of manuscripts throughout the Middle Ages, during the sixteenth century, miniatures such as the lifelike depiction of a stag beetle (*Lucanus cervus*) by Albrecht Dürer (1471–1528) and a female southern hawker dragonfly (*Aeshna cyanea*) in Plate 53 of the album

59 DaCosta Kaufmann (1993, 44) suggests that these depictions made their appearance perhaps in reference to the habit of pilgrims to collect real insects between the pages of their prayer books. On insect depictions in the margins of manuscripts, see also Neri 2011 and Lammertse 2022.

60 On the development of miniature painting in natural history, see Swan 2005, 72–79. On the technique itself, see Colding 1953, 17–19.

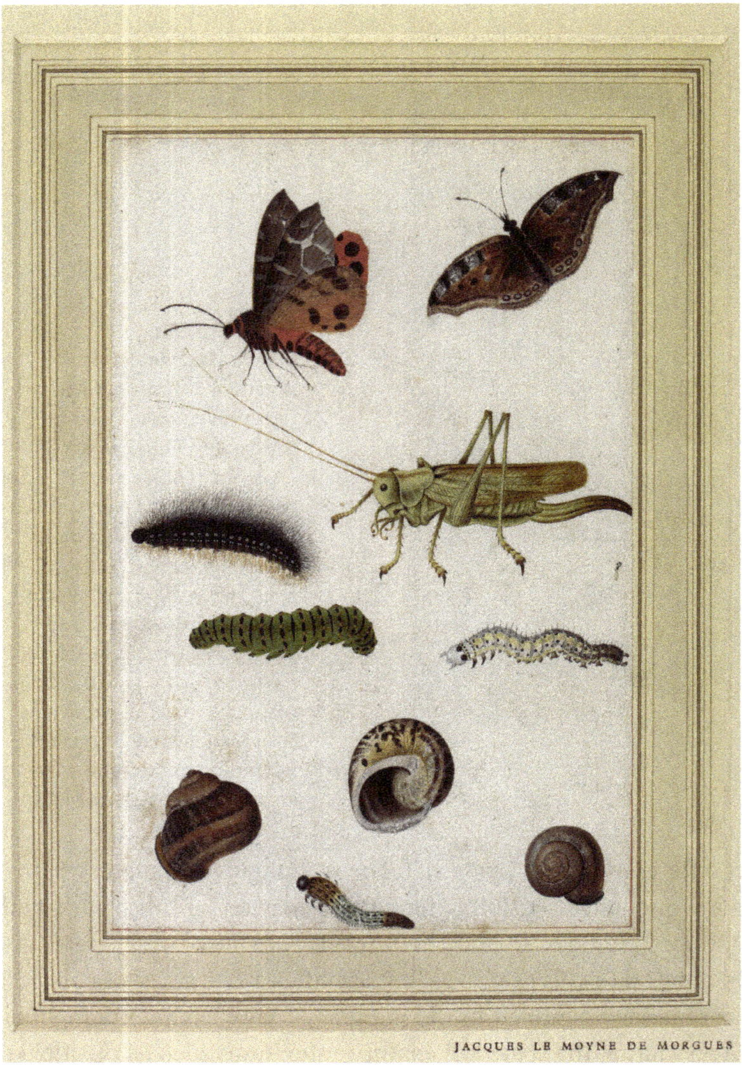

FIGURE 1.2 Jacques le Moyne de Morgues, *Studies of Insects and Shells*, c.1585, watercolor and body color on paper, 212 × 140 mm, London, British Museum, 1962,0714.1.2

Ignis by Joris Hoefnagel (1542–1601) (Fig. 2.2) indisputably established insect illustrations as individual artworks.[61] Many artists followed in Dürer's and

61 Albrecht Dürer, *Stag Beetle*, 1505, watercolor and body color on paper, 141 × 114 mm, Los Angeles, CA, J. Paul Getty Museum Collection, inv. no. 83.GC.214: https://www.getty

Hoefnagel's footsteps. Around 1585, for example, the French miniature painter Jacques le Moyne de Morgues (*c.*1532–1588) finely detailed a grasshopper with green and brownish pigments, and painted with minute brushwork the hair of a black caterpillar on a sheet of paper (Fig. 1.2).[62] Likewise, in her *Libro delle Miniature*, the Italian miniaturist Giovanna Garzoni (1600–1670) accurately portrayed on vellum a bowl of red cherries with a violet carpenter bee (*Xylocopa violacea*) flying rightwards (Fig. 1.3).

Paper as a painting support allowed much sough-after colored images of insects to circulate among connoisseurs and collectors of *naturalia*, which were in turn replicated in prints or copied by the hand of other artists.[63] For instance, the draftsman, printer, and glass painter Pieter Holsteyn II (1614–1673)

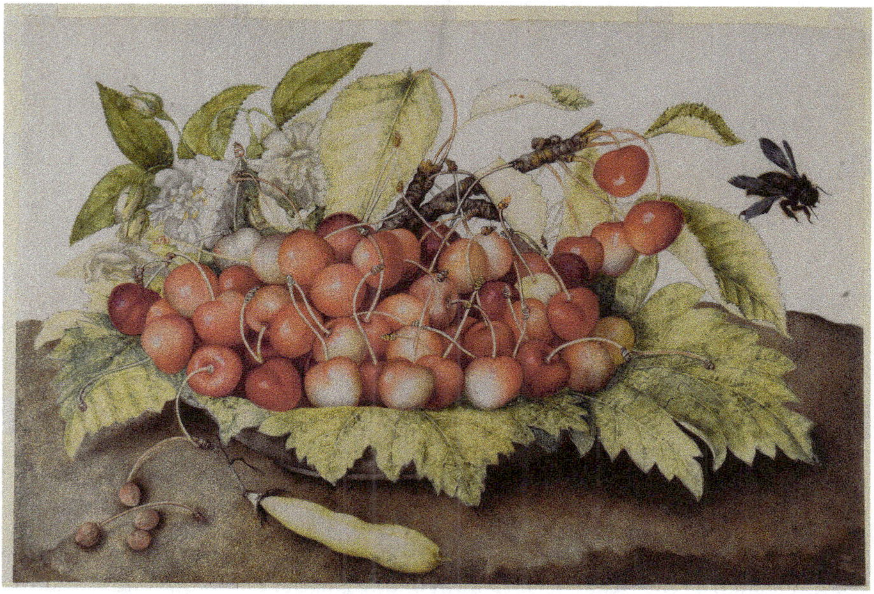

FIGURE 1.3 Giovanna Garzoni, *Plate of Cherries with Roses and a Carpenter Bee*, *c.*1655/1662, watercolor and body color on vellum, 245 × 375 mm, Poggio a Caiano, Museo della Natura Morta. 1890, 4779

.edu/art/collection/object/103QS6. Joris Hoefnagel, *Stag Beetle*, in: (*Ignis*), Plate 5, watercolor and gouache, with oval border in gold, on vellum, 143 × 184 mm, Washington, National Gallery of Art, inv. no. 1987.20.5.6: https://www.nga.gov/collection/art-object-page.69724.html. On the various copies of this image, see Koreny 1988, 112–126, Neri 2011, 6–11, and Bass 2019, 233. On Hoefnagel and for further literature, see Kimberly Schenck and Stacey Sell's essay in this volume.

62 On Le Moyne de Morgues, see Hulton 1962; 1977 and Sloan 2007.

63 See Neri 2011, esp. Ch. 2.

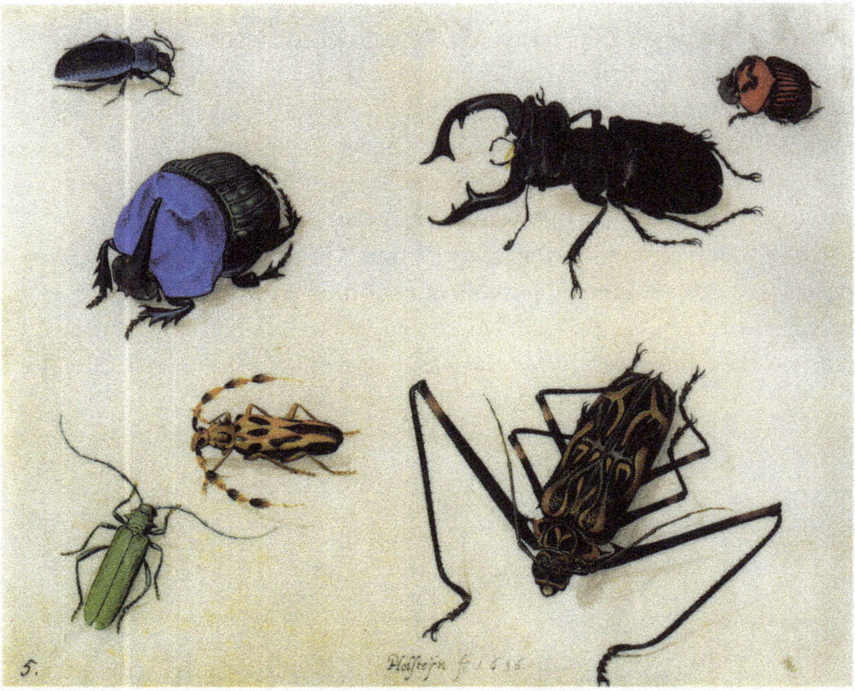

FIGURE 1.4 Pieter Holsteyn II, *Stag Beetle, Arlequin, Other Beetles*, 1636, gouache, black chalk, grey wash on parchment, 163 × 206 mm, private collection. Photo Collection RKD – Nederlands Instituut voor Kunstgeschiedenis, The Hague HTTPS://RKD.NL/IMAGES/30866

created a large body of natural history illustrations.[64] Among these, a gouache miniature on parchment dated 1636 depicts several species of beetles. Next to the European stag beetle, we also find exotic beetles imported from the Dutch colonies, such as the tropical Harlequin beetle (*Acrocinus longimanus*) and the Amazonian scavenger dung-beetle (*Coprophanaeus lancifer*, Fig. 1.4).[65] The stag beetle could be a reference to Dürer, but, despite the accurate delineation, Holsteyn did not represent its characteristic purplish coloration and simply

64 His father, Pieter Holsteyn I (1587–1662), also depicted insects.

65 Merian justified the inclusion of the image of the Harlequin beetle in a plate of her *Metamorphosis insectorum surinamensium* as follows: "The beautiful black beetle decorated with red and yellow flecks shown resting on the fruit was added on account of its rarity to complete and decorate the engraving." Merian 1705, 1: 11. Translation from Neri 2011, 172.

rendered it in black. Holsteyn's scavenger dung-beetle was subsequently copied in the same posture and in varying shades of blue in other miniatures.[66]

Oil on copper plates was another suitable medium for the representation of the shapes and colors of insects with fine brushwork, not only because it allowed artists to modulate minute particulars and to work with transparent superimposed glazes, but also because it allowed to imitate their luster and glow.[67] Like miniatures on paper and vellum, these small-sized painted copper plates were meant to be taken in the hand and observed closely by collectors, who were mostly interested in natural history subjects. The Flemish painter Jan van Kessel I (1626–1679) is known to have favored this technique to paint insects. The resulting images were judged to be "drawn with such power, colored so well, and highlighted with such mastery" that they became highly sought after by art connoisseurs.[68] Van Kessel was so skilled in the depiction of insects that he playfully staged caterpillars and worms in the form of his signature.[69] In an oil painting on copper, now in Cambridge, the artist represented a selection of insects such as the bumblebee (genus *Bombus*) and the speckled yellow moth (*Pseudopanthera macularia*) together with flowers, using the natural coloration of insects and blossoms to create a harmonious chromatic arrangement (Fig. 1.5).

Another medium related to the depiction of insects is oil still life painting on canvases and panels. Insects were a significant component of flower pieces by Flemish and Dutch painters in particular.[70] In this context, it is interesting to note that still-life artists mostly favored the colors of species endemic to Europe, since only a few paintings feature tropical species like the miniatures

---

66 See, for instance, Jan Augustin van der Goes (1671–after 1698), *Blue beetle*, gouache and body color on parchment, 60x80 mm, Amsterdam, Rijksmuseum, inv. no. RP-T-1884-A-330H: http://hdl.handle.net/10934/RM0001.COLLECT.50664 and anonymous artist, *Studies of a Blue Beetle and Insects*, gouache and watercolor, 100 × 132 mm, New York, Metropolitan Museum of Art, inv. no. 2008.427: https://www.metmuseum.org/art/collection/search/385463. We are thankful to David Edmonds and Joan Greer for pointing out the latter example to us.

67 Christine Göttler (2021, 245) discusses the shiny properties of oil painting on prepared metallic supports.

68 "zo fix getêkent, zo braaf gekoloreert, en zo meesterlijk getoetst" Weyerman 1729, 4:208–209. English translation by Marije Osnabrugge.

69 See Baadj 2016, 73–74. For in-depth analysis of Van Kessel's technique, see Don 2020.

70 Flemish and Dutch artists included insects in their still lifes, particularly in their floral compositions, at an early stage, but Italian still life painters, such as Paolo Porpora (1617–1673) and Giovanni Stanchi (*c.*1608–after 1686), only did so in the second half of the seventeenth century. On the insect species represented in still life paintings, see Dicke 2000; Dicke 2004, esp. 11. For recent publications about Dutch floral still life paintings, see Segal and Alen 2020 as well as Suchtelen 2022.

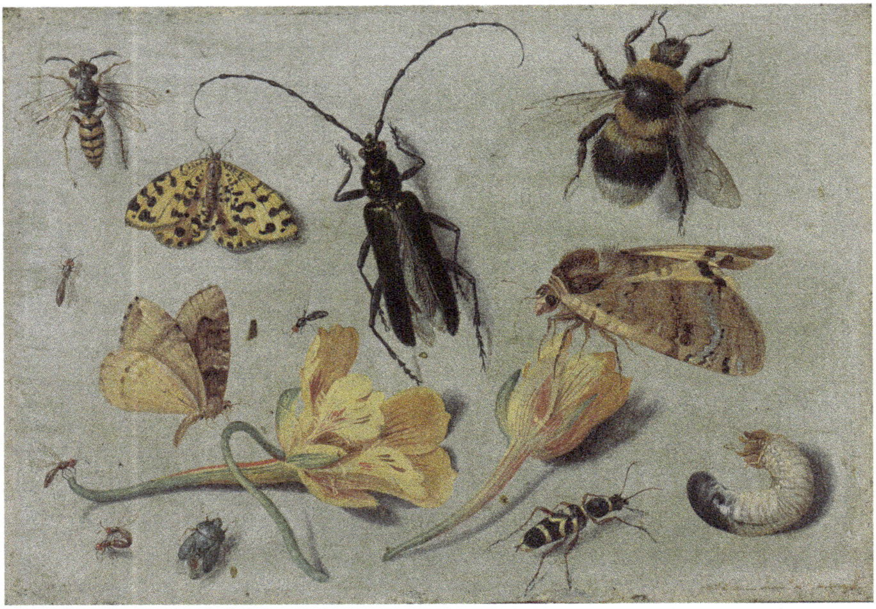

FIGURE 1.5 Jan van Kessel I, *Insects*, no date, oil on copper, 88 × 127 mm, Cambridge, Fitzwilliam Museum, 506
PHOTOGRAPH © THE FITZWILLIAM MUSEUM, UNIVERSITY OF CAMBRIDGE. HTTPS://DATA.FITZMUSEUM.CAM.AC.UK/ID/OBJECT/1613

presented above.[71] The Dutch painter Abraham Mignon (1640–1679), for instance, hid many accurately colored caterpillars, butterflies, ants, and other tiny insects in his sumptuous bouquets. A particularly stunning example of this is *Flowers in a metal vase* (1670) from the Mauritshuis in The Hague (Fig. 1.6), in which a lavishly colored plethora of flowers complements the accurate coloration of a bright green and orange larva of the pale tussock (*Calliteara pudibunda*), the blue and red patterns on the wings of the peacock butterfly (*Aglais io*), as well as the vibrant green metallic hue of the rose chafer (*Cetonia aurata*).

Insects, as we have shown above, were especially admired for their shimmering and iridescent colors. In seventeenth-century optical theories, iridescent colors were referred to as "apparent colors," that is, as transient chromatic phenomena that materialized in the rainbow, in nacre, on the skin of chameleons, on the exoskeleton of beetles, and on the wings of birds and butterflies.[72] Early modern artists attempted to imitate apparent colors with different techniques. Painters commonly made use of shell gold and metal leaves, sometimes with

71 Segal and Alen 2020, 118.
72 Leonhard 2014, 186.

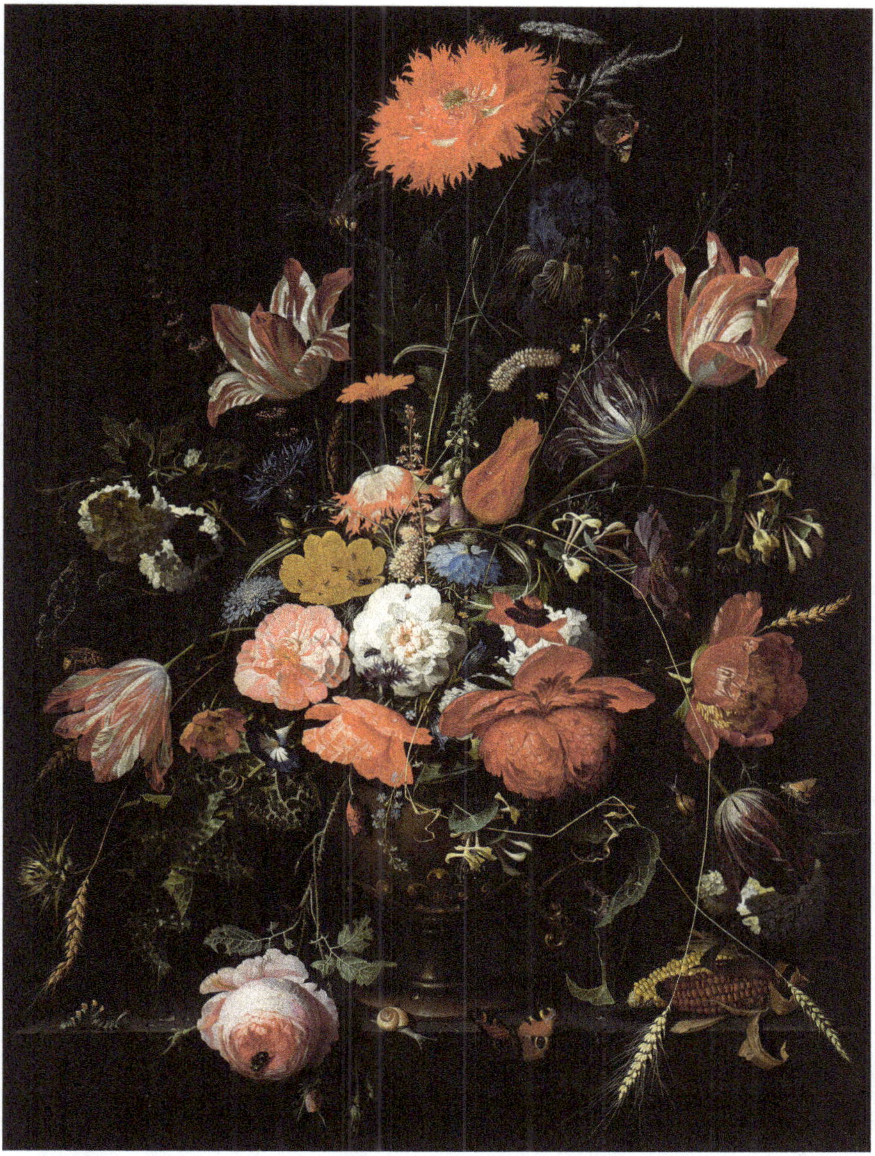

FIGURE 1.6 Abraham Mignon, *Flowers in a Metal Vase*, 1670, oil on canvas, 90 × 72.5 cm, The Hague, Mauritshuis, 111
HTTPS://WWW.MAURITSHUIS.NL/ONTDEK-COLLECTIE/KUNSTWERKEN/111-BLOEMEN-IN-EEN-METALEN-VAAS/

a transparent glaze, as the French miniaturist Catherine Perrot (d. 1690) in *Les leçons royales* (1686) recommended to paint a butterfly with shell silver.[73] The use of these materials can be observed in several media and with variable painting supports. Drawings on paper with metal details appear in the oeuvres of Felix Platter (1536–1614), Le Moyne de Morgues, Pieter Withoos (1655–1692), and Merian's daughter Johanna Helena Herolt (1668–1723).[74] In their oil paintings, Magdalena van der Hecken (1615?–after 1635) and Otto Marseus van Schrieck (*c.*1620–1678) adorned with gold, either in leaves or in powder, the wings of butterflies and the bodies of flies.[75] More interestingly still, silver and gold were quite frequently employed in the process of hand-coloring the plates of printed books. In Plate LXXI of a copy of *Insectorum Angliæ Naturalis Historia* (1731) by Eleazar Albin (active 1690–1742) and William Derham (1657–1735), the colors of the moth marked C at the upper left part of the image are rendered with a brown glazing over a gold leaf (Fig. 1.7).[76] The use of metal fragments certainly echoes some descriptions of insects by entomologists – as mentioned above, Moffet and Merian, among others, marveled at the silver and gold colors of the winged insects they studied.

73 Perrot 1686, 27. On shell gold, see Kirby et al. 2003. On metal leaves, see Richter 2013. The word shell refers to the receptacle in which the pigments were stored. On the use of gold and silver in general, see Morgan 2016.

74 On Platter, see Egmond 2017, 214. In the folio *French Marigold and Butterfly* (inv. no. 1962,0714.1.29) by Le Moyne de Morgues and included in the album from the British Museum shown in Fig. 1.2, silver has been detected as well, but it could also be oxidized white pigment. It is very likely that the shiny dots of the exotic species in the gouache on paper by Pieter Withoos (1655–1692) entitled *South-American and Asian butterflies* and now in the particular collection Boersma-Paulussen; See Boersma 2021, 133. Herolt's watercolor a species of morpho butterfly (genus Morpho) is held at the Herzog Anton Ulrich-Museum in Braunschweig, inv. no. z-06523. We thank Katharina Schmidt-Loske for pointing out to us this watercolor by Johanna Helena Herolt. Hoefnagel used the same technique – see Seehafer 2021 and Kimberly Schenck and Stacey Sell's essay in this volume.

75 One of the two insect studies by Van der Hecken on display at the Kunsthaus in Zurich contains gold shell pigments: Magdalena van der Hecken, *Study of Two Butterflies and Six Insects*, oil on panel, 5.2 × 13.5 cm, Zurich, Kunsthaus, inv. D.2016/0036: https://collection.kunsthaus.ch/de/collection/item/7213/. The two oil paintings were intended to decorate a cabinet, as were Jan van Kessel's above-mentioned depictions of insects on copper plates. See Büttner, Hoffmann, and Gloor 2015, 151–153 and 181. On Marseus, see Steensma 1999, 58 and the contribution of V.E. Mandrij in this volume.

76 We have observed the use of metal powder in other books held by the Staatsbibliothek Berlin, such as Carl Alexander Clerck's (1709–1765) *Icones insectorum rariorums* (1759), call no. 4 ° Ls 3647 R, Plate 17; Esper 1777, call no. 4 ° Lt 5085-1/1, Plate 43; Domenico Maria Leone Cirillo's (1739–1799) *Entomologiae Neapolitanae specimen primum* (1787), call no. gr. 2 ° 12431 R, Plate 1.

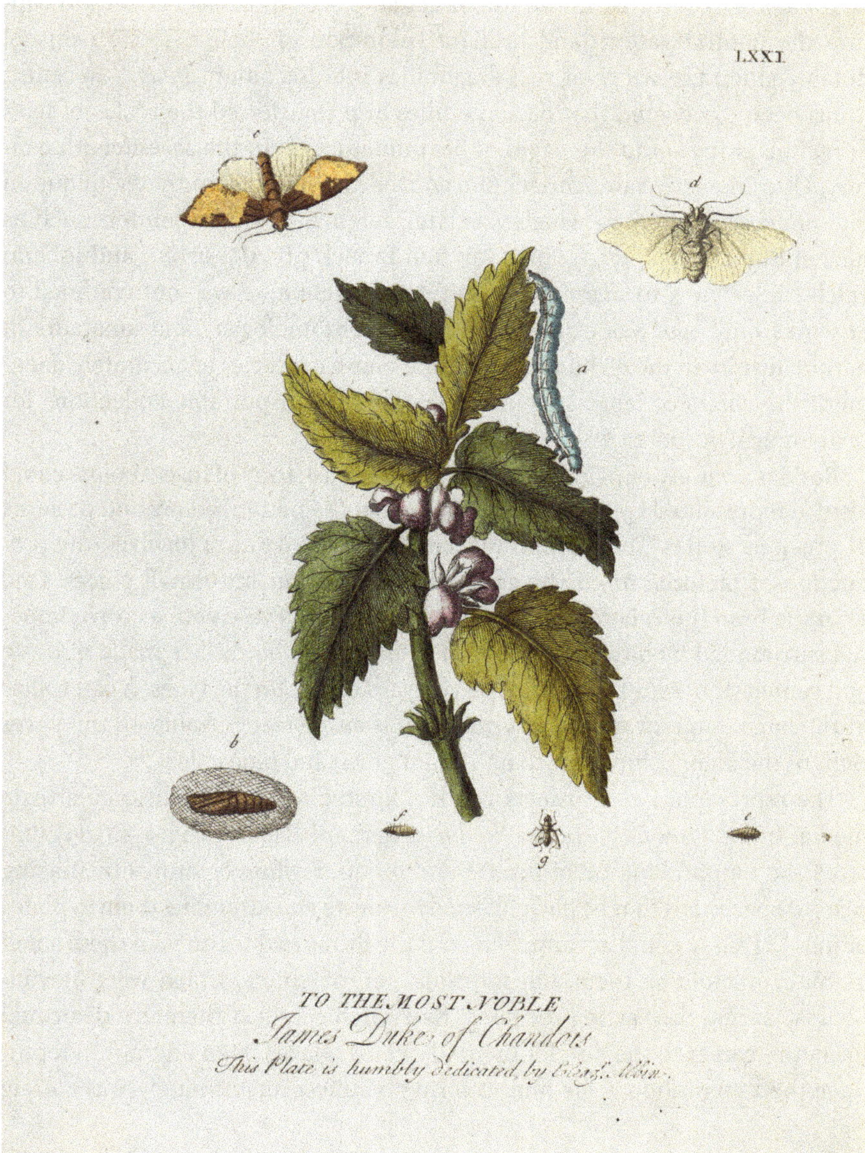

FIGURE 1.7 Eleazar Albin, Plate 71 showing an unidentified species of sallow moth with golden wings, in: Eleazar Albin and William Derham, *Insectorum Angliæ Naturalis Historia*, 1731, quarto, Staatsbibliothek zu Berlin – Preußischer Kulturbesitz, Abteilung Handschriften und Historische Drucke, 4″ Lt 12255<a> : R

Some artists resorted to the use of actual pieces of insects' bodies to capture the natural pattern and brilliant coloration of their wings. Hoefnagel notably glued the wings of real dragonflies into his album *Ignis*.[77] Recently, it has been discovered that he sometimes also transferred the scales of lepidopterans' wings onto the pages of his miniatures.[78] In the seventeenth century, Otto Marseus van Schrieck introduced the lepidochromy technique in his *sottobosco* paintings, which was later adopted by other painters such as Rachel Ruysch (1664-1750), Elias van den Broeck (1650/51–1708), and Johann Falch (1687–1727), to name but a few.[79] This technique was not confined to artworks only, but was extensively used by entomologists and amateurs in natural history in the eighteenth and nineteenth centuries to accurately document the colors of lepidopterans and to build transportable collections for taxonomical purposes.[80]

Besides a variety of painting techniques, the vast array of insect colors could also be represented by other means, including the natural colors and patterns of precious stones. The Florentine technique of *pietre dure* involved the production of pictorial motifs by assembling stones cut into small pieces. One example from the Palazzo Pitti in Florence displays a vase with a crown imperial surrounded by birds and grapes (Fig. 1.8). Here, butterflies made of agate and veined stones flutter around the vineyard or rest on the vines. A caterpillar of the large emperor moth (*Saturnia pyri*) is easily recognizable on the lower right of the composition due to its vibrant green and blue colors.[81]

The representation of insects was not limited to two-dimensional artistic media. In his famous *rustiques figulines*, Bernard Palissy (*c*.1510–*c*.1589) imitated the natural biotope of insects and other crawling creatures by making casts of specimens that he had collected in nature and arranging them in plates or jugs.[82] Palissy could accurately record the forms and textures of these small animals by molding them, but not their natural colors, which were inevitably lost during the casting process. The French ceramist therefore dedicated his entire career to the development of enamel glazes in an attempt to reproduce their coloration – for him, coloring his lifecasts accurately was closely

77 Bass 2019, 232.
78 See Schenck and Sell 2022 as well as their contribution to the present collection.
79 On Marseus van Schrieck and his followers, see Beier 1987, Seelig 2017, and Leonhard 2020. On the neologism 'lepidochromy,' see V.E. Mandrij's essay in this volume.
80 On lepidochromy in natural history and oil painting, see V.E. Mandrij's essay in this volume.
81 Gallori and Wolf (2015, 245–250) connected the representations of caterpillars in this artwork to drawings by the Italian master Jacopo Ligozzi (1547–1627), who is known for his accurate natural history watercolors.
82 Felfe 2015, 44–45.

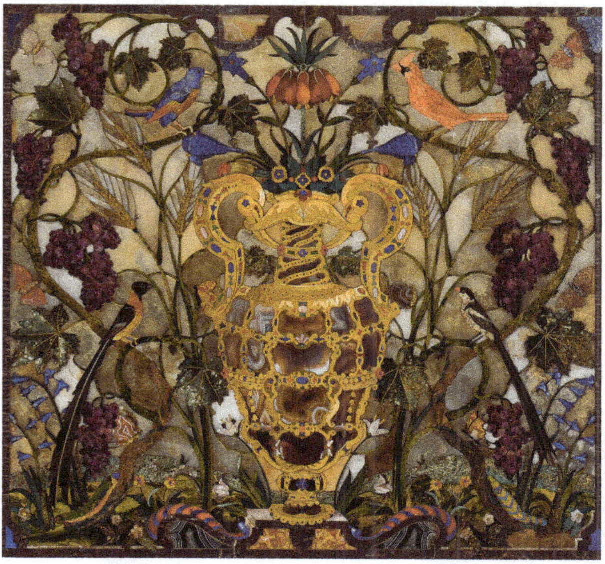

FIGURE 1.8 Giovan Battista Sassi and Jan Flach after Bernardino Poccetti, detail of *Table top decorated with two vases of flowers with birds, grapevine and ear of wheat*, c.1603–1610, *pietre dure* Florence, Palazzo Pitti, Galleria Palatina, OdA1911, 1512

intertwined with his philosophical and empirical approach to the understanding of nature.[83]

3 Taxonomy and Colors

The interconnectedness of color standardization practices and the investigation of insect life has been highlighted by Stefanie Jovanovic-Kruspel in her recent essay "The Art of Painting Butterflies."[84] That the standardization of colors occupied the minds of early modern natural historians in general and of entomologists in particular is not only corroborated by a range of historical studies, but also documented first-hand by said entomologists themselves. In his *De animalibus insectis libri septem* (1602), the first book on insects in Europe, Aldrovandi, explicitly acknowledged the significance of colors for the field of entomology and stressed his view that the act of producing colored images of insects was a joint effort between "the most erudite philosopher, who precisely

83 See Shell 2004, 20–24.
84 Jovanovic-Kruspel 2021, 471–472.

distinguishes and describes the differences of colors, and the most meticulous painter, who paints each part of such small and much-speckled bodies."[85]

While the images of insects published by Aldrovandi were left black and white, they were trumpeted in the book title as illustrations "ad vivum expressis," namely "closely modelled after life." Nonetheless, the original drawings – the so-called *tavole acquerellate* – were fully colored by his painters and marked with chromatic annotations by Aldrovandi, just as described in the previously quoted sentence.[86] Why it was so important for insects to be depicted in color Aldrovandi did not reveal, yet color was clearly regarded as the most astounding characteristic of these little animals.

The German physician August Christian Kühn (1773–1808) critically argued that most of the insect collections he had seen had been put together in order to please the eye with beautiful colors, much like decks of colorful playing cards that were not intended for actual playing.[87] The oil painter Benjamin Wilkes (*fl.* 1740–1750) introduced his *One Hundred and Twenty Copper-Plates of English Moths and Butterflies* (1749) with an interesting observation regarding the color of lepidopterans: Wilkes maintained that at his first meeting with members of the Aurelian Society, he "saw such Specimens of Nature's admirable Skill in the Disposition, Arrangement, and contrasting of Colours" that he was inspired to strive for the right color selection and arrangement for his own paintings.[88]

Insect colors were soon recognized as a taxonomical trait, and therefore either meticulously described with words or represented through hand-colored images. In *Der Raupen wunderbare Verwandelung*, Merian, who did not always color the illustrations in her printed works, highlighted color terms she used to describe insect specimens with bold type both in the text and in the register, thus providing a sort of rudimentary color vocabulary for entomologists.[89] For a time, uncolored illustrations complemented by descriptions predominated. The entomologist René-Antoine Ferchault de Réaumur (1683–1757) explained in *Memoires pour servir à l'histoire des insects* (1734) that plates in black and white were preferable for at least two reasons: They were less expensive, and they did not deceive the observer about the "colors specific to the insects," as

85 "doctissimum saepè Philosophum requirant qui exactè colorum diversitatem discernat, & describat; pictoremque diligentissimum, qui singula membra tantillis in corpusculis tantoperè maculatis dipingat," Aldrovandi 1602, 3. On Aldrovandi's cooperation with painters, see Olmi 1987.

86 The digitized *tavole* are available at http://aldrovandi.dfc.unibo.it/pinakesweb/main.asp. See also the reproductions of some plates in Neri 2011, 35, 39–40 and Egmond 2017, 211.

87 Kühn 1783, 1.

88 Wilkes 1749, iii. On the Aurelian Society, see Salmon, Marren, and Harley 2000.

89 Etheridge 2021, 102.

did plates painted with haste.[90] Strictly opposed to colored illustrations in entomological works was Swammerdam, who was mainly interested in the anatomy of insects and was not prepared to countenance the "defacement" of his images with colors.[91]

In many cases, though, color descriptions in books on insects were accompanied by detailed colored illustrations. This is surely the case with the series of booklets on insects entitled *Der monatlich herausgegebenen Insecten-Belustigung* (1746–1761) and authored by Rösel von Rosenhof. In the first volume, he acknowledged that the peculiar iridescence of certain butterfly wings could not be imitated by the "dead colors" of painters.[92] Nonetheless, the miniaturist was pleased to inform the reader that he had painted the colors of butterflies as naturally and as similarly to the original as possible.[93]

In some instances, authors simply deemed chromatic descriptions superfluous on the grounds that their specimens were "truly represented" and "painted with the utmost Exactness."[94] While this was the exception rather than the rule, Goedaert was one such example – indeed, the naturalist Martin Lister (1639–1712) complained in his translation of Goedaert's *Metamorphosis naturalis* (1662) that "I cou'd have wished the author had taken the care of describing to us the paintings and colour of the insects, which he delineated with his Pencill."[95] Instead, Goedaert had opted to hand-color the plates in his book with the help of his family specifically for buyers who were willing to pay more.[96]

---

90 "couleurs propres aux insectes," Réaumur 1734, 1:54.

91 "bekladden/commaculare," Swammerdam 1737, 66. See also Giulia Simonini's essay in this volume. Swammerdam may have been acquainted with a similar statement by Willem Goeree (1635–1711) in *Verlichtery-konst*, where he argued that "to color prints is to corrupt prints" ("printen beverven is printen bederven," Goeree 1668, 4; translation from Stijnman and Savage 2015, ix). Despite this utterance, Swammerdam played with strong contrasts by contouring the illustration of the insect with a dark halo. Similarly, Hooke used a black background around the insect to enhance the visibility of white details on the original specimen. See Bertoloni Meli 2010, 415–416.

92 "todten Farben," Rösel von Rosenhof 1746, 1:102.

93 Rösel von Rosenhof 1746, 1:102.

94 Wilkes 1749, v.

95 Goedaert and Lister 1682, 48. See Greven and Stoll 2012, 240; Etheridge 2021, 36. Lister stressed this point very clearly in the first page of his foreword, where he wrote that Goedart probably had never "aimed at any things more, then to please and gratifie his Fancy, and the excellent skill he had in Limning. And this appears from the almost totall neglect of descriptions, which he had sufficient opportunity to have well performed; and the many references to the Colours in his Designs, which Calcography, or Printing could not represent," Goedaert and Lister 1682, i. See Greven and Stoll 2012, 235.

96 Goedings 2019, 24. On the hand-coloring of plates, see Giulia Simonini's contribution to this volume.

Similarly, in his aforementioned work, Wilkes stated that "the common Method of amusing the Reader with long Accounts of the Colours and Markings of each Fly is avoided in this Work, as entirely needless, when the Objects are truly represented before his eyes."[97] The naturalist and inventor Jacob Christian Schäffer (1718–1790) likewise presented the plates in his *Icones insectorum circa Ratisbonam* (1766) in full color and with a meagre description. His goal was to cover all vital information about the studied insects through colored images, for which he proposed a method to normalize a set of colorants and their mixtures.[98]

Elsewhere, the complete opposite held true, with great emphasis being placed on chromatic descriptions while complementing images, even black-and-white ones, were conspicuously absent. For example, images were rejected by Carl Linnaeus (1707–1778) and his student Johan Christian Fabricius (1745–1808), who, however, referenced illustrations in other entomological works for the sake of clarity. Geoffrey E. Hancock explains that this practice was "a combination of inability to pay artists" and of "the delays inherent in producing plates."[99] Indeed, there is a strong likelihood that many entomologists had nothing against illustrations, but were simply unable to afford them. Together with the "almost numberless" colors in the insect realm and the absence of a fixed terminology to describe them, the inability to furnish entomological books with (colored) illustrations spurred many scholars to propose strategies to normalize the color vocabulary of entomologists and natural historians.[100] In *Entomologia Carniolica* (1763), Giovanni Antonio Scopoli (1723–1788) provided a catalog of forty-four color terms that he partly obtained by spinning a color top painted with different pigments and that were partly based on chromatic similitude with everyday objects.[101] Schiffermüller found Scopoli's method impractical. In his announcement of a catalog of butterflies gathered in the surroundings of Vienna, he proclaimed that he had found an instrument "which, in effect, is meant to replace the lack of hand coloring, remedy any ambiguity in the descriptions, and establish a general terminology of colors."[102] This instrument was published in *Versuch eines Farbensystems*

97 Wilkes 1749, v.

98 See Giulia Simonini's essay in this volume as well as Nickelsen 2006, 168.

99 Hancock 2015, 156. See also Beth Tobin's essay in this volume.

100 Schiffermüller 1771, 2; Simonini 2024.

101 Karliczek 2018, 37–38.

102 "welches allenfalls den Abgang des Illuminirens ersetzen, jeder Undeutlichkeit in den Beschreibungen abhelfen, und überhaupts den Begriff von den Farben festsetzen soll," Denis and Schiffermüller 1775, 9. On the absence of images in Schiffermüller's work, see Stefanie Jovanovich-Kruspel and Harald Bruckner's essay in this volume.

(1771), where Schiffermüller provided an incomplete system intended to systematize blue color terms by means of color samples.[103]

Color charts were sometimes enclosed in entomological works to avoid terminological ambiguity. In *An Exposition of English Insects* (1776/1780), Moses Harris supplied a colored plate entitled "Scheme of Colours" (Fig. 1.9) together with an annex of seventy-two color terms, which, according to Harris, were "little known but to painters."[104] The scheme was intended to "assist the conception of the reader, and to give some idea of each meant by the terms in the Catalogue."[105] Towards the end of the eighteenth century, the naturalist Franz de Paula von Schrank (1747–1835) deemed color samples to be vital sources of reference when it came to normalizing the chromatic terminology used for the description of insects.[106] At the beginning of the nineteenth century, the French entomologist Pierre-André Latreille (1762–1833) considered the topic so relevant that he explicitly discussed it in the first volume of his *Histoire naturelle générale et particulière des crustacés et des insectes* (1802), where he stated that "almost all specific distinctions of insects are based, in whole or in part, on their colors and arrangement."[107] To describe the colors of insects, Latreille recommended the use of Jean Baptiste Lamarck's (1744–1829) "chromometric ladder" over previous systems like the ones proposed by Scopoli and Harris.[108] Such tools were still considered useful and relevant in the second half of the nineteenth century, when Julius Müller included in his *Terminologia entomologica* (1860) a color chart with 136 nuances taken from "the most reliable examples in nature" that served to illustrate the color terms used in entomology.[109]

The use of color charts to accurately replicate the colors of insects in drawings and printed plates was apparently an established practice. For example, there is reason to assume that a color chart was employed by the Flemish printmaker Jacques de Gheyn II (1565–1629) when he started experimenting with oil and miniature painting. Intriguingly, it has been suggested that the chart in question was used to execute the illustrations in the *Lugt Album*, which,

103 See Jovanovic-Kruspel and Harald Bruckner's essay in this volume.

104 Harris 1776, iv.

105 Harris 1776, v.

106 de Paula von Schrank 1796, 2.

107 "presque toutes les distinctions spécifiques des insectes sont fondées, en tout ou en partie, sur leurs couleurs et leur disposition." Latreille 1802, 1:316.

108 "échelle chromométrique" Latreille 1802, 1:333. On Lamarck's chromometric ladder, see Boskamp 2016; Kuehni 2016.

109 "den verlässlichsten Beispielen aus der Natur" Müller 1860, vi. For a reproduction of the color chart, see Karliczek 2016, 391.

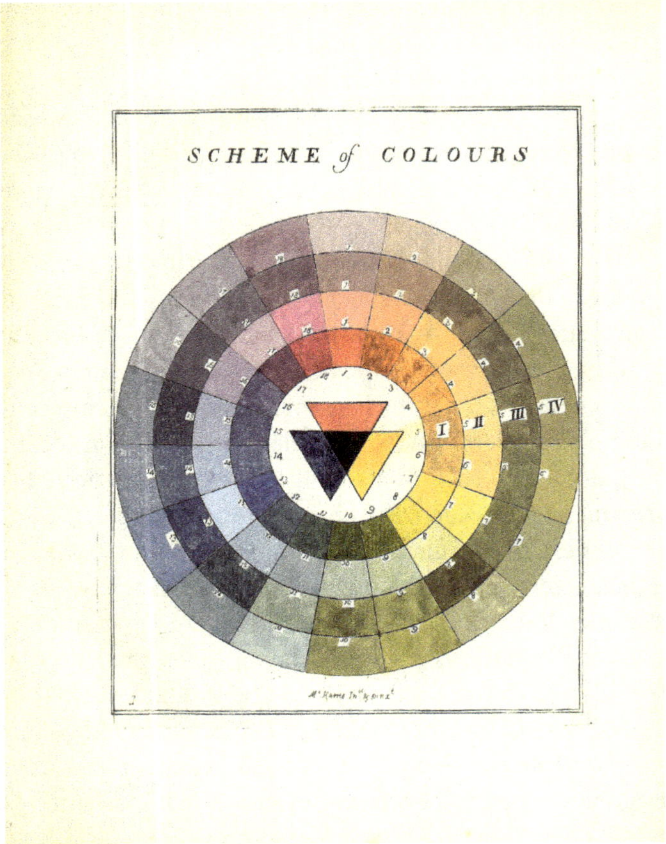

FIGURE 1.9 Moses Harris, *Scheme of Colours*, in: Moses Harris, *An Exposition of English Insects*, 1776, quarto, Washington, Smithsonian Libraries, Cullman Library Rare Books, QL482.G7H315 1776. Public Domain Mark 1.0 HTTPS://ARCHIVE.ORG/DETAILS/EXPOSITIONOFENGL00HARR/PAGE/8/MODE/2UP

among other motifs, depict insects.[110] Other known examples are the chart prototype published by Schäffer in 1769 and the extensive color notes complemented with color samples that the natural historian William Jones (1745–1818) used to produce his famous *Icones* (1780–1810), a collection of 1,292 watercolor images of Lepidoptera originally bound into seven volumes (Fig. 1.10).[111]

110 Swan 2005, 74. See also Giulia Simonini's essay in this volume.

111 On Schäffer, see Giulia Simonini's essay in this volume. On Jones, see Vane-Wright 2021, 22–23.

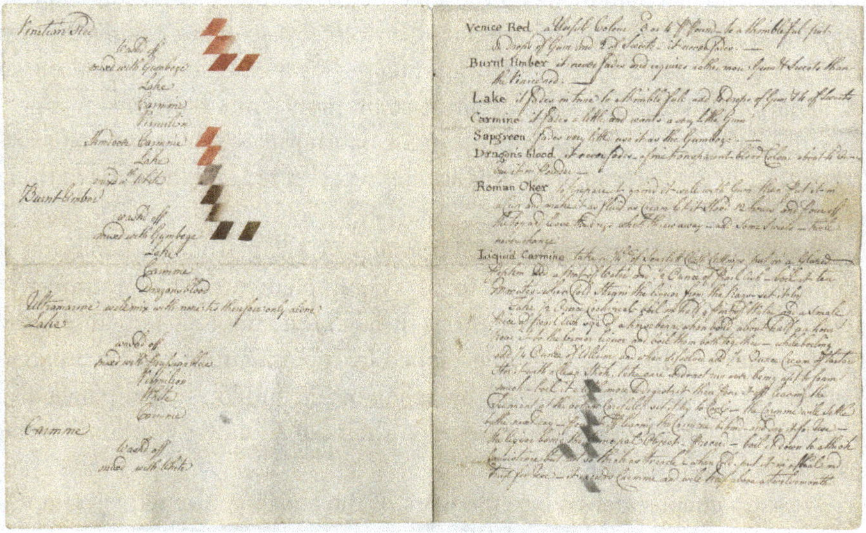

FIGURE 1.10 William Jones, Paint samples and mixes used for his *Icones*, no date, Oxford, University Museum of Natural History, WJ/B/3/0/001, fol. 3

4 Content of the Volume

In the above, we have sketched out the manifold entanglements of insects and colors in the early modern and modern world of artists and natural historians. We are fully aware that this is by no means an exhaustive reconstruction. However, we are confident that the six papers assembled in this volume provide further insights into this fascinating subject, especially with regard to the order of Lepidoptera.

In the second essay of this volume, Kimberly Schenck and Stacey Sell present the first in-depth technical analysis of Hoefnagel's insect miniatures in the album *Ignis* (National Gallery Museum, Washington), describing how the Flemish artist integrated insects' body parts – such as butterfly scales and dragonfly wings – into his artworks, and how he used color glazes over metal gilding to reproduce the iridescence of beetles and flies, so challenging to depict through paint.

The closely related third essay deals with the transfer of lepidopteran scales on paper and canvas between the seventeenth and early nineteenth century. Introducing the terms 'lepidochromy,' 'lepidochromes,' and 'lepidochromist,' V.E. Mandrij explores Marseus's motives for the use of butterfly scales in oil paintings by examining the Dutch painter's *sottoboschi* under microscope and by replicating lepidochromy recipes.

In the fourth chapter, Kay Etheridge discusses the biology of color in lepidopterans along with the attendant historical discourses, taking the caterpillars of the red admiral (*Vanessa atalanta*) as a case study. Based on an explanation of the various functions of coloration from a biological perspective, the author investigates how the artist-naturalists who noticed color discrepancies in the caterpillars of this species represented and thematized these chromatic variations.

Chapter five, authored by Giulia Simonini, investigates the hand-coloring of entomological plates. The author's focus is the implementation of a strategy to standardize the coloring of illustrations through the sampling and numbering of pigments and their mixtures developed by the miniature painter Stephan Loibel (1725–1775) and described by Schäffer, his employer. The paper also discusses points of comparison between their collaborative work and other painting-by-number techniques.

The sixth chapter delves into the work of the abovementioned Jesuit entomologists Denis and Schiffermüller, who classified many European species of lepidopterans in the eighteenth century. Stefanie Jovanovic-Kruspel and Harald Bruckner discuss the use of colors in their collaboratively published *magnum opus*, the *Wiener Verzeichnis* (1775/76), in relation to Schiffermüller's color theory, while also highlighting the iconographic impact of the latter's watercolors of caterpillars. An appendix lists the modern scientific names of the species depicted in Schiffermüller's watercolors to make them accessible for entomological research.

Finally, Beth Fowkes Tobin thematizes the problem of the faded colors of taxidermized North American insects, which led to misleading information about their coloration being published when they were employed as models for entomological illustrations. Conversely, *The Natural History of the Rarer Lepidopterous Insects of Georgia* (1797), authored by the English illustrator John Abbot (1751–1840) working in North America and the London-based naturalist James Edward Smith (1759–1828), is discussed as an exemplary model for the reliable coloring of entomological plates. Furthermore, the author points out how the successful collaboration between Abbot and Smith contributed to the establishment of the study of the larval stages of lepidopterans for taxonomical purposes.

The essays in this collection not only provide six individual case studies on key aspects of the topic – in conjunction, they mirror the fruitful interdisciplinarity of the early days of entomology. Moreover, the present volume clearly illustrates how the depiction, description, and documentation of insect coloration in words and images between the sixteenth and the eighteenth century invites further research.

Bibliography

Aguilar, Jacques d'. 2006. *Histoire de l'entomologie*. Paris: Delachaux and Niestlé.

Aldrovandi, Ulisse. 1602. *De animalibus insectis libri septem, cum singulorum* ... Bologna: G.B. Bellagamba.

Aloi, Giovanni. 2007a. "Insect Poetics." *Antennae. The Journal of Nature in Visual Culture*, no. 3B (Autumn).

Aloi, Giovanni. 2007b. "Insect Poetics." *Antennae. The Journal of Nature in Visual Culture*, no. 2.1 (Autumn).

Aristotle. 1961. *Parts of Animals*. Translated by E.S. Forster and A.L. Peck. Loeb Classical Library. Cambridge, Mass: Harvard University press.

Baadj, Nadia. 2016. *Jan van Kessel I (1626–79): Crafting a Natural History of Art in Early Modern Antwerp*. London: Harvey Miller Publishers.

Ball, Clara Sue. 1966. "The Apiarium: An Early Example of Microscopic Study." *Proceedings of the Oklaoma Academy of Science*, 148–151.

Barnard, Peter C., 2018. "Bat-Fowlers, Pooters and Cyanide Jars: A Historical Overview of Insect Collecting and Preservation." In *Naturalists in the Field: Collecting, Recording and Preserving the Natural World from the Fifteenth to the Twenty-First Century*, Arthur MacGregor (ed.). [Emergence of Natural History 2]. Leiden: Brill, 646–85. https://doi.org/10.1163/9789004323841_023.

Bass, Marisa Anne. 2019. *Insect Artifice: Nature and Art in the Dutch Revolt*. Princeton and Oxford: Princeton University Press.

Beier, Bodo. 1987. "'Contre-Epreuves' in der barocken Stillebenmalerei." *Maltechnik/ Restauro* 1 (93): 35–39.

Bertoloni Meli, Domenico. 2010. "The Representation of Insects in the Seventeenth Century: A Comparative Approach." *Annals of Science* 67 (3): 405–429. https://doi.org/10.1080/00033790.2010.488920.

Beurs, Willem. 1692. *De groote waereld in 't kleen geschildert, of schilderagtig tafereel van 's weerelds schilderyen. kortelijk vervat in ses boeken. verklarende de hooftverwen, haare verscheide mengelingen in oly en der zelver gebruik*. Amsterdam: J. and G. Janssonius van Waesberge.

Bodenheimer, Friedrich S. 1928. *Materialien zur Geschichte der Entomologie bis Linné*. Vol. 1. Berlin: W. Junk.

Boersma, Albert. 2021. *Ander licht op Withoos: drie generaties Withoos*. Amersfoort: Uitgeverij Boiten boekprojecten.

Bodson, Liliane. 1983. "The Beginnings of Entomology in Ancient Greece." *The Classical Outlook* 61 (1): 3–6.

Boskamp, Ulrike. 2016. "Jean-Baptiste de Lamarcks vergessene Chemie, Physik und Naturgeschichte der Farbe." In *Die Farben der Klassik: Wissenschaft – Ästhetik – Literatur*,

Martin Dönike, Jutta Müller-Tamm, and Friedrich Steinle (eds.). [Schriftenreihe des Zentrums für Klassikforschung 3]. Göttingen: Wallstein Verlag, 205–230.

Brez, Jaques. 1791. *La flore des insectophiles: précédée d'un discours sur l'utilite des insectes et de l'etude de l'insectologie*. Utrecht: B. Wild and J. Altheer.

Butler Greenfield, Amy. 2005. *A Perfect Red: Empire, Espionage, and the Quest for the Color of Desire*. New York: HarperCollins.

Büttner, Philippe, Claire Hoffmann, and Lukas Gloor, eds. 2015. *Ein goldenes Zeitalter: Holländische und flämische Gemälde aus einer privaten Sammlung*. Exhibition Catalogue. Zurich, Kunsthaus. Zurich: Kunsthaus Zürich; Scheidegger & Spiess.

Cambefort, Yves. 2007. "Entomologie et mélancolie. Quelques aspects du symbolisme des insectes dans l'art européean du XIVe au XXIe siècle." In *Le symbolisme des animaux: l'animal, clef de voûte de la relation entre l'homme et la nature? = Animal symbolism: animals, keystone in the relationship between man and nature?*, Edmond Dounias, Elisabeth Motte-Florac, and Margaret Dunham (eds.). Paris: IRD Éditions, 393–323.

Cardon, Dominique. 2007. *Natural Dyes: Sources, Tradition, Technology and Science*. London: Archetype Publications.

Cerami, Cristina. 2018. "Toward a New Criterion of the Scale of Being in Aristotle's Generation of Animals from Part II – The Principles of Animal Generation Reconsidered." In *Aristotle's Generation of Animals: A Critical Guide*, David Lefebvre and Andrea Falcon (eds.). Cambridge: Cambridge University Press, 130–149. https://doi.org/10.1017/9781316459386.

Chapman, Allan. 2005. *England's Leonardo: Robert Hooke and the Seventeenth Century Scientific Revolution*. Bristol: Institute of Physics Publishing.

Cobb, Matthew. 2006. *The Egg & Sperm Race: The Seventeenth-Century Scientists Who Unravelled the Secrets of Sex, Life and Growth*. London: The Free Press.

Colding, Torben Holck. 1953. *Aspects of Miniature Painting: Its Origins and Development*. Copenhagen: E. Munksgaard.

Cramer, Pieter, and Caspar Stoll. 1779. *De uitlandsche kapellen voorkomende in de drie waereld-deelen Asia, Africa En America: Onder deszelfs opzigt allen naar het leven getekend, in het koper gebragt en met natuurlyke koleuren afgetekend*. Amsterdam, Utrecht: J. Baalde and B. Wild.

DaCosta Kaufmann, Thomas (ed.). 1993. *The Mastery of Nature: Aspects of Art, Science, and Humanism in the Renaissance*. Princeton: Princeton University Press.

Daston, Lorraine, and Peter Galison. 2007. *Objectivity*. Cambridge, New York: Zone Books.

Denis, Michael, and Ignaz Schiffermüller. 1775. *Ankündung eines systematischen Werkes von den Schmetterlingen der Wienergegend, herausgegeben von einigen Lehrern am k.k. Theresianum*. Vienna: A. Bernardi.

Denis, Michael, and Ignaz Schiffermüller. 1776. *Systematisches Verzeichniß der Schmetterlinge der Wienergegend*. Vienna: A. Bernardi.

Dicke, Marcel. 2000. "Insects in Western Art. American Entomologist." *American Entomologist* 46 (4): 228–237. https://doi.org/10.1093/ae/46.4.228.

Dicke, Marcel. 2004. "From Venice to Fabre: Insects in Western Art." *Proceedings of the Netherlands Entomological Society Meeting* 15: 9–14.

Doherty, Meghan C. 2012. "Discovering the 'True Form:' Hooke's Micrographia and the Visual Vocabulary of Engraved Portraits." *Notes and Records: The Royal Society Journal of the History of Science* 66 (3): 211–234. https://doi.org/10.1098/rsnr.2012.0031.

Don, Anna. 2020. "'They Are so True to Life and Nature': The Technique and Treatment of Jan van Kessel the Elder's Insects." *Hamilton Kerr Institute Bulletin* 8: 7–13.

Egmond, Florike. 2017. *Eye for Detail: Images of Plants and Animals in Art and Science, 1500–1630*. London: Reaktion Books.

Egmond, Florike. 2020. "The 'ad Vivum' Conundrum Eyewitnessing and the Artful Representation of Naturalia in Sixteenth-Century Natural Science." In *Zeigen – Überzeugen – Beweisen: Methoden der Wissensproduktion in Kunstliteratur, Kennerschaft und Sammlungspraxis der Frühen Neuzeit*, Elisabeth Oy-Marra and Irina Schmiedel (eds.). Heidelberg: arthistoricum.net, 33–62.

Ernst, Jean-Jacques, and Jacques-Louis-Florentin Engramelle. 1779a. *Papillons d'Europe, peints d'après nature: premiere partie. chenilles, crisalides, & papillons de jour*. Paris: Delaguette.

Ernst, Jean-Jacques, and Jacques-Louis-Florentin Engramelle. 1779b. *Insectes d'Europe, peints d'après nature, gravès et coloréz: seconde partie. chenilles, crisalides et papillons de jour*. Paris: Delaguette.

Esper, Eugen Johann Christoph. 1777. *Die Schmetterlinge in Abbildungen nach der Natur mit Beschreibungen: Europäische Gattungen*. Vol. 1. Erlangen: W. Walthers.

Essig, Edward Oliver. 1936. "A Sketch History of Entomology." *Osiris* 2: 80–123.

Etheridge, Kay. 2021. *The Flowering of Ecology: Maria Sibylla Merian's Caterpillar Book*. [Emergence of Natural History 3]. Leiden: Brill.

Felfe, Robert. 2015. *Naturform und bildnerische Prozesse: Elemente einer Wissensgeschichte in der Kunst des 16.und 17. Jahrhunderts*. [Actus et Imago: Berliner Schriften für Bildaktforschung und Verkörperungsphilosophie, Vol. XIII]. Berlin; Boston: De Gruyter.

Gallori, Corinna Tania, and Gerhard Wolf. 2015. "Tre serpi, tre vedove e alcune piante: I disegni 'inimitabili' di Jacopo Ligozzi e le loro copie o traduzioni tra i progetti di Ulisse Aldrovandi e le pietre dure." *Mitteilungen des Kunsthistorischen Institutes in Florenz* 57 (2): 212–252. https://doi.org/10.11588/mkhi.2015.2.66741.

Geisthardt, Michael. 1990. "Die Gerningsche Insektensammlung im Landesmuseum Wiesbaden. Ein Beitrag zur Geschichte der Entomologie." *Mitteilungen des internationalen entomologischen Vereins* 15 (1/2): 29–39.

Goedaert, Johannes. 1662a. *Metamorphosis et historia naturalis insectorum*. Middelburg: J.

Goedaert, Johannes. 1662b. *Metamorphosis naturali: ofte historische beschryvinghe van den oirspronk, aerd, eygenschappen en de vreemde veranderinghem der wormen, rupsen, maeden, vliegen, witjens, byen motten ende dierghelijcke dierkens meer: niet uyt eenighe boecken, maer aileenelijck door eygen ervarentheyd uytgevonden, beschreven, en de na de konst afgeteyckent*. Middelburg: J. Fierens.

Goedaert, Johannes, Martin Lister (ed.). 1682. *Johannes Godartius of Insects*. York: J. White.

Goedings, Truusje. 2019. "De werking van zwart en wit. Kleur in Goedaerts *Metamorphosis naturalis*." *Tijd-Schrift. Erfgoedpraktijk in Vlaanderen* 9 (3): 20–37.

Goeree, Willem. 1668. "Verlichtery-Konst." In *Inleydinge tot de al-gemeene Teycken-Konst*. Middelburg: W. Goeree.

Göttler, Christine. 2021. "Yellow, Vermilion, and Gold: Colour in Karel van Mander's Schilder-Boeck." In *Materialized Identities in Early Modern Culture, 1450–1750: Objects, Affects, Effects*, Susanna Burghartz, Lucas Burkart, Christine Göttler, and Ulinka Rublack (eds.). [Visual and Material Culture, 1300–1700]. Amsterdam: Amsterdam University Press, 233–280.

Greven, Hartmut, and Silke Stoll. 2012. "Ein Band von Goedaerts niederländischer Ausgabe der „Metamorphosis Naturalis" – Eine entomologische Kostbarkeit aus der Bibliothek des „Aquazoo/ Löbbecke-Museum Düsseldorf" (A Volume of Goedaert's Dutch Edition of the 'Metamorphosis Naturalis' – an Entomological Treasure from the Library of the 'Aquazoo/ Löbbecke-Museum Düsseldorf')." *Entomologie heute* 24 (January): 217–263.

Hancock, E. Geoffrey. 2015. "The Shaping Role of Johann Christian Fabricius: William Hunter's Insect Collection and Entomology in Eighteenth-Century London." In *William Hunter's World: The Art and Science of Eighteenth-Century Collecting*, E. Geoffrey Hancock, Nick Pearce, and Mungo Campbell (eds.). [The Histories of Material Culture and Collecting, 1700–1950]. Farnham: Ashgate, 151–164.

Harris, Moses. 1776. *An Exposition of English Insects: With Curious Observations and Remarks, Wherein Each Insect Is Particularly Described; Its Parts and Properties Considered; the Different Sexes Distinguished, and the Natural History Faithfully Related: The Whole Illustrated with Copper Plates*. London: Printed for the author.

Hermens, Erma. 2022. "Crawly Creatures, Forest Floors and Butterfly Wings Otto Marseus van Schrieck and Sottobosco." In *Crawly Creatures. Little Animals in Art and Science*, Jan de Hond, Eric Jorink, and Hans Mulder (eds.). Exhibition Catalogue. Amsterdam, Rijksmuseum. 30 September 2022–15 January 2023. Amsterdam: Rijksmuseum, 139–149.

Hollingsworth, Cristopher. 2005. *Poetics of the Hive: Insect Metaphor in Literature*. Iowa City: University of Iowa Press.

Hond, Jan de. 2022. "Crawly Creatures from Abhorrence to Amazement." In *Crawly Creatures. Little Animals in Art and Science*, Jan de Hond, Eric Jorink, and Hans Mulder (eds.). Exhibition Catalogue. Amsterdam, Rijksmuseum. 30 September 2022–15 January 2023. Amsterdam: Rijksmuseum, 7–24.

Hond, Jan de, Eric Jorink, and Hans Mulder (eds.). 2022. *Crawly Creatures. Little Animals in Art and Science*. Exhibition Catalogue. Amsterdam, Rijksmuseum. 30 September 2022–15 January 2023. Amsterdam: Rijksmuseum.

Hoogstraten, Samuel van. 1678. *Inleydinge tot de hooge schoole der schilderkonst: Anders de zichtbaere werelt*. Rotterdam: F. van Hoogstraeten.

Hoogstraten, Samuel van, Celeste Brusati, and Jaap Jacobs. 2021. *Samuel van Hoogstraten's Introduction to the Academy of Painting, or, The Visible World*. Texts & Documents. Los Angeles: Getty Research Institute.

Hooke, Robert. 1665. *Micrographia, or Some Physiological Descriptions of Minute Bodies Made by Magnifying Glasses, with Observations and Inquiries Thereupon*. London: J. Martyn and J. Allestry.

Hulton, Paul. 1962. "An Album of Plant Drawings by Jacques Le Moyne de Morgues." *The British Museum Quarterly* 26 (1/2): 37–39. https://doi.org/10.2307/4422769.

Hulton, Paul. 1977. *The Work of Jacques Le Moyne de Morgues. 1: Foreword, Catalogue and Introductory Studies*. London: British Museum Publishing.

Hünniger, Dominik. 2018. "Nets, Labels and Boards: Materiality and Natural History Practices in Continental European Manuals on Insect Collecting 1688–1776." In *Naturalists in the Field: Collecting, Recording and Preserving the Natural World from the Fifteenth to the Twenty-First Century*, Arthur MacGregor (ed.). [Emergence of Natural History 2]. Leiden: Brill, 686–705. https://doi.org/10.1163/9789004323841_024.

Jarvis, Charles E. 2018. "'Take with You a Small Spudd or Trowell': James Petiver's Directions for Collecting Natural Curiosities." In *Naturalists in the Field: Collecting, Recording and Preserving the Natural World from the Fifteenth to the Twenty-First Century*, Arthur MacGregor (ed.). [Emergence of Natural History 2]. Leiden: Brill, 212–239. https://doi.org/10.1163/9789004323841_009.

Jervis, Mark A. 2013. "Robert Hooke's Micrographia: An Entomologist's Perspective." *Journal of Natural History* 47 (39–40): 2531–2573. https://doi.org/10.1080/00222933.2013.780270.

Jorink, Eric. 2007. "Between Emblematics and the 'Argument from Design': The Representation of Insects in the Dutch Republic." In *Early Modern Zoology: The Construction of Animals in Science, Literature and the Visual Arts*, Karl Alfred Engelbert Enenkel and Paul J. Smith (eds.). [Intersections 7]. Leiden: Brill, 147–175.

Jorink, Eric. 2010. *Reading the Book of Nature in the Dutch Golden Age, 1575–1715*. [Brill's Studies in Intellectual History 191]. Leiden: Brill.

Jorink, Eric. 2018. "Insects, Philosophy and the Microscope." In *Worlds of Natural History*, Helen Anne Curry, Nicholas Jardine, James Andrew Secord, and Emma C. Spary (eds.). Cambridge; New York: University of Cambridge, 131–147.

Jovanovic-Kruspel, Stefanie. 2021. “The Art of Painting Butterflies.” In *Iconotypes: A Compendium of Butterflies and Moths, Jones' Icones Complete*, Oxford University Museum (ed.). Oakland: University of California Press, 462–473.

Karliczek, André. 2016. “Katalog.” In *Farre. Farbstandards in den frühen Wissenschaften*, edited by André Karliczek and Andreas Schwarz. Jena: Salana, 283–523.

Karliczek, André. 2018. “Zur Herausbildung von Farbstandards in den frühen Wissenschaften des 18. Jahrhunderts.” *Ferrum – Farben der Technik – Technik der Farben / Colors in Technology – Technology of Colors* 90: 36–49. http://doi.org/10.5169/seals-787106.

Kirby, Jo. 1977. “A Spectrophotometric Method for the Identification of Lake Pigment Dyestuffs.” *National Gallery Technical Bulletin* 1: 35–45.

Kirby, Jo. 2003. *Varnish*. Grove Dictionary of Art Online. Oxford: Oxford University Press. https://doi.org/10.1093/gao/9781884446054.article.T087987.

Kirby, Jo. 2015. “‘One of the Most Beautiful Reds': Cochineal in European Painting.” In *A Red Like No Other: How Cochineal Colored the World: An Epic Story of Art, Culture, Science, and Trade*, Carmella Padilla, Barbara C. Anderson, and Blair Clark (eds.). New York: Santa Fe: Skira/Rizzoli Museum of International Folk Art, 174–189.

Kirby, Jo, Maarten van Bommel, and André Verhecken. 2014. *Natural Colorants for Dyeing and Lake Pigments: Practical Recipes and Their Historical Sources*. London: Archetype Publications in association with The National Gallery, Cultural Heritage Agency, IRPA KIK, Doerner Institut, Charisma Charisma.

Kirby, Jo, P.T. Craddock, Jonathan Stephenson, A. Broderick, Malcolm Green, and Nicholas Pickwoad. 2003. “Gilding.” In *Oxford Art Online*. Oxford University Press. https://doi.org/10.1093/gao/9781884446054.article.T032215.

Kirby, Jo, Marika Spring, and Catherine Higgitt. 2005. “The Technology of Red Lake Pigment Manufacture: Study of the Dyestuff Substrate.” *National Gallery Technical Bulletin* 26: 71–87.

Kirby, Jo, Marika Spring, and Catherine Higgitt. 2007. “The Technology of Eighteenth- and Nineteenth-Century Red Lake Pigments.” *The National Gallery Technical Bulletin* 28 (January): 69–96.

Kuehni, Rolf G. 2016. “Lamarck's ‘Chromometric' Color Ladder.” In *Farre. Farbstandards in den frühen Wissenschaften.*, edited by André Karliczek and Andreas Schwarz. Jena: Salana, 65–75.

Kühn, August Christian. 1783. *Kurze Anleitung Insecten zu sammlen*. Eisenach: Wittekind.

Kusukawa, Sachiko. 2011. “Patron's Review. The Role of Images in the Development of Renaissance Natural History.” *Archives of Natural History* 38 (2): 189–213. https://doi.org/10.3366/anh.2011.0028.

L'Admiral, Jacob. 1740. *Naauwkeurige waarneemingen van veele gestaltverwisselende gekorvene diertjes: in omtrent 30 Jaaren, zo in Vrankryk, als in Engeland en Holland, by een vergaaderd, na het leeven geschilderd, en nu in't koper gebragt*. Amsterdam: F. Changuion, S. van Esveldt.

Lammertse, Friso. 2022. "With Fresh Eyes Renaissance Artists and Crawly Creatures." In *Crawly Creatures. Little Animals in Art and Science*, Jan de Hond, Eric Jorink, and Hans Mulder (eds.). Exhibition Catalogue. Amsterdam, Rijksmuseum. 30 September 2022–15 January 2023. Amsterdam: Rijksmuseum, 39–54.

Latreille, Pierre-André. 1802. *Histoire naturelle, générale et particulière des crustacés et des insectes. ouvrage faisant suite aux œuvres de Leclerc de Buffon, et partie du cours complet d'histoire naturelle rédigé par C.S. Sonnini. principes élémentaires. Tome Premier*. Edited by Charles-Nicolas-Sigisbert Sonnini. Vol. 1. Paris: F. Dufart.

Lawson, Ian. 2016. "Crafting the Microworld: How Robert Hooke Constructed Knowledge about Small Things." *Notes and Records: The Royal Society Journal of the History of Science* 70 (1): 23–44. https://doi.org/10.1098/rsnr.2015.0057.

Leonhard, Karin. 2014. "Larve und Mimikry: Etymologisches und Entomologisches im 17. Jahrhundert." In *Maske, Maskerade und die Kunst der Verstellung: vom Barock bis zur Moderne*, Christiane Kruse (ed.). [Wolfenbütteler Arbeiten zur Barockforschung 52]. Wiesbaden: Harrossowitz, 173–192.

Leonhard, Karin. 2017. "Verlorene Farben – Gewonnene Einsichten: Systematisches Sehen in der Mikroskopie Antoni van Leeuwenhoeks." In *Gesprächs-Stoff Farbe: Beiträge aus Wissenschaft, Kunst und Gesellschaft*, Konrad Scheurmann and André Karliczek (eds.). Cologne: Böhlau Verlag, 182–195.

Leonhard, Karin. 2020. *The Fertile Ground of Painting. 17th-Century Still Lifes and Nature Pieces*. London, Turnhout: Harvey Muller Publishers.

Marren, Peter. 2017. *Rainbow Dust: Three Centuries of Butterfly Delight, with a New Preface*. Chicago: University of Chicago Press.

Merian, Maria Sibylla. 1679. *Der Raupen wunderbare Verwandelung, und sonderbare Blumen-nahrung*. Vol. 1. Nuremberg: Graff; Funk.

Merian, Maria Sibylla. 1705. *Metamorphosis insectorum surinamensium. Ofte verandering der surinaamsche insecten. waar in de surinaamsche rupsen en wormen met alle des zelfs veranderingen na het leven afgebeeld en beschreeven worden, zynde elk geplaast op die gewassen, bloemen en vruchten, daar sy op gevonden zyn; waar in ookde generatie der kikvorschen, wonderbaare padden, hagedissen, slangen, spinnen en mieren werden ver toond en beschreeven, alles in america na het leven en levensgroote geschildert en beschreeven*. Amsterdam: M.S. Merian; G. Valck.

Moffett, Thomas. 1658. "The Theater of Insects: Or, Lesser Living Creatures as Bees, Flies, Caterpillars, Spiders, Worms, &c. a Most Elaborate Work by Tho. Mouffet, Doctor in Physick." In *The History of Four-Footed Beasts and Serpents. Describing at Large Their Ture and Lively Figure, Their Several Names, Conditions, Kinds, Virtuees ... Collected out of the Writings of Conradus Gesner and Other Authors*. Translated by Edward Topsell. London: E. Cotes.

Müller, Julius. 1860. *Terminologia entomologica: ein Handbuch sowohl für den angehenden Entomologen als auch für den Fachmann: nach dem neuesten Standpunkte dieser Wissenschaft*. Brünn: J. Müller.

Neri, Janice. 2011. *The Insect and the Image: Visualizing Nature in Early Modern Europe, 1500–1700*. Minneapolis: University of Minnesota Press.

Nickelsen, Kärin. 2006. *Draughtsmen, Botanists and Nature: The Construction of Eighteenth-Century Botanical Illustrations*. Archimedes. [New Studies in the History and Philosophy of Science and Technology 15]. Dordrecht: Springer Science & Business Media.

Ogilvie, Brian W. 2008. "Nature's Bible: Insects in Seventeenth-Century European Art and Science." *Tidsskrijft for Kulturforkning* 7 (3): 5–21.

Ogilvie, Brian W. 2014. "Order of Insects. Insect Species and Metamorphosis between Renaissance and Enlightenment." In *The Life Sciences in Early Modern Philosophy*, Ohad Nachtomy and Justin E.H. Smith (eds.). Oxford; New York: Oxford University Press, 222–245. https://doi.org/10.1093/acprof:oso/9780199987313.003.0012.

Ogilvie, Brian W. 2020. "Maxima in Minimis Animalibus: Insects in Natural Theology and Physico-Theology." In *Physico-Theology: Religion and Science in Europe, 1650–1750*, Ann Blair and Kaspar von Greyerz (eds.). [Medicine, Science, and Religion in Historical Context]. Baltimore: Johns Hopkins University Press, 171–182.

Olmi, Giuseppe. 1987. "Ulisse Aldrovandi and the Bolognese Painters in the Second Half of the 16th Century." In *Emilian Painting of the 16th and 17th Centuries. A Symposium, National Gallery of Art, Washington*. Bologna: Nuova Alfa, 63–74.

Onaga, Lisa and Hünniger, Dominik. 2024. "Introduction: Expanded Perspectives on Tiny Animals as Epistemic Agents." *Isis* 115 (1): 126–130. https://doi.org/10.1086/728965.

Pater, Ellen, and Eric Jorink. 2022. "The View through a Microscope Insects Up Close and Personal." In *Crawly Creatures. Little Animals in Art and Science*, Jan de Hond, Eric Jorink, and Hans Mulder (eds.). Exhibition Catalogue. Amsterdam, Rijksmuseum. 30 September 2022–15 January 2023. Amsterdam: Rijksmuseum, 127–136.

Paula von Schrank, Franz de (ed.). 1796. *Sammlung naturhistorischer und physikalischer Aufsätze*. Nuremberg: Raspe.

Perrot, Catherine. 1686. *Les leçons royales ou la manière de peindre en mignature les fleurs & les oiseaux, par l'explication des livres de fleurs & d'oiseaux de feu Nicolas Robert fleuriste. Composées par Damoiselle Catherine Perrot, peintre académiste, femme de M' C. Horry notaire apostolique de l'archevêché de Paris. Dédiées à Madame La Dauphine*. Paris: J.B. Nego.

Plinius Secundus, Gaius. 2013. *Naturkunde: Buch XI: Zoologie: Insekten: Vergleichende Anatomie; lateinisch – deutsch*. Roderich König and Joachim Hopp (eds.). Sammlung Tusculum. Berlin: De Gruyter.

Ratcliffe, Brett. 2006. "Scarab Beetles in Human Culture." *Coleopterists Bulletin* 60 (mo5):85–101.

Raven, Charles. 1942. *John Ray, Naturalist His Life and Works*. New York: Cambridge University Press. https://doi.org/10.1017/CBO9780511694066.

Ray, John, and Martin Lister. 1710. *Historia insectorum. Opus posthumum jussu Regiae Societatis Londinensis editum. Cui subjungitur appendix de scarabaeis britannicis, Autore M. Lister*. London: A. & J. Churchill.

Réaumur, René-Antoine Ferchault de. 1734. *Memoires pour servir a l'histoire des insectes. Tome prèmier sur les chenilles et les papillons*. Vol. 1. Paris: De l'imprimerie royale.

Reynolds, Stuart. 2019. "Cooking up the Perfect Insect: Aristotle's Transformational Idea about the Complete Metamorphosis of Insects." *Philosophical Transactions of the Royal Society Biology* 374 (1783): 20190074. https://doi.org/10.1098/rstb.2019.0074.

Richter, Mark. 2013. "Coloured Glazes on Metal Leaf from the Baroque and Rococo Period." In *Lüsterfassungen in Barock und Rokoko: Coloured Glazes on Metal Leaf from the Baroque and Rococo Period*, edited by Erwin Emmerling, Michael Kühlenthal, and Mark Richter. [Studien aus dem Institut für Baugeschichte, Kunstgeschichte, Restaurierung mit Architekturmuseum/Technische Universität München, Fakultät für Architektur Restaurierung]. Munich: Siegl, 33–38.

Roemer, Bert van de, Florence Pieters, Hans Mulder, Kay Etheridge, and Marieke van Delft, (eds.). 2022. *Maria Sibylla Merian: Changing the Nature of Art and Science*. Tielt: Uitg. Lannoo N.V.

Roos, Anna Marie. 2016. "The Saline Chymistry of Color in Seventeenth-Century English Natural History." In *Early Modern Color Worlds*, Tawrin Baker, Sven Durpé, Sachiko Kusukawa, and Karin Leonhard (eds.), Leiden: Brill, 274–300. https://doi.org/10.1163/9789004316607_012.

Roque, Georges. 2021. *La cochenille, de la teinture à la peinture: une histoire matérielle de la couleur*. [Art et artistes]. Paris: Gallimard.

Rösel von Rosenhof, August Johann. 1746. *Der monatlich herausgegebenen Insecten-Belustigung in welchem die in sechs Classen eingetheilte Papilionen mit ihrem Ursprung, Verwandlung und allen wunderbaren Eigenschafften, aus eigener Erfahrung beschrieben und in sauber illuminirten Kupfern, nach dem Leben abgebildet, vorgestellet werden von August Johann Rösel, Miniatur-Mahlern: Nebst einer Vorrede, in welcher von dem Nutzen derer Insecten gehandelt, was sie seyen gezeiget, und von der Eintheilung dererselben Nachricht gegeben wird*. Christian Friedrich Carl Kleemann and August Johann Rösel von Rosenhof (eds.). Vol. 1. Nuremberg: J.J. Fleischmann.

Roth, Michel, Magdalena Bushart, and Martin Sonnabend (eds.). 2017. *Maria Sibylla Merian und die Tradition des Blumenbildes von der Renaissance bis zur Romantik*. Exhibition Catalog (Kupferstichkabinett, Berlin, 07.04.2017–02.07.2017, Berlin; Städel Museum, 11.10.2017–14.01.2018, Frankfurt Am Main). Munich: Hirmer.

Salmon, Michael A., Peter Marren, and Basil Harley. 2000. *The Aurelian Legacy: British Butterflies and Their Collectors*. Berkeley; Los Angeles: University of California Press.

Schenck, Kimberly, and Stacey Sell. 2022. "Insect Transfer Prints in Joris Hoefnagel's Ignis from the Four Elements." *Master Drawings* 60 (2): 161–170.

Schenk, Franziska, and Andrew Parker. 2011. "Iridescent Color: From Nature to the Painter's Palette." *Leonardo* 44 (2): 108–115. https://doi.org/10.1162/LEON_a_00114.

Schenk, Franziska, Bodo D. Wilts, and Doekele G. Stavenga. 2013. "The Japanese Jewel Beetle: A Painter's Challenge." *Bioinspiration & Biomimetics* 8 (4): 045002. https://doi.org/10.1088/1748-3182/8/4/045002.

Schiffermüller, Ignaz. 1771. *Versuch eines Farbensystems.* Vienna: A. Bernardi.

Schmidt-Loske, Katharina. 2007. *Die Tierwelt der Maria Sibylla Merian (1647–1717): Arten, Beschreibungen und Illustrationen.* [Acta Biohistorica 10]. Marburg/Lahn: Basilisken-Presse.

Schmidt-Loske, Katharina. 2010. "Historical Sketch Maria Sibylla Merian – Metamorphosis of Insects." *Deutsche Entomologische Zeitschrift* 57 (1): 5–10. https://doi.org/10.1002/mmnd.201000001.

Schmidt-Loske, Katharina. 2021. "Maria Sibylla Merian: A Woman's Pioneering Work in Entomology." In *Women and the Art and Science of Collecting in Eighteenth-Century Europe*, Arlene Leis and Kacie Wells (eds.). [The Histories of Material Culture and Collecting, 1700–1950]. New York: Routledge, 61–77.

Seehafer, Michèle. 2021. "Shimmering Virtue: Joris Hoefnagel and the Uses of Shell Gold in the Early Modern Period." In *Materialized Identities in Early Modern Culture, 1450–1750: Objects, Affects, Effects*, Susanna Burghartz, Lucas Burkart, Christine Göttler, and Ulinka Rublack (Eds.). [Visual and Material Culture, 1300–1700]. Amsterdam: Amsterdam University Press, 281–321.

Seelig, Gero (ed.). 2017. *Medusa's Menagerie. Otto Marseus van Schrieck and the Scholars.* Exhibition Catalogue Medusa's Menagerie, Staatliches Museum Schwerin 7 July–15 October 2017, Rijksmuseum Twenthe, Enschede 5 November 2017–11 March 2018. Munich: Hirmer.

Segal, Sam, and Klara Alen. 2020. *Dutch and Flemish Flower Pieces: Paintings, Drawings and Prints up to the Nineteenth Century.* 2 vols. Leiden: Brill.

Sepp, Jan Christiaan. 1762. *Beschouwing der wonderen Gods, in de minstgeachte schepzelen: of Nederlandsche insecten.* Amsterdam: J.C. Sepp.

Simonini, Giulia. 2024. *Color Charts in 18th-Century Europe: Natural, Pigmentary, and Trichromatic.* Heidelberg: arthistoricum.net.

Sloan, Kim. 2007. *A New World: England's First View of America.* Chapel Hill: The University of North Carolina Press.

Steensma, Susanna. 1999. *Otto Marseus van Schrieck: Leben und Werk.* Hildesheim: G. Olms.

Stijnman, Ad, and Elizabeth Savage (eds.). 2015. *Printing Colour 1400–1700: History, Techniques, Functions and Receptions.* [Library of the Written Word 41]. Leiden: Brill.

Stork, Nigel E. 2018. "How Many Species of Insects and Other Terrestrial Arthropods Are There on Earth?" *Annual Review of Entomology* 63 (1): 31–45. https://doi.org/10.1146/annurev-ento-020117-043348.

Suchtelen, Arianne van (ed.). 2022. *In Full Bloom Ariane van Suchtelen Fred G. Meijer Erik A. de Jong Epco Runia Charlotte Rulkens Marya Albrecht*. Zwolle: Waanders Publishers.

Swammerdam, Jan. 1669. *Historia Insectorum Generalis, ofte agemene verhandeling der bloeelose Dierkens*. Utrecht: Merinardus van Dreunen.

Swammerdam, Jan. 1737. *Bybel der Natuure of Historie der insecten/Biblia naturae sive, Historia insectorum*. Herman Boerhaave (ed.), translated in Latin by Hieronymus David Gaubius. Leiden: Severinus and Vander Aa.

Swan, Claudia. 2005. *Art, Science, and Witchcraft in Early Modern Holland: Jacques de Gheyn II (1565–1629)*. Cambridge: Cambridge University Press.

Thomas, Jacqueline M.C. 2003. *"Insects" in Oral Literature and Traditions*. Peeters Publishers.

Vane-Wright, Richard I. 2021. "Introduction. William Jones & the Birth of the Icones." In *Iconotypes. A Compendium of Butterflies & Moths or: Jones's Icones Complete. An Enhanced Facsimile*, Oxford University Museum (ed.). Oakland: University of California Press, 11–35.

Wettengl, Kurt (ed.). 2013. *Maria Sibylla Merian: 1647–1717. Künstlerin und Naturforscherin*. Exhibition Catalogue. Historischen Museums in Frankfurt am Main. Ostfildern: Hatje Cantz.

Weyerman, Jacob Campo. 1729. *De levens-beschryvingen der nederlandsche konst-schilders en konst-schilderessen*. Vol. 4. Dordrecht: Blussé.

Wiersma, Lisa. 2020. "'Colouring' – Material Depiction in Flemish and Dutch Baroque Art Theory." *Art and Perception* 8 (3–4): 243–265. https://doi.org/10.1163/22134913-bja10005.

Wilkes, Benjamin. 1749. *One Hundred and Twenty Copper-Plates of English Moths and Butterflies Representing Their Changes into the Caterpillar, Chrysalis, and Fly States, and the Plants, Flowers, and Fruits, Whereon They Feed. Coloured with Great Wxactness from the Subjects Themselves. With a Natural History of the Moths and Butterflies, Describing the Method of Managing, Preserving, and Feeding Them*. London: B. White.

Witkowski, Nicolas. 2007. *Papillonnages: une histoire culturelle du papillon*. Paris: Seuil.

CHAPTER 2

Insect Color in Joris Hoefnagel's *Ignis*

Kimberly Schenck and Stacey Sell

1 Introduction

Joris Hoefnagel's (1542–1600) *Ignis: Animalia Rationalia et Insecta* has long been recognized as a landmark in the early study and artistic depiction of insects. The present paper examines the role of Hoefnagel's materials and techniques in his meticulous portrayal of insect color in this volume of miniatures. A careful examination of individual insects provides a glimpse into the creative techniques the Flemish artist employed to describe the complexities of color and texture.[1] A consideration of the findings in the context of sixteenth-century natural history illustration as well as Hoefnagel's cabinet miniatures and his other manuscripts offers insights into some of the original purposes and meanings of this complex work.

Ignis is part of a set of four volumes of animal miniatures executed in watercolor on parchment, now known as *The Four Elements*.[2] In many ways, *Ignis* diverges from its companion volumes. The other three books – *Terra*, containing images of creatures that walk the earth, *Aqua*, with its water animals, and *Aier*, filled with depictions of birds – draw heavily upon images from other sources, especially the natural history treatises of Conrad Gessner

1 Plates were studied in various lighting conditions (specular, raking, normal, and transmitted), in ultraviolet light, and using a microscope. Characterization of the pigments was made primarily using a microscope. National Gallery of Art conservation scientist Kathryn Morales identified some pigments and metals using x-ray fluorescence analysis (XRF). XRF analysis was conducted using Röntec (now Bruker) ArtTAX and Bruker M6 Jetstream µXRF spectrometers. Using an adapted Nikon D7100 camera, infrared photographs were taken to assist in determining the possible presence of underdrawings.

2 For *The Four Elements* and their extensive bibliography, see Hendrix 1984, Vignau-Wilberg 2017, cat. no. A6, and Bass 2019. Although most of the plates from the series are bound into the four volumes now at the National Gallery of Art, Washington (acc. no. 1987.20.5-8), loose sheets have been dispersed to other collections. See Vignau-Wilberg 2017, 110–130. Images of all intact sheets are available at https://www.nga.gov/collection/collection-search.html. Many of the animals in the four volumes are numbered in colored inks like those used in the inscriptions, suggesting that Hoefnagel made or planned a key describing or naming the numbered animals, but no such text has survived. An article on materiality in *Ignis* by Nancy K. Turner was forthcoming but unavailable at the time of writing (Turner 2023).

 | DOI:10.1163/9789004684553_003

(1516–1565).[3] Most of the insects in *Ignis*, created before the publication of the earliest treatises on insects by Ulisse Aldrovandi (1522–1605) and Thomas Moffet (1553–1604), appear instead to be closely studied from specimens and depicted at approximately life size. In all four volumes, textual references to sources such as Erasmus, classical literature, and the Bible elaborate upon the emblematic significance of the subjects. While *Ignis* maintains this tradition with regard to some insects, such as bees, most of the inscriptions in the volume reflect more generally upon the wonder and variety of God's creation.[4] At the same time, the extraordinary detail of the watercolors shifts the focus to a more direct engagement with the creatures themselves.

Despite the voluminous scholarship devoted to the series, some aspects of its creation remain in dispute, including the dates of its execution. Hoefnagel spent his early adulthood working in the family business as a merchant of luxury goods, and his first known artworks were topographical drawings made during his travels to France and Spain. He also spent time in London and his native Antwerp, where he probably began to make miniatures in the early 1570s. He left Antwerp forever after the Spanish Fury, traveling to Italy before settling down to a career as a court artist, first to the Dukes of Bavaria and ultimately to Emperor Rudolf II (1552–1612). Writing a few years after Hoefnagel's death, art historian Karel van Mander (1548–1606) stated that *The Four Elements* had been commissioned by Rudolf.[5] However, dates scattered throughout the volumes range from 1575 to 1582, indicating that Hoefnagel began work long before entering the Emperor's service in 1591.[6] The present paper, supported

3 For Hoefnagel's sources, see Hendrix 1984, 338–352; Dreyer 1986, 115–144; Rikken 2016, 32–44.

4 Typical of dozens of examples is the passage on the page facing Plate 6, "They shall speak of the magnificence of the glory of thy holiness: and shall tell thy wondrous works" ("Magnificentiam gloriæ sanctitatis tuæ loquentur, et mirabilia tua narrabunt." Psalms 144:5). Hoefnagel's reasons for grouping insects under the element of fire remains unresolved. For discussions of the subject, see Hendrix 1995, 373–390, Bass 2019, 236, Cambefort 2006, 212–213, and Carroll 2020, 1017. Hoefnagel's interest in the emblematic and religious implications of insects was far from unusual and persisted well into the seventeenth century; see, for instance, Jorink 2007, 147–175.

5 Van Mander and Miedema 1994, 1: 310–311, fol. 263r.

6 Although the latest date in the series has been read as 1589 (Bass 2019, 12), microscopic examination reveals a date of 1580: the last numeral is a zero beside a minute patch of lighter brown. Thea Vignau-Wilberg (2017, 101) maintains that the series was made for Rudolf II during the 1590s and posits that the dates commemorate events rather than the creation of the miniatures. Hendrix (1984, 39–41) reviews earlier scholarship and concludes that the series was begun during the 1570s. For a more recent argument supporting this view, see Bass 2019, 11–14 and Bass 2020, 127–129. In addition, Marrigje Rikken (2016, especially 40–47) has plausibly demonstrated *The Four Elements*' roots in the community around Abraham Ortelius (1527–1598) in Antwerp, which Hoefnagel left permanently in 1577.

by an ongoing technical examination of all four volumes, maintains that Hoefnagel worked on the series for years as a personal project, beginning in the 1570s.[7] Closely tied to discussions of dating are varying ideas about Hoefnagel's reasons for creating the series. Most recently, Marisa Anne Bass has explored the deeply personal significance of the series to the artist.[8] A multifaceted work, *The Four Elements* combines aspects of emblem book, commonplace book, private painted natural history collection, and reference book of motifs.[9] Hoefnagel's extraordinarily detailed and closely observed depictions of insect color relate directly to the last two of these functions.

2 *Ignis* in Context

Like the other volumes in the series, *Ignis* participates in many of the conventions of natural history imagery found in albums owned by naturalists and other collectors, suggesting that Hoefnagel viewed his miniatures in part within this context.[10] Butterflies, for instance, generally appear in dorsal views, side views, or both, positioned to provide clear images of the markings on their wings. Individuals rarely overlap and are shown close to or at life size on a blank background, providing unobstructed views of their contours.

7 *Ignis* features a range of artistic styles, suggesting its execution stretched over a period of years. Hoefnagel continued to add inscriptions for some time, as demonstrated by ink that transferred to facing interleaving pages while still wet. The interleaving sheets themselves feature watermarks dating from the late 1550s (castle with two towers, centered on a chain line, similar to Briquet 1985, nos. 15952 and 15953).

8 Bass 2019.

9 Vignau-Wilberg (2017, 100; 1994, 11) dismisses the idea that *Ignis* was used as a model book, stating that none of the insects were "drawn directly from life onto the vellum" and suggesting that Hoefnagel copied them into *Ignis* from a different model book, now lost. Elsewhere, she notes that Hoefnagel would not have worked directly on parchment because options for correcting mistakes were limited (Vignau-Wilberg 2007b, 225–226). However, corrections and changes are visible throughout the entire volume (see below, note 27). Conversely, Hendrix (1984, 64) states that Hoefnagel used the four volumes as a pattern book to supply elements for his later ornamental projects. Bass (2019, 192) concludes that Hoefnagel used the volumes in part as model books, "a source and stimulus for his inventive process." For the distinctions between the earlier traditions of model and pattern books, see Scheller 1995, 6.

10 For conventions in depictions of insects in natural history, see Neri 2011, 27–60. As Bass (2019, 192) has noted, Hoefnagel's goals diverged from those of naturalists such as Gessner and Aldrovandi. *The Four Elements* compares more closely to a "private museum of images" than to collections of drawings to be reproduced in natural history treatises for wider dissemination of knowledge. For early naturalists in Hoefnagel's social circles, see Vignau-Wilberg 2007b, 234–236, and Bass 2019, 235.

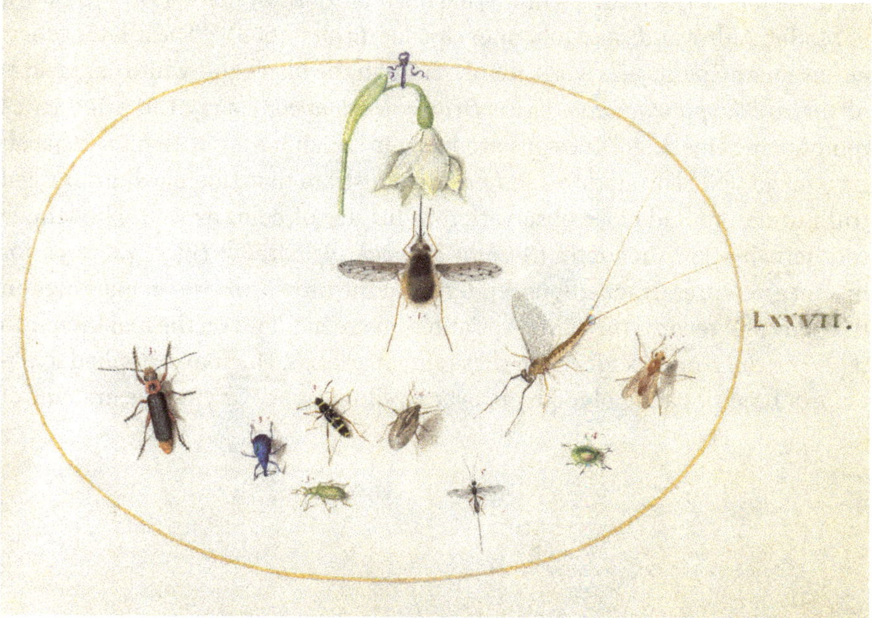

FIGURE 2.1 Joris Hoefnagel, *Dotted Bee Fly with a White Flower, a Mayfly, a Blue Weevil, and Other Insects*, in: *Ignis*, Plate 77, c.1575/1590, watercolor and gold, with oval border in gold, on parchment, c.143 × 184 mm (leaf), Washington, National Gallery of Art, Gift of Mrs. Lessing J. Rosenwald, 1987.20.5.78

Although he occasionally depicted insect behavior, as with the hummingbird hawkmoth (*Macroglossum stellaturum*) shown sucking nectar from a flower in Plate 77 (Fig. 2.1), the painter generally focused on their physical appearance, an emphasis shared with collectors of natural history images. In common practice with many of these collections, Hoefnagel organized his insects into loose categories such as butterflies and moths, beetles, grasshoppers, flies, and bees, though some pages break this pattern by mixing different creatures.[11] Hoefnagel's insects are unusual for their time in their wealth of accurate detail, which earned the admiration of seventeenth-century naturalists. Jan Swammerdam (1637–1680), for instance, expressed his appreciation of "the late

11 See Egmond 2017, 49–67. In keeping with the practices of some early naturalists, Hoefnagel generally separated caterpillars from butterflies. Only six are pictured in *Ignis* (Plates 20, 22, 34, 49, 50, and 51). In *The Four Elements*, they are mostly assigned to *Terra*, while water insects appear in *Aqua*.

accurate James Hoefnagel," comparing his renditions of insects with the work of Moffet, Aldrovandi, and Johannes Goedaert (1617–1668).[12] These later generations knew Hoefnagel's work mainly through the black-and-white engravings of the *Archetypa studiaque patris Georgii Hoefnagelii* (1592), the print series reproducing Hoefnagel's designs produced in collaboration with his son, Jacob (1575–1630).[13] The miniatures in *Ignis* demonstrate that Hoefnagel prioritized truth to nature and close observation in his use of color as well. The female southern hawker dragonfly (*Aeshna cyanea*) in Plate 53 (Figs. 2.2, 2.3), for instance, features a meticulous rendering of the brown and green markings on the body and records the subtle variations in green found on the real insect. To replicate the complex greens in these tiny markings, Hoefnagel applied malachite, a bright opaque blue-green, over a yellow-green, varying the amount of

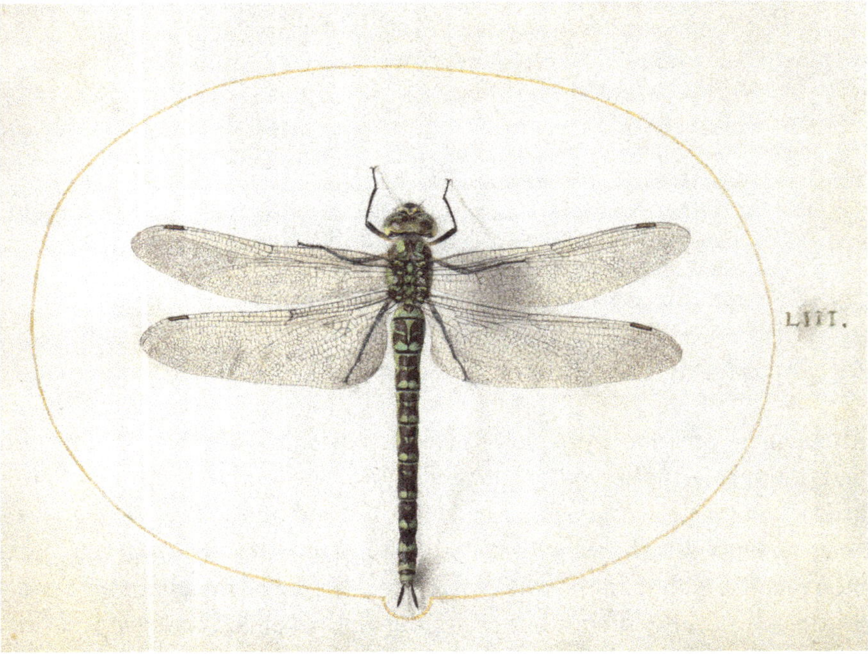

FIGURE 2.2 Joris Hoefnagel, *Southern Hawker Dragonfly*, in: *Ignis*, Plate 53, *c*.1575/1590, watercolor, with oval border in gold, on parchment, *c*.143 × 184 mm (leaf), Washington, National Gallery of Art, Gift of Mrs. Lessing J. Rosenwald, 1987.20.5.54

12 Swammerdam 1758, 24; see also 93, 95, 118, 122–124, 143.

13 Other engravers were likely involved as well. For the series, see Vignau-Wilberg 1994, 7–93.

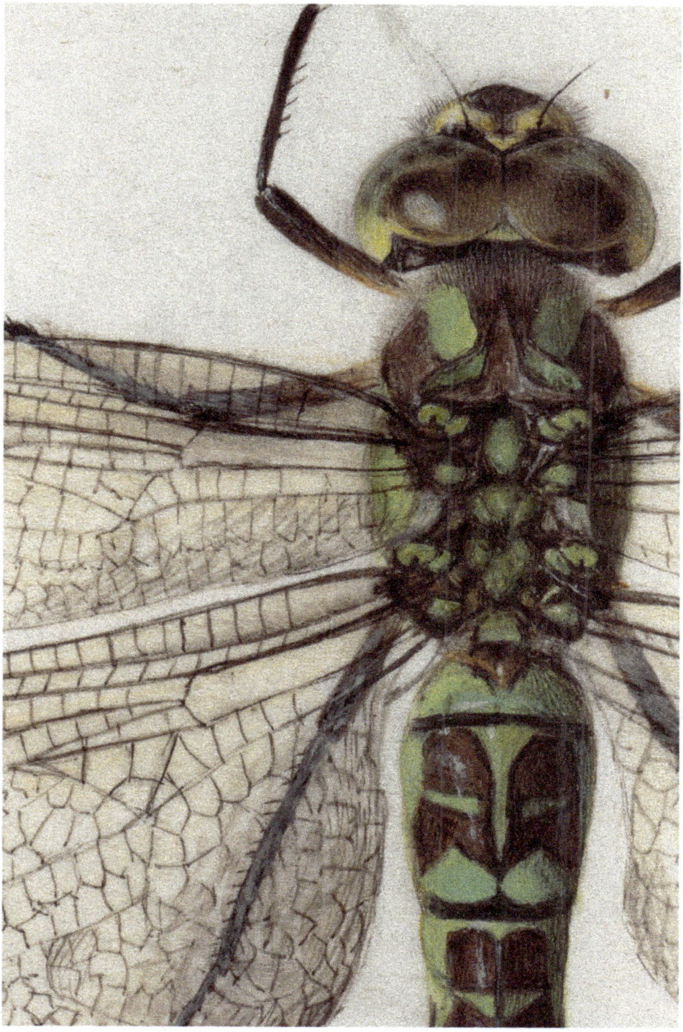

FIGURE 2.3 Joris Hoefnagel, body of a female southern hawker dragonfly (*Aeshna cyanea*), detail of Fig. 2.2, *c.*1575/1590, watercolor, with oval border in gold, on parchment, *c.*143 × 184 mm (leaf), Washington, National Gallery of Art, Gift of Mrs. Lessing J. Rosenwald, 1987.20.5.54

each from spot to spot and adding touches of yellow, a third green, and white. The miniatures in *Ignis* also demonstrate an extraordinary sensitivity to the effects of light and texture on insect color. In the same dragonfly, Hoefnagel used a pale, glossy brown wash to convey the color and texture of the translucent wings, but he also scumbled opaque white over the legs to mark the

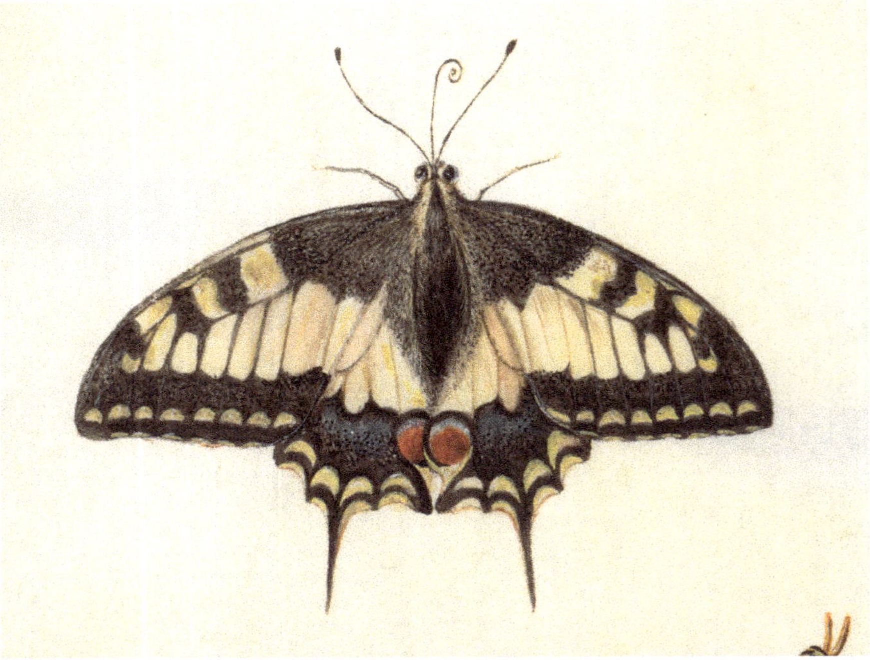

FIGURE 2.4 Circle of Giorgio Liberale, Swallowtail (*Papillo machaon*), detail of *49 Insekten*, before 1580, watercolor on parchment, 640 × 860 mm (leaf), Vienna, Österreichische Nationalbibliothek, Cod. Ser. N. 2669, fol. 95r

change in their color as seen through the wings. A complex interweaving of minute brushstrokes of browns, yellows, green, blue, and white conveys the luminosity and roundness of the insect's eyes. Hoefnagel was not alone among his contemporaries in his detailed approach to natural history illustration.[14] The meticulous nature studies of Giorgio Liberale (1527–before 1580) and his circle, executed in watercolor on parchment like the miniatures in *Ignis*, present an especially relevant comparison: In fact, they were once attributed to Hoefnagel himself.[15] Even within this tradition, however, Hoefnagel's

14 For "high-definition naturalism," with earlier bibliography, see Egmond 2017, especially 88–163.

15 This series was commissioned by Ferdinand II of Tyrol (1529–1595) and kept at his *Kunstkammer* at Ambras. It is now in the collection of the Österreichische Nationalbibliothek in Vienna, Cod. Ser. N. 2669 and is available online at https://www.onb.ac.at/. See Weiler 2011, especially 18–51 and 130–144. Ferdinand II commissioned a missal from Hoefnagel during the artist's stay in Munich (Vignau-Wilberg 2017, cat. no. A2), but there

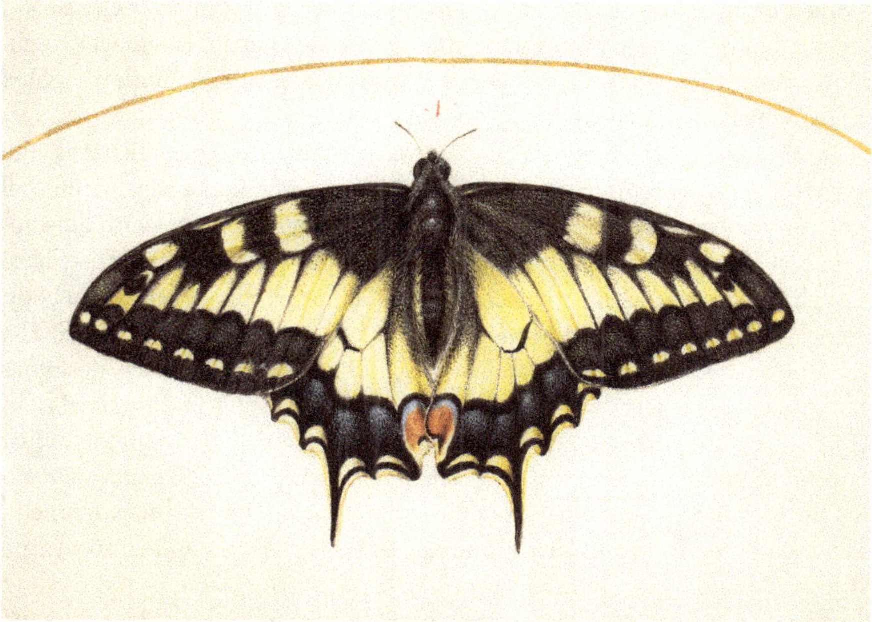

FIGURE 2.5 Joris Hoefnagel, Swallowtail Butterfly (*Papillo machaon*), detail of *Swallowtail Butterfly*, in: *Ignis*, Plate 7, *c.*1575/1590, watercolor, with oval border in gold, on parchment, *c.*143 × 184 mm, Washington, National Gallery of Art, Gift of Mrs. Lessing J. Rosenwald, 1987.20.5.8
© NATIONAL GALLERY OF ART WASHINGTON. PHOTO CREDITS: RICARDO BLANC. HTTPS://WWW.NGA.GOV/COLLECTION/ART-OBJECT-PAGE.69746.HTML

attention to details of insect color and his expertise at handling watercolor is exceptional. At first glance, Liberale's swallowtail (*Papilio machaon*) (Fig. 2.4) appears almost as closely observed as Hoefnagel's version in Plate 7 (Fig. 2.5).[16] Liberale, however, painted the coloration at the base of the forewings by dotting black onto a layer of yellow, while Hoefnagel correctly observed that the real butterfly features yellow scales scattered on a black layer. Similarly, Hoefnagel accurately depicted the bright blue areas of the hindwings by applying an opaque blue paint to a black background, gradually increasing the space between the minute strokes to create a smooth transition to black. Liberale

is no firm evidence that Hoefnagel visited Ambras or had access to the watercolors. Focusing on the fish by both artists, Hendrix (1984, 142–146) concludes that they represent two different traditions.

16 In contrast, the swallowtail in Plate 13 is less closely observed and handled differently, a discrepancy to be addressed in a future publication on stylistic variations within *Ignis*.

worked in the opposite manner by dotting dense black paint over circles of a less opaque blue, which caused the color to recede visually. Hoefnagel's technique more successfully describes the appearance of the real butterfly's blue scales, which seem to float on a field of black.

In addition to their minute detail and carefully described textures, the insects in *Ignis* are remarkable for their lifelike presence on the page. Hoefnagel was certainly aware of the artistic tradition of depicting insects with illusionistic cast shadows, but he took this convention to new heights.[17] His elaborate shadows vary in strength according to their proximity to limbs and body. Some provide important information about the insects that cast them. The shadow of the crane fly (*Tipulidae* family) in Plate 58, for instance, describes the curve of the insect's body, which is not otherwise visible in this dorsal view. Hoefnagel sometimes added minute touches of opaque white watercolor just beyond the contours of the insect, a subtle and nearly invisible means of making the creature appear to rise from the creamy background of the parchment. The blue bottle fly (*Calliphora vomitoria*) in Plate 78 (Fig. 2.6) features white strokes along the right wing between the legs, covering some of the shadow. As a result, the front of the fly appears to rise higher from the surface than the rear end, mimicking the stance of a real fly. In addition to these illusionistic touches, Hoefnagel frequently exploited the interaction between the colors of his painted insects and the light in the viewer's space. Wings, shiny eyes, and beetle shells throughout the volume are coated with a glossy gum, which catches the light as the book is handled and pages turned.[18] For translucent wings, he sometimes used glossy gum tinted pale brown or gray, rather than simply applying diluted watercolor.[19] Hoefnagel supported these techniques with the animated poses and arrangements of many of the insects. As others have observed, the slightly asymmetrical placements of some of the most

17 For Hoefnagel's complicated relationship to the tradition of illusionism in illuminated manuscripts, see DaCosta Kaufmann and Kaufmann 1991, 43–64. Insects with illusionistic shadows are also common in albums of natural history images: among others, see those in the Album Gessner-Platter (such as Egmond 2018, 316, plate III C23, fol. 127), the *Libri picturati A16–30* (de Koning et al. 2006, 148, A21.003), and the Aldrovandi's *Tavole acquerellate* (such as Tavole 007 Animali, fol. 74).

18 In many cases, a single page demonstrates a variety of approaches: in Plate 19, for instance, Hoefnagel added glossy touches to the eyes of the black-veined white butterfly, the shell of insect no. 6, the wings of insect no. 5, and the wings and body of no. 2. For the green beetle (no. 3), the artist mixed green paint with a substantial amount of gum to create the shiny wings, body, and antennae. In Plate 21, gum is present in the wings of nos. 2, 5, 8, 9 and the body of beetle no. 6.

19 For example, the bee (no. 4) in Plate 20 and the flies in Plate 47 have glossy wings made with lightly pigmented gum.

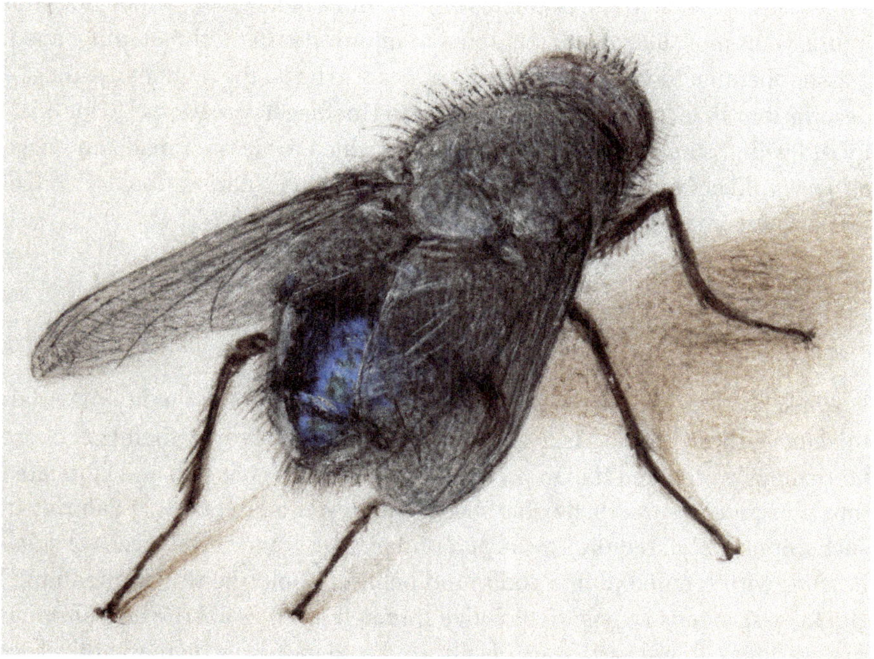

FIGURE 2.6 Joris Hoefnagel, a blue bottle fly (*Calliphora vomitoria*), detail of *Ten Insects, Including a Blue Fly*, in: *Ignis*, Plate 78, *c.*1575/1590, watercolor, with oval border in gold, on parchment, *c.*143 × 184 mm (leaf), Washington, National Gallery of Art, Gift of Mrs. Lessing J. Rosenwald, 1987.20.5.79

elaborately painted insects in the volume impart a feeling of movement.[20] The illusionism of Hoefnagel's insects provides a display of his artistic virtuosity and secures his place in a long line of artists stretching back to Zeuxis. In fact, Hoefnagel explicitly compared his own talents to those of the Greek painter, whose convincingly rendered grapes fooled the birds.[21] This intensely lively presence may also have lent Hoefnagel's insects value and meaning in their role as part of a painted natural history collection. Gessner, for instance, placed special merit on images that provoked the same reaction from the viewer as the

20 See, for instance, Bass 2019, 231 and 235–236.

21 The inscription in Hoefnagel's hand appears in his decorations for Georg Bocskay's (died 1575) *Schriftmusterbuch* (Model Book of Calligraphy), Vienna, Kunsthistorisches Museum, Kunstkammer, inv. no. KK 975 (see Vignau-Wilberg 2017, cat. no. A3 and Hendrix and Vignau-Wilberg 1992, 7). For an examination of the tradition of the illusionistic painted fly in Renaissance art, see Chastel 1984, especially 20.

subject itself. He cited the examples of renditions of a spider, "which does not injure, yet is most like one that is about to injure and that is threatening" and a grasshopper that "does not jump, but it appears to be on the point of jumping" – descriptions that could be readily applied to Hoefnagel's creations.[22] This quality of liveliness and its effect on the viewer, which to Gessner made an image *ad vivum*, likely had special implications for painted natural history collections, where images substituted for animals or specimens.[23]

3 Hoefnagel's Materials and Painting Technique

In *Ignis*, Hoefnagel relied mainly on conventional but high-quality materials to achieve these effects. As a support for his detailed work, Hoefnagel chose meticulously prepared Italian parchment, made from goat skin and fabricated into translucent, smooth, flexible sheets with few imperfections.[24] Painting in such minute detail required evenly ground pigments and various sizes of small brushes with tapered points and round bellies to hold the watery medium.[25] Hoefnagel's extensive palette included minerals, such as azurite and ultramarine; various colored earths; artificially produced inorganic pigments, like lead white; and lakes made by precipitating dyes from plants and scale insects onto

22 "Non laedit araneus, simillimus tamen laesuro et minanti[:] non salit locusta, sed iam iam salitura videtur." Kusukawa (2014, 356) cites and translates this passage from a letter from Gessner to Georgius Fabricius (1516–1571).

23 See also the account of Jacomo Antonio Buoni (1527–1587), a visitor to Aldrovandi's collection in 1571, who marveled at natural objects portrayed so vividly that the eye could barely discern between model and depiction, in Swan 2005, 72.

24 Abigail Quandt, conservator of rare books and manuscripts at the Walters Art Museum, Baltimore, identified the parchment as goat. For a guide to identifying skins in parchment, see Moog 2021, 419–431. Archival records from Hoefnagel's period in Munich indicate that he ordered parchment from Rome (Vignau-Wilberg 1985, 115–116). Although the choice of parchment as a support recalls its use in illuminated manuscripts, it was also used for nature studies, ranging from the works of Albrecht Dürer (1471–1528) and Hans Hoffmann (*c.*1545/50–1591/2) (see, for instance, Koreny 1988, nos. 22 and 25), Jacopo Ligozzi (see Conigliello 1991, 22–29), Liberale (see Weiler 2011, 130–135), and the nature albums of Hans Bol (1534–1593), which served as an important model for *The Four Elements* (see Hendrix 1984, 40–49 and Rikken 2016, 37–40). For a more modest example, see the album of simple bird miniatures on parchment in Haag 2015, cat. no. 4.15.

25 While the idea of painters of miniatures using single-hair brushes is widespread, they actually require brushes composed of several hairs that form a point. For a discussion of capillary action in watercolor painting, see Mysels 1981, 22–27.

semi-transparent white bases.[26] He used these pigments alone or sometimes in mixtures, for example combining yellow lake and indigo to produce a transparent green. Hoefnagel's courtly connections and his proximity to artistic circles and major trade centers enabled him to obtain pigments of fine quality and color. Throughout *Ignis*, for instance, he used a finely ground azurite of exceptionally strong blue.

Hoefnagel's development of the insects on the parchment varies. Suggestions of underdrawings in dry drawing materials, such as graphite or metalpoint (a thin metal rod inserted into a stylus), are found below the watercolors in some insects. For example, traces of lightly drawn marks outline the tips of the wings and legs of a grasshopper of the genus *Acrida* in Plate 45. In other cases, Hoefnagel first sketched in the shape and parts of insects with pale strokes of dilute watercolor that he could easily blot from the parchment by selectively applying tiny amounts of water. Evidence of these initial watercolor outlines are found throughout *Ignis*, either as light strokes extending beyond an insect's perimeter or pigment residues left from their partial removal with water. Upon close inspection, numerous plates in *Ignis* bear traces of corrections from lifting unwanted brushstrokes using water or by covering mistakes with touches of opaque white watercolor.[27] In a few cases, such as the grasshoppers in Plates 50–52, insects are relatively loosely sketched with rapidly painted brushstrokes, contrasting with the more detailed approach of the rest of the volume. This range of approaches and the numerous corrections suggest

26 Pigments include carbon-based blacks, various earths, organic brown, yellow lake, lead-tin yellow, red lakes, purple lake, indigo, azurite, malachite, verdigris, and inorganic reds. Discolored white paint indicates the presence of lead white; a less opaque, calcium-based white is used in other areas. For a list of pigments and materials available in Germany at the time, see Burmester et al. 2005, 44–48. For trade of pigments in Antwerp, see Vermeylen 2010, 356–365. Although some of Hoefnagel's materials were expensive, he worked with diluted colors on such a minute scale that he required extremely small amounts of any given material.

27 Only a few corrections are noted here. In Plate 19, a touch of white watercolor covers adjustments to the end of the abdomen of insect number 4; in Plate 37, several spiders are altered; in Plate 46, white covers watercolor changes to the head and underside of the green grasshopper; in Plate 51, a mistake is corrected near the head of the pink grasshopper and white is noted at the lower edge of the blue-winged grasshopper; in Plate 52, the position of the bottom grasshopper has been changed; and in Plate 65, the end of the cricket shows evidence of being reduced in size. Even the seemingly perfect dragonfly in Plate 53 bears traces of correction with white. Comparison of an infrared photograph of the bottom grasshopper in Plate 45 with an image taken in normal light reveals that Hoefnagel changed the size and shape of the eye and the location of the indentation in the upper body from his original underdrawing to the finished insect.

that Hoefnagel saw *Ignis* as a personal record of his observations, not as a perfected final product for a patron.[28]

Hoefnagel possessed an unusual command of the qualities of his pigments, selecting them not only for their hues but also for their degrees of transparency or opacity, tinting strengths, and particle characteristics. He used at least two yellows and two blues with differing opacities, for instance, for the swallowtail in Plate 7 (Fig. 2.5). He selected an opaque pigment, likely lead-tin yellow, for laying in the yellow of the wings and for the scattered scales on the black background of the upper wings, but switched to a transparent yellow lake of a deeper hue to glaze areas of the wings, evoking the glowing colors seen in nature.[29] The blue spots on the black background of the lower wings were painted with a mineral-based blue pigment, probably azurite, mixed with white, while the head was tinted with indigo, a more transparent and duller blue. Hoefnagel painted the wings of a blue bottle fly in Plate 78 (Fig. 2.6) with numerous strokes of a dilute, gritty black paint but drew in the veins with smooth lines of a different black pigment composed of very fine particles. He also may have added depth to the fly's hairs on its body by combining strokes of a cool black with a more transparent, warm-toned pigment, either dark brown or weak black.

Hoefnagel used these materials to address complex color effects. While contemporary naturalists struggled to describe iridescent and metallic insects, Hoefnagel experimented with a variety of techniques to depict these effects. Sometimes he relied on paint alone. As he did throughout the volume, Hoefnagel considered his pigments' other properties as well as hue when selecting his paints, highlighting an oil beetle (*Meloe* genus) in Plate 43 (Fig. 2.7, top left insect) with a glossy, transparent purple to impart shine and to intensify the deep blue-black of the insect's body. Although Moffet described the oil beetle as "betwixt black and dark blew shining," images in contemporary natural history collections generally depict it as solid black.[30] In the more

28 Other works by Hoefnagel, either examined in person or in high resolution images, do not exhibit the same broad range of approaches to developing the insects and the number of corrections.

29 Light is transmitted through transparent colorants and is reflected off the parchment back through the transparent layer, making the color appear more luminous than when light is reflected off opaque paints. Hoefnagel innately understood the impact of light on the human perception of color and balanced his use of opaque and transparent pigments accordingly.

30 Moffet 1658, 1016. See the similar beetle depicted by Liberale or a follower without any attempt to evoke iridescence (Vienna, Österreichische Nationalbibliothek, Cod. Ser. N. 2669, fol. 94v). Another oil beetle in Aldrovandi's *Tavole acquerellate*, labeled "scarabus

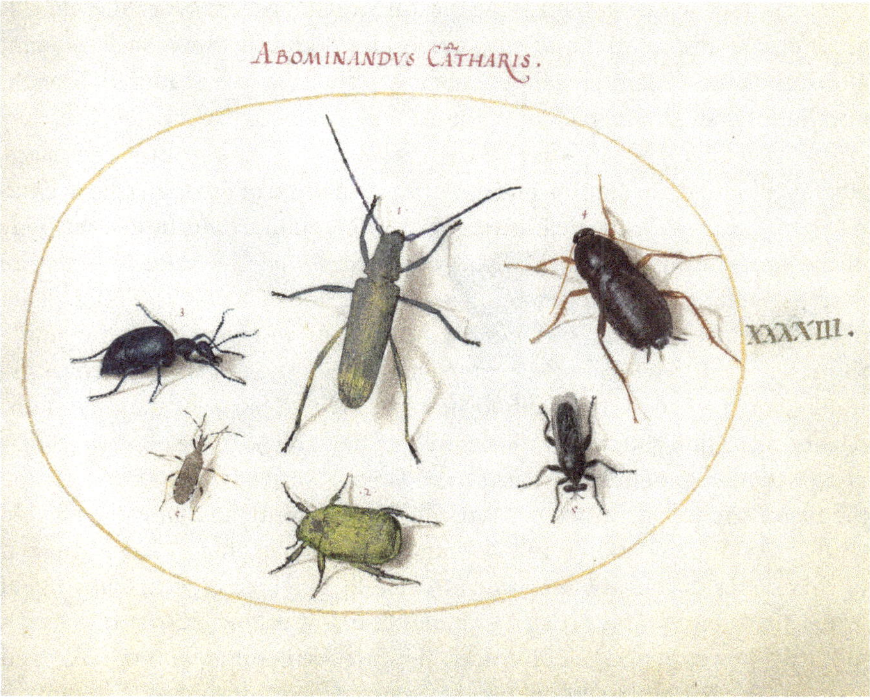

FIGURE 2.7 Joris Hoefnagel, *Spanish Fly, Oil Beetle, Tansy Beetle(?), Cockroach, Leaf-Footed Bug, and Other Insects*, in: *Ignis*, Plate 43, c.1575/1590, watercolor and gold leaf, with oval border in gold, on parchment, c.143 × 184 mm (leaf), Washington, National Gallery of Art, Gift of Mrs. Lessing J. Rosenwald, 1987.20.5.44

colorful blue bottle fly mentioned above (Fig. 2.6), Hoefnagel supplemented the range of black tones with the suggestion of iridescence in the abdomen by dotting vivid green malachite and light blue on top of a thickly applied, vibrant ultramarine.

4 Silver and Gold

Hoefnagel also used metallic materials to capture both color and texture. In keeping with his true-to-life use of color in *Ignis*, he used silver and gold descriptively rather than ornamentally; that is, he used metals to convey the

terrestris taurus cantharus," is painted solid black (Tavole 007 Animali, fol. 80; image available at http://aldrovandi.dfc.unibo.it/pinakesweb/main.asp).

sheen or glow he saw in nature. He most frequently used shell gold (gold leaf ground to a fine powder and suspended in a liquid medium, such as plant gums, and used as paint).[31] Minute touches of shell gold are scattered throughout the volume. Although many of these are not obvious to the naked eye, they create a metallic glint in some of the insects as the viewer turns the pages. Hoefnagel varied the dilution and application method of the shell gold according to his purpose. When delineating the shining golden hairs beside the body of the small tortoiseshell butterfly (*Aglais urticae*) in Plate 8, for instance, he used the point of a small brush to create clear lines that replicate the gleam of the hairs in sunlight.[32] Elsewhere, attempting to evoke the more subtle shimmer of butterfly scales, he painted a dilute solution of shell gold over the watercolors he had already laid down or mixed gold particles into his colors, resulting in a fine glitter.[33] In the central beetle in Plate 33, he painted strokes of a glossy, transparent reddish brown on top of a wash of shell gold to describe the insect's rounded striped exoskeleton. He frequently combined shell gold with orpiment, a sparkling yellow-orange mineral. Hoefnagel likely chose it to counteract the slightly grayish tint of shell gold, which occurs when gold leaf is ground into fine particles. Orpiment's bright yellow lent warmth to the reflective power of the gold, resulting in a precise combination of color and texture.[34] He used this combination in varying proportions. A small amount of

---

31 Hoefnagel, in his *Allegory for Abraham Ortelius* (Antwerp, Museum Plantin-Moretus, Stedelijk Prentenkabinet, acc. no. PK.OT.00535; Vignau-Wilberg 2017, cat. no. C16), pictures shells of paint including gold, along with other tools used by a cartographer or artist. For a discussion of the historic use of shells, see Howard 2006. For shell gold in Hoefnagel's other work and a review of recipes, see Seehafer 2021, 281–321. For a broader description of the history of the use of gold and other metals in manuscripts, see Morgan 2016, 193–199 and Turner 2017–2018, 81–96.

32 For other examples of Hoefnagel applying individual strokes of shell gold to butterflies, see the mourning cloaks (*Nymphalis antiopa*) in Plate 6, the small tortoiseshell (*Aglais urticae*) and red admiral (*Vanessa atalanta*) in Plate 11, and the red admiral in Plate 13. For verification of gold in Plate 8, see analysis report, Morales 2022a.

33 For examples of overall application of shell gold, see the top butterflies in Plate 9 and the dragonfly in Plate 13. Hoefnagel may have used ground mother-of-pearl to convey shimmer or iridescence in butterfly wings in his cabinet miniatures (Schmückle 2017, 267). This technique has not been noted in *Ignis*.

34 Although the miniaturist Nicholas Hilliard (1547–1619) recommended against the use of orpiment because of its toxicity and unpleasant smell, he used it in a similar fashion, placing a layer of orpiment under a layer of shell gold (see, for instance, Fiorillo et al. 2021, 1177–1178). In addition to being poisonous due to its arsenic content, orpiment can be incompatible with pigments containing copper and lead (Fitzhugh 1997, 52). The authors painted samples of shell gold combined with orpiment onto parchment and paper to better understand why Hoefnagel may have chosen the mixture. When orpiment was added

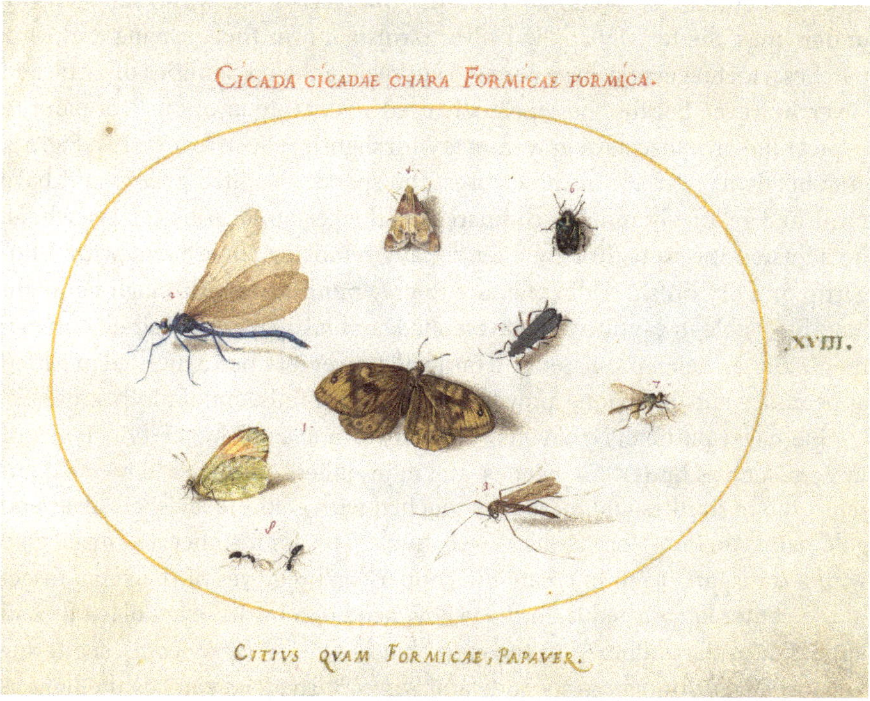

FIGURE 2.8 Joris Hoefnagel, *Two Butterflies and a Moth with a Dragonfly, Two Ants, and Four Other Insects*, in: *Ignis*, Plate 18, *c.*1575/1590, watercolor, gold and silver, with oval border in gold, on parchment, *c.*143 × 184 mm (leaf), Washington, National Gallery of Art, Gift of Mrs. Lessing J. Rosenwald, 1987.20.5.19

orpiment was added to shell gold for the hairs in the small tortoiseshell butterfly mentioned above (Plate 8). In broader areas such as butterfly wings, it lends a subtle glow to the entire insect (Plate 18, Fig. 2.8, central butterfly).[35]

Hoefnagel also used shell silver. As with shell gold, his handling of silver varied according to the task. He sometimes applied it in a thick, flat layer, as in the spots of two fritillary butterflies (tribe *Argynnini*) (Plates 10 and 12). Although

to gold watercolor (23.75 karat), the samples appeared warmer in tone than when the gold paint was used alone.

35 For verification of gold and orpiment in Plates 8 and Plate 18, see analysis report, Morales 2022a and 2022c. Hoefnagel mixed gold with orpiment in the painted lady butterfly in Plate 18, but he also used orpiment without gold (for example, in spider no. 1 in Plate 38).

this was certainly an attempt to capture the pearlescent white spots on the underwings, the heavily applied silver tarnished over time, leaving dark gray patches. Hoefnagel may have come to appreciate the possibilities of tarnished silver, however, because he repeatedly used a very thin layer of silver paint to depict blue-gray insects or gray insects with slightly reflective textures (Fig. 2.8, gray beetle at right).[36] In these cases, the sparsely applied gray would have tarnished relatively quickly compared to thicker applications.[37] Elsewhere, he adjusted the wet silver by intentionally removing some paint with a fingertip or a dry brush, adding texture to gray moth wings. Although he might have been able to capture the precise shade of bluish gray he desired with pigments alone, the manipulated and tarnished silver retained a mottled or fritted appearance and an opaque, powdery shimmer reminiscent of moth scales.

Hoefnagel particularly favored shell silver and gold for certain types of insects, such as butterflies, beetles, and damselflies: Approximately one quarter of the butterflies and moths are touched with gold. He lavished silver and gold paints on humbler insects as well. In Plate 60, for instance, he mixed gold with a transparent red to create the coppery colored eyes of the small insect in the center left, dotted it on the yellow head of a brown-eyed black fly and applied it to the abdomen of the green bottle fly (*Lucilia sericata*), scattering areas of silver throughout its body and wings.[38] The gold catches the light as the angle changes, while the darkened silver particles shape the glossy green upper body.

In addition to metallic paints, Hoefnagel experimented with highly reflective glazed gold leaf. In all three cases in *Ignis*, he used it to depict iridescent green beetle exoskeletons. Gold leaf appears in the jewel beetle (family Buprestidae) in Plate 42 and in the Spanish fly (*Lytta vesicatoria*) and tansy beetle (*Chrysolina graminis*) in Plate 43. The application of gold leaf was a more involved process than painting with shell gold: The artist coated the area with glue or glair (beaten egg white) and then pressed the leaf into place while the surface was still tacky. He then burnished the leaf, a process that increased its luster and brilliance, but slightly deformed the parchment. For all three

36 See, for instance, the top and lower right moths in Plate 32, the bluish gray moth in Plate 49, and the top right insect in Plate 79. For verification of silver in Plate 18, see analysis report, Morales 2022c.

37 The authors painted shell silver sparingly on top of black watercolors and placed part of the samples next to a wool felt, which contains sulfur, to induce tarnishing of the metal. When first applied, the silver appeared glistening light gray. As it tarnished, the silver oxidized to a darker gray with less metallic gleam. However, it retained a glittery, textured appearance when the samples were moved in the light.

38 For verification of gold and silver, see analysis report Morales 2022f.

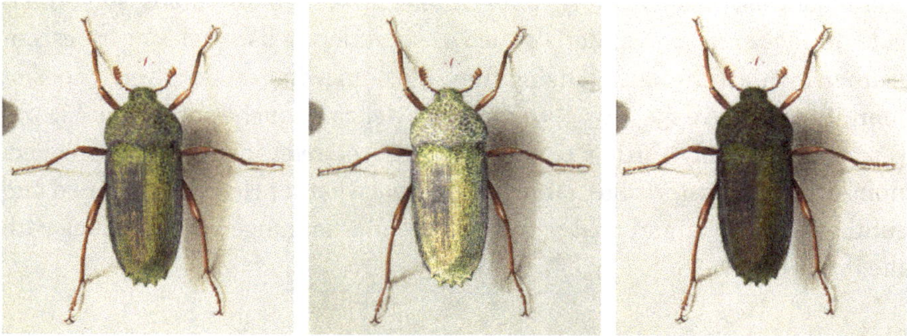

FIGURE 2.9 Joris Hoefnagel, A jewel beetle (family Buprestidae), detail of *Jewel Beetle with a Plant Gall(?) and a Flower*, in: *Ignis*, Plate 42, *c.*1575/1590, watercolor over gold leaf, with oval border in gold, on parchment, *c.*143 × 184 mm (leaf), Washington, National Gallery of Art, Gift of Mrs. Lessing J. Rosenwald, 1987.20.5.43 PHOTO CREDITS: KURT HEUMILLER. HTTPS://WWW.NGA.GOV/COLLECTION/ART-OBJECT-PAGE.69706.HTML

beetles, Hoefnagel selected a leaf of gold combined with silver, a combination that lent a cool, faintly green tint to the gold.[39] He then added strokes of watercolor, attempting to reproduce both the colors and the textures of the specimens before him. The jewel beetle in Plate 42 features glazes of transparent blue, green, and purple, communicating its more varied array of colors, while the color range is more limited in the emerald-green Spanish fly in Plate 43 (Fig. 2.7, center top insect). Hoefnagel dotted watercolor onto the jewel beetle's thorax, replicating its pitted appearance, but applied long strokes of watercolors to the Spanish fly's abdomen to imitate the vertical striations of its shell. Perhaps more so than any other images in the volume, these beetles rely upon the movement of light over the page for their intended effect, as the gold reflects the shifting light and different colors predominate according to the angle of light (see, for example, Fig. 2.9).

Hoefnagel was not alone in choosing metallic materials to convey effects he saw in nature. The use of silver to replicate the shimmering white spots of fritillaries, for example, was common in contemporary natural history images. Moffet described the complex coloration of the spots in the real insect as a

39 See analysis report, Morales 2022d and 2022e. *Zwischgold*, a layered foil in which gold leaf and silver leaf were hammered together, was lighter in color and a quarter of the cost of pure gold leaf (Burmester 2005, 46).

combination of pearls and silver.[40] Other artists used silver, and sometimes gold, to render natural materials such as shimmering fish scales or iridescent feathers, with techniques ranging from the somewhat crude silver stripes of John White (active 1585–1593) to the more delicate touches of Jacopo Ligozzi (1547–1627).[41] Only further technical studies of relevant artists and collections of natural history illustrations will reveal whether Hoefnagel's varied and subtle use of silver and gold represents an unusual range of techniques with these materials.

5 Insect Inclusions

Perhaps the most unusual materials in *Ignis* are the parts of real insects. As many have noted, Hoefnagel used real dragonfly wings, adhering them to the painted bodies of four dragonflies in three plates.[42] Although the process of attaching the wings to the parchment was relatively straightforward, it required planning as he painted the bodies and the cast shadows before adding the wings. Traces of blind stylus marks around the wings in three of the dragonflies indicate that he established their positions in advance, allowing him to paint the insect and his shadow with the precise location of the wings already decided.[43] The fragile wings quickly began to flake from the page, as revealed by Hoefnagel's watercolor inpainting – as seen, for example, along the tops of the wings in Plate 47 and lower portions of the wings in both insects in Plate 54 (Fig. 2.10). The wings continued to crumble and fall from the page long after Hoefnagel's time, leaving gaps where indentations of the wings' veins into the parchment can be seen in raking light.

40 *Ignis*, Plates 10 and 12. See Moffet 1658, 969. For silver paint on fritillaries, see anonymous depictions in Aldrovandi's *Tavole acquerellate*, Tavole 007 Animali, pl. 83; Rudolf II's collection (Österreichische Nationalbibliothek, Vienna, Cod Min 42, fol. 135); and the album of animal drawings from the Giardino dei Semplici, Pisa, now at the Biblioteca Universitaria, Pisa, carta 154r (IT-PI112MS.HORTUS PISANUS.514).

41 For White, see Bescoby et al. 2007, 20–21. For Ligozzi's use of silver and gold, see Alberghina et al. 2014, 314–315. A gold chrysalis appears in the Felix Platter albums. See Egmond 2017, 214.

42 Plates 47, 54, and 63. Using magnification and lighting from different angles, the authors readily determined that the dragonfly wings were glued on top of the painted bodies and shadows.

43 Plate 47, the dragonfly on the lower left of Plate 54, and Plate 63.

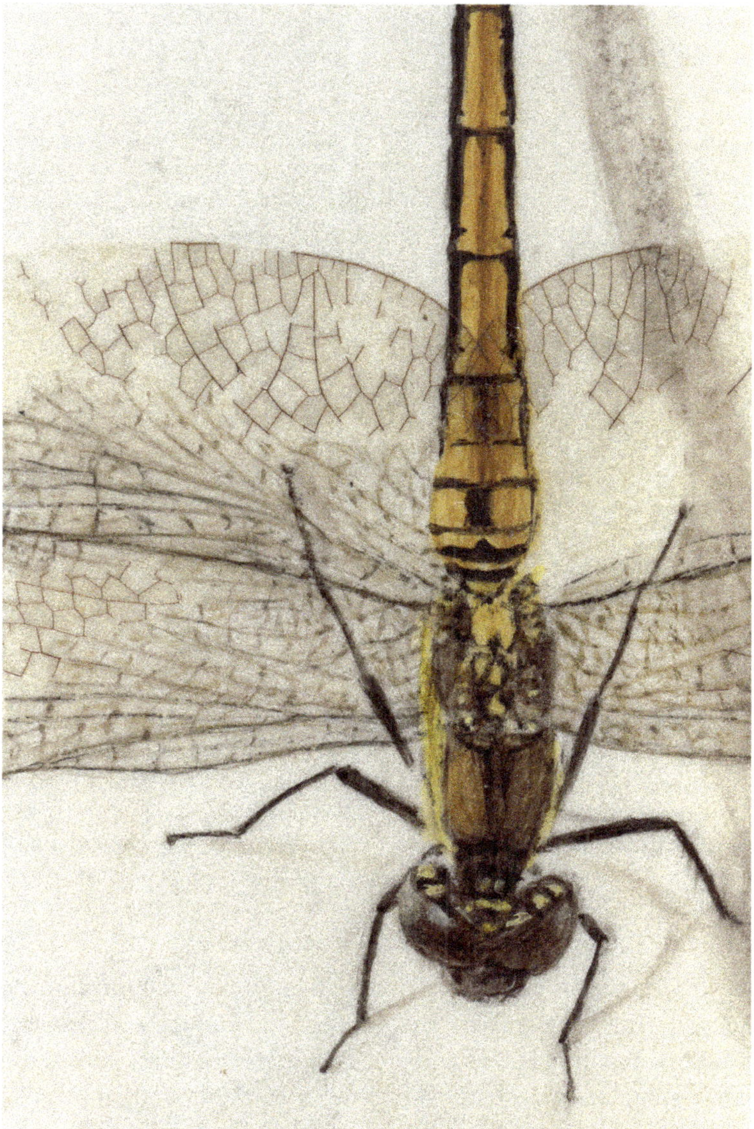

FIGURE 2.10 Joris Hoefnagel, A dragonfly with flaked wings, detail of *Hairy Dragonfly and Two Darters*, in: *Ignis*, Plate 54, *c.*1575/1590, watercolor and dragonfly wings, with oval border in gold, on parchment, *c.*143 × 184 mm (leaf), Washington, National Gallery of Art, Gift of Mrs. Lessing J. Rosenwald, 1987.20.5.55 © NATIONAL GALLERY OF ART WASHINGTON. PHOTO CREDITS: KIMBERLY SCHENCK. HTTPS://WWW.NGA.GOV/COLLECTION/ART-OBJECT-PAGE.69719.HTML

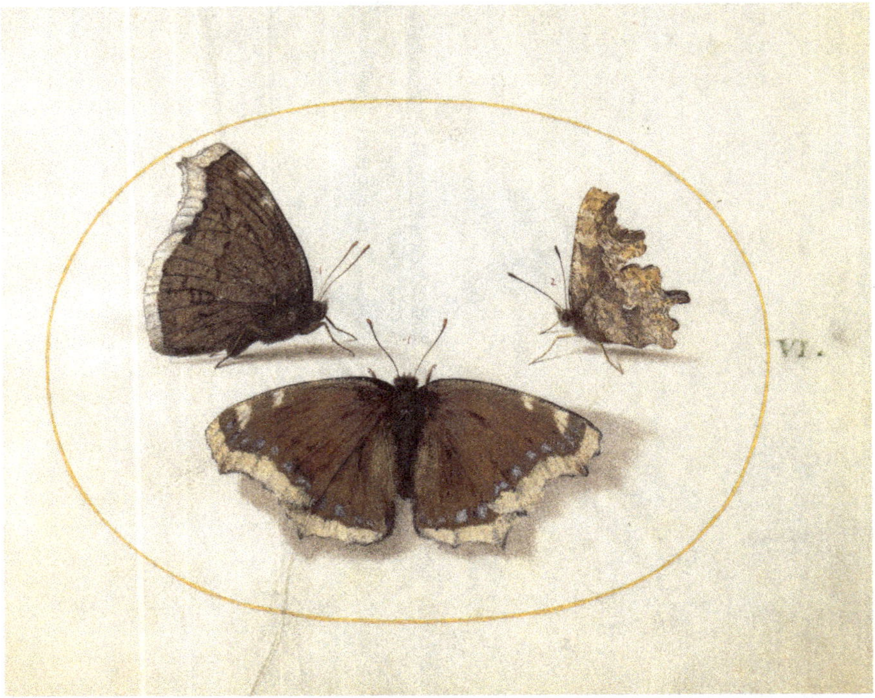

FIGURE 2.11 Joris Hoefnagel, *Two Views of a Mourning Cloak Butterfly* (*Camberwell Beauty*) *with a Comma Butterfly*, in: *Ignis*, Plate 6, *c.*1575/1590, watercolor and gold, with oval border in gold, on parchment, *c.*143 × 184 mm (leaf), Washington, National Gallery of Art, Gift of Mrs. Lessing J. Rosenwald, 1987.20.5.7

More recently, lepidochromy has been identified in over one third of the depictions of butterflies and moths on the pages of *Ignis*.[44] The dates of Hoefnagel's earliest experiments with the technique are unknown and all the pages featuring lepidochromy are undated. He was certainly using it by 1589, when he incorporated butterfly transfers into a dated cabinet miniature.[45] Plate 6 (Fig. 2.11) may represent an early effort: Hoefnagel applied excessive

44 For a comprehensive discussion of the technique and its process, see V.E. Mandrij's chapter in this volume. Lepidochromy appears in *Ignis* in Plates 6, 8–10, 15, 16, 23, 25, and 28–32. Other examples may exist within the pages that were separated from the volume. See, for example, Vignau-Wilberg 2017, cat. nos. A6c and A6d (Staatliche Museen zu Berlin, Kupferstichkabinett, acc. nos. KdZ 4810 and 4812). For a detailed discussion of lepidochromy in *Ignis*, see Schenck and Sell 2022, 161–170.

45 See below, note 57.

pressure and moisture when making the transfer, wrinkling and distorting the parchment. Otherwise, his technique varies little throughout the volume. He consistently used a single transfer of the lepidopterous scales directly onto the parchment. He occasionally established the position of the wings by laying the dead butterfly on the surface and outlining the wings with fine, pale gray lines made with goldpoint (a thin gold rod inserted into a stylus).[46] He then separated the wings from the body and made the lepidochrome by pressing the wings onto the surface and removing the membranes. To protect the scales from being rubbed away, he coated the lepidochromes with gum or a similar substance.[47] When it was complete, he painted the body and retouched the lepidochrome, filling in areas where groups of scales had failed to transfer or had done so unsatisfactorily, as in the blue spots of the small tortoiseshell (Plate 8).[48] Because the colors in lepidochromes tend to appear more muted than they do on a butterfly wing, he deepened and enhanced colors where appropriate: The brown hairstreak butterfly (*Thecia betulae*) in Plate 10, for example, is extensively overpainted to create a close approximation of the many shades of brown and orange visible in the live butterfly. Minute dabs of black paint indicate that he also tried to emphasize the lines of brown or black scales on either side of the white stripes to create a more accurate depiction of the butterfly's markings. The task of delineating such a tiny area of black proved too delicate even for his skills, and he abandoned the effort. In most cases, however, Hoefnagel let the scales dominate even though this resulted in a slightly muted version of the wing colors. This suggests that the authenticity of the butterfly materials was more important to him than precisely replicating their coloring. The mourning cloak butterflies (*Nymphalis antiopa*) in Plate 6 (Fig. 2.11, left and bottom butterflies), for example, lack the rich maroon brown of the real insects, while the markings in spotted butterflies such as the painted

46 See analysis report, Morales 2022a. Hoefnagel's selection of gold point for this purpose is an unusual but practical choice: a stylus fitted with a narrow metal wire would be better suited than black chalk or graphite for the delicate task of tracing around a butterfly wing. Not all printed butterflies show evidence of outlines. When Plate 10 is viewed in ultraviolet light, faint fluorescence is visible far beyond the brighter fluorescence lying close to the butterflies, indicating Hoefnagel applied the adhesive over broad areas and wiped away the excess outside the wings when he finished. Conversely, the adhesive used to print the tortoiseshell butterfly in Plate 8 fluoresces only inside the guidelines.

47 The thin coating appears glossy when viewed at an oblique angle except where paint is applied on top of it, dulling the shine.

48 For the complications of transferring blue structural scales, see Mandrij's chapter in this volume.

lady (*Vanessa cardui*) in Plate 8 are discernible but less distinct than in the real butterfly.

Because Hoefnagel made his lepidochromes directly onto the parchment, he was able to use each wing only once and required two sets of wings if he wanted to show the ventral and dorsal sides of the same species, as in Plate 6 (Fig. 2.11, left and bottom butterflies). Nearly all the lepidochromes show the butterflies and moths from the top or the side, providing a clear view of a single side of the wings. Occasionally, however, he experimented with a more complicated pose, showing a butterfly from the side with its wings slightly open and a portion of the dorsal side of the wing visible. In most of these, he painted in the portion of dorsal wing when he added the body. In one instance, the clouded yellow (*Colias croceus*) in Plate 9, he transferred what appears to be a sliver of wing as well. Instead of attempting to manipulate a narrow slice of wing, he kept both sets of wings closed and adjoined after removing the body and then pressed them onto the tacky gum. Imprints of veins in most of the transfers confirm Hoefnagel's consistent use of a single-transfer technique (Fig. 2.12).[49] He had little reason to experiment with a double transfer. The powdery substance covering the wings was generally believed at this time to be a type of dust rather than flat scales. Therefore, he would have had no desire to display a particular side of the scale or any idea that each side might present a different appearance.[50] Hoefnagel may have observed, however, that this "dust" was arranged into horizontal rows on the wings: Although difficult to see without powerful magnification on real butterfly wings, the row formation is visible under low magnification in some lepidochromes where the scales have transferred sparsely, and the artist sometimes imitated the effect with horizontal striations in his painted butterflies.[51] For Hoefnagel, the patterns of the veins produced by lepidochromy could be as important as the scales' colors: In

49 For a detailed explanation of the single and double transfer techniques, see Mandrij's chapter in this volume.

50 Moffet (1658, 958) describes the Phalena, "a kind of Butterfly flying in the night ... its wings powdered or sprinkled as it were with a fine kind of ashes or dust ..." Seventeenth-century advances in microscopy allowed naturalists to discern rows of scales, which they sometimes identified as feathers. See Power 1664, 7–8: "that small meal and dust of their wings (which sticks to your fingers when you catch them) is all small little feathers, which grow out of their wings; and you may plainly see the twills by which they stick to the wings; and the holes in the wings, out of which they were pluck'd ... which shews how vastly they were mistaken, that held this mealy dust to be an exudation of atoms out of their wings." See also Hooke 1665, 194–195.

51 Using dots and dashes of watercolor, Hoefnagel imitated individual scales and linear groupings in entirely painted butterflies: for example, he painted the gray markings on the upper wing of the clouded yellow (*Colias croceus*) in Plate 14 with small horizontal

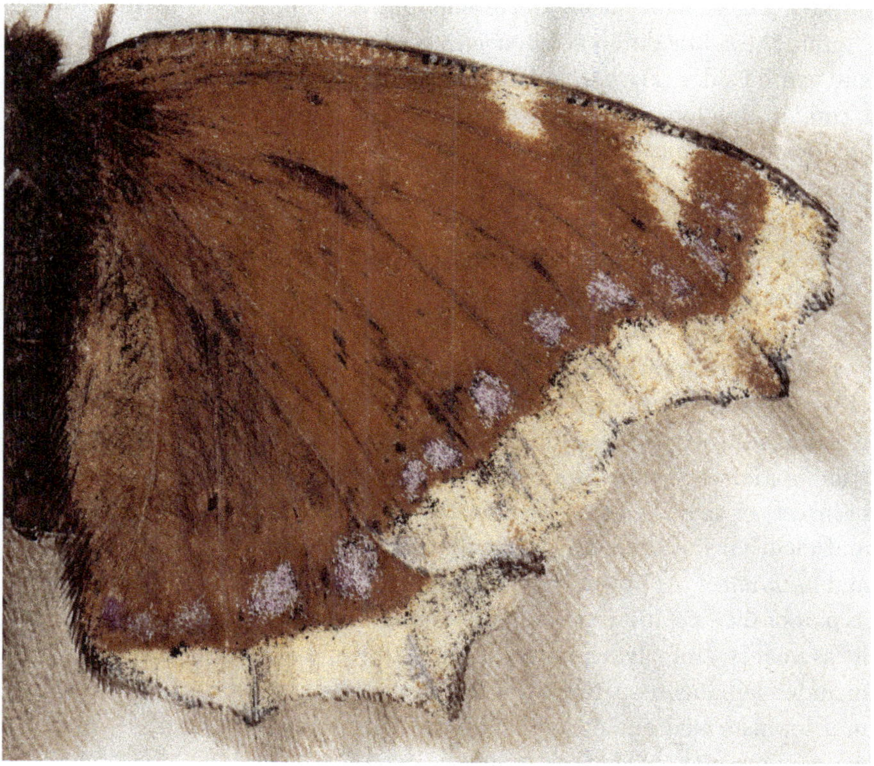

FIGURE 2.12 Joris Hoefnagel, right wings of a mourning cloak butterfly (*Nymphalis antiopa*), detail of Fig. 2.11, *c*.1575/1590, lepidochrome, watercolor, and gold, with oval border in gold, on parchment, *c*.143 × 184 mm (leaf), Washington, National Gallery of Art, Gift of Mrs. Lessing J. Rosenwald, 1987.20.5.7

the black-veined white butterfly (*Aporia crataegi*) in Plate 16, he overpainted much of the butterfly with white paint, but he faithfully traced the impressions of the veins with black, possibly planning to refer to them later in other work.[52]

The dragonfly wings and lepidochromes are likely related, though not identical, in purpose and meaning. Despite the differences in color intensity between the lepidochromes and the insects themselves, Hoefnagel may have appreciated these insect inclusions in part for their accuracy and proximity

---

dashes that correspond to the direction of the rows of scales on the wing. Traces of lepidochromes were not detected in this plate.

52 A similar butterfly appears (in reverse) in the *Archetypa Studiaque*, Part 1, Plate 3 and in the *Mira Calligraphiae*, fol. 40 (see below, note 67).

to the living insects. The lepidochromes of the mourning cloaks, for instance, capture the subtle difference in color between the brown scales on the dorsal and ventral sides (Fig. 2.11, left and bottom butterflies). Lepidochromy served a similar function to the nature prints that had been included in some collections of botanical illustrations for over a century, although of course the processes differ from each other.[53] Early collectors apparently sometimes preserved butterflies by pressing them in books, a practice that put the fragile wings at risk of crumbling.[54] The lepidochromes in *Ignis*, however, remain legible and intact. Lepidochromy presented a more durable alternative to including whole wings or insects.

Hoefnagel's attention to detail in these composite images – such as his careful retouching of the lepidochromes and the delicate painted shadows cast by real dragonfly wings – resulted in an almost seamless blend of art and nature, a combination especially appealing to erudite collectors. Much like nature casts of insects or lizards, life masks, or feather mosaics, Hoefnagel's lepidochromes and inclusions of dragonfly wings blurred the boundary between *artificialia* and *naturalia*.[55] As objects that required physical contact with their subjects as part of their creation, the butterfly transfers may have had particular authority as images, embodying one meaning of the term *ad vivum*.[56] Hoefnagel also included lepidochromy in his cabinet miniatures, suggesting that their appeal to a sophisticated audience outside the community of naturalists formed a major component of their desirability as well.[57]

---

53 Reeds 2006, 203–237. Nature printing requires the use of ink or another substance, whereas in lepidochromy the scales themselves act as the medium. Some creators of nature prints enhanced the images with missing information to make them more accurate, much as Hoefnagel enhanced the lepidochromes (Reeds 2006, 230). Marjorie Shelley (2020, 33–34) also links Hoefnagel's use of insect parts to natural history collecting practices.

54 Barnard 2018, 646–651; Jarvis 2018, 218 and 223. A single butterfly was discovered pressed inside Thomas Moffet's manuscript for *Insectorum sive Minimorum Animalium Theatrum* (1634), now in the British Library, Sloane MS 4104; see Thomson 2012, 103.

55 See, for instance, Swan 2005, 93–94. For life casts, see Smith and Beentjes 2010, 140–144. For the reception of feather mosaics in Europe, see Russo 2015, 23–63.

56 For this sense of the term "*ad vivum*," see remarks on Girolamo Cardano (1501–1576) in Kusukawa 2019, 107–109.

57 Lepidochromes are present in two cabinet miniatures at the Metropolitan Museum of Art, New York: *Still Life with Flowers, a Snail, and Insects* (acc. no. 2008.110; for this work, see Vignau-Wilberg 2017, cat. no. C4 and Vignau-Wilberg 2007a) and *Insects and the Head of a Wind God* (acc. no. 63.200.4; Vignau-Wilberg 2017, cat. no. C19b). Marjorie Shelley (2020, 33) identified insect inclusions in the latter but interpreted them as adhered wings rather than transfers. Lepidochromes likely appear in other cabinet miniatures, such as *Arrangement of Flowers in a Vase with Insects* (Oxford, Ashmolean Museum, acc. no. WA1951.49, Vignau-Wilberg 2017, cat. no. C18); *Still Life with Flowers and Insects*

The lepidochromes had a practical use as models for Hoefnagel's other work. Sometimes he simply copied them: He adapted the hummingbird hawkmoth (*Macroglossum stellaturum*) in Plate 23, for instance, in a cabinet miniature.[58] However, the lepidochromes in *Ignis* also served as models for lepidopterans in the *Archetypa studiaque*, where the engraved versions are identical in size within a millimeter or two to their sources. Reversing the engravings to account for the printing process and superimposing them on the originals demonstrates that their primary contours match as well, indicating that they were traced for reproduction in the print series.[59] The presence of lepidochromy in several of the butterflies that were used in this way establishes beyond doubt that the versions in *Ignis* are the original sources and that Hoefnagel used at least some of the insects in the volume as models for later work. Insects in *Ignis* not made with the lepidochromy technique, such as grasshoppers, dragonflies, and beetles, are also reproduced in nearly identical size in *Archetypa*, suggesting that they, too, may have been traced from the volume. For instance, one plate from the *Archetypa* (Plate 10, Part 2; Fig. 2.13) features both a blue underwing moth (*Catocala fraxini*) copied from a lepidochrome in *Ignis* (Plate 25), as well as a mayfly (*Ephemeroptera* order) and a long-headed grasshopper copied from entirely painted versions (Plates 44 and 45).[60] While Hoefnagel certainly had other model books, *Ignis* served in part as a repository of accurate depictions of insects, to be copied precisely or adapted in a variety of ways for other projects.[61]

Similarly, the extraordinarily complex and delicate arrangements of cells in the dragonfly wings presented a challenge even to Hoefnagel's command of detail, as demonstrated in Plate 54, where two dragonflies with real wings

---

(Middelburg, Zeeuws Museum, acc. no. M98-072-01, Vignau-Wilberg 2017, cat. no. C5), but this has not yet been confirmed.

58 The moth appears in the lower right of *Insects and the Head of a Wind God* (see above, note 57).

59 Both mourning cloak butterflies in Plate 6 appear in the *Archetypa* (Part 2, Plates 9 and 11) and closely match the original lepidochromes, as do the orange tip butterfly (*Anthocharis cardamines*) in Plate 8 (*Archetypa* Part 2, Plate 9) and blue underwing moth (*Catocala fraxini*) in Plate 25 (*Archetypa* Part 2, Plate 10). Other transfers used in the *Archetypa* include the *comma butterfly* (*Polygonia c-album*) (Plate 6 and *Archetypa* Part 2, Plate 7) and the hummingbird hawkmoth (Plate 23 and *Archetypa* Part 1, Plate 5).

60 Dimensions of prominent features in all three primary insects in the print measured within 2 mm of those in the paintings. When superimposing an image of the reversed transfer over an image of the engraving in Adobe Photoshop, the contours match closely. Although finer lines, such as the veining in the moth, do not directly align, the general stance and features of the insects clearly correspond. Even the bald patch on the blue underwing pictured in *Ignis* (Plate 25) is replicated in the *Archetypa*.

61 Some motifs appear repeatedly in Hoefnagel's work but are absent from *Ignis*, indicating that he had other model books. See, for instance, Hendrix and Vignau-Wilberg 1992, 44.

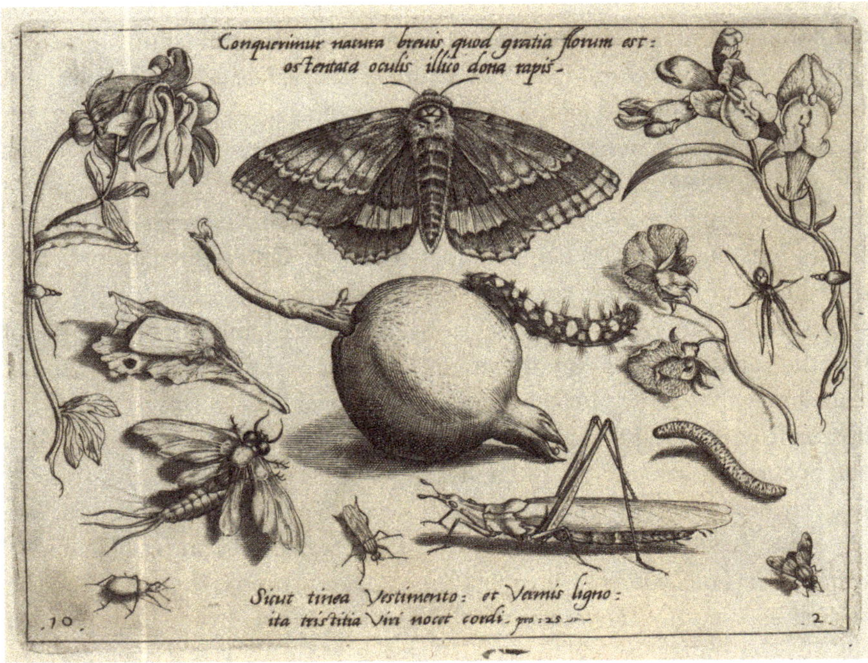

FIGURE 2.13 Jacob Hoefnagel, Composition of insects, spiders, flowers, and fruits, in: *Archetype studiaque patris Georgii Hoefnagelii*, 1592, Plate 10 of Part 2, engraving on laid paper, Washington, National Gallery of Art, Gift of Mrs. Lessing J. Rosenwald, 1987.20.9.24
© NATIONAL GALLERY OF ART WASHINGTON. PHOTO CREDITS: TRICIA ZIGMUND. HTTPS://WWW.NGA.GOV/COLLECTION/ART-OBJECT-PAGE.224577.HTML

are juxtaposed with one entirely painted example: In the latter, Hoefnagel only approximated the wing pattern, relying on tiny flecks of gray watercolor to give the impression of rows of cells. As Bass has observed, the juxtaposition poses a deliberate comparison between the hand of the artist and that of God or nature.[62] The applied wings may have had practical implications as well. Gluing real wings to the painted bodies allowed him to retain the most accurate version possible for later reference. His careful study of the color and translucency of wings is also reflected in his varied treatment of legs and bodies as they pass beneath wings in his entirely painted dragonflies: While Hoefnagel applied white paint on parts of the legs in Plate 53, elsewhere in *Ignis* he diluted the watercolor with water or lightly tamped the wet paint with a fingertip, leaving behind only traces of color to evoke the muted colors of leg

62 Bass 2019, 233.

and body beneath the translucent wings.[63] The dragonflies with the adhered wings, then, may have contributed to the experience and insight Hoefnagel needed to create the southern hawker dragonfly in Plate 53 (Fig. 2.2) as well as the many other painted dragonflies and damselflies that appear throughout his manuscripts and cabinet miniatures.

6 Conclusion

While most of the miniatures in *Ignis* are highly finished, a few may have served as experiments, which shaped Hoefnagel's techniques when he approached the same insects elsewhere in his work. He may have decided that the rapid deterioration of the dragonfly wings, for example, made them an unsuitable material, as they are not recorded in other works.[64] His attempts to depict iridescence or pearlescence with metallic materials in *Ignis* varied in success, which may have influenced his choices in other projects. In the miniatures he provided for Georg Bocskay's (died 1575) calligraphy in the *Mira Calligraphiae Monumenta*, a project he undertook during the 1590s for Rudolf II, his practices deviated from those in *Ignis*.[65] In the *Mira Calligraphiae*, he chose a non-tarnishing material for the spots in the fritillary and a new approach for two metallic green beetles.[66]

Hoefnagel's treatment of insect color in *Ignis* sometimes contrasts with his depictions of the same creatures elsewhere, reflecting the varying purposes of different projects. In the *Mira Calligraphiae*, he frequently simplified both form and color when depicting insects found in *Ignis*, generally without describing the effects of texture and light on color. The brown hairstreak butterfly (*Thecia betulae*) in Folio 32, for example, lacks the gradations in oranges and browns that Hoefnagel added to the lepidochrome in Plate 10 in *Ignis*. The white stripes are enlarged and simplified, allowing the artist space to add the

63 Hoefnagel used subtractive watercolor techniques throughout the volume. For example, he diminished the strength of some shadows by lifting wet paint with a finger or dry brush.

64 Vignau-Wilberg (2017, 180) points out applied wings in the crane fly at the lower right of *Insects and the Head of a Wind God* (see above, note 57). However, these are painted with tinted gum, which has now cracked, and no insect wings are present. The possible pasted wing Vignau-Wilberg notes in *Vanitas* (London, British Museum, acc. no. 1997,0712.56) appears in images to be painted.

65 Los Angeles, J. Paul Getty Museum Collection, ac. no. Ms. 20 (86.MV.527). See Hendrix and Vignau-Wilberg 1992; Vignau-Wilberg 2017, cat. no. A4. Images of every page are available at https://www.getty.edu/art/collection/object/103RWD.

66 Los Angeles, J. Paul Getty Museum Collection, ac. No. Ms. 20 (86.MV.527), Fol. 47 and 129.

dark lines he abandoned in *Ignis*, the hindwing is abstracted into a more regular curve, the white body is changed to a tan that harmonizes with the wings, and the facial features are exaggerated and anthropomorphized.[67] While based on the knowledge Hoefnagel derived from studies like the ones in *Ignis*, many of the insects in the *Mira Calligraphiae* are imaginary. The colorful dragonfly in Folio 76 (Fig. 2.14), for example, remains unidentified, though its markings vaguely recall those of the southern hawker in Plate 53 (Fig. 2.2). In the *Missale Romanum*, he also repeated many of the insects found in *Ignis*. Here, the insects vary in size from miniature versions just a few millimeters long to approximately life-sized, and Hoefnagel adjusted his modus operandi accordingly, simplifying the tiniest versions while treating the larger depictions with a level of detail similar to that seen in *Ignis*.[68] His approach often deviated from the convincing accuracy of *Ignis* when depicting the same insects in his cabinet miniatures as well. The versions of insects repeated in these paintings are often simplified, created with fewer and broader brushstrokes and a more restricted range of pigments.[69] Often, if Hoefnagel created multiple versions of the same insect, the most thorough and faithful treatment appears in *Ignis*, suggesting that the volume may have been the ultimate source of many of these motifs – a possibility strengthened by his unquestionable use of some of the lepidochromes in the volume as sources in his other work. Retaining the most accurate and detailed versions of the insects would have left Hoefnagel free to adjust, simplify, or copy them precisely, according to the needs of the project and tastes of the patron.

---

67 Similarly, the veins of the black veined white butterfly that Hoefnagel traced in the lepidochrome in Plate 10 are simplified in the *Mira Calligraphiae* (Folio 40). The black thickening near the top of the wing is transformed into a more regular circle, the overall wing color darkened to differentiate it from the parchment, and the black antennae decorated with a striped pattern.

68 Collection of the Österreichische Nationalbibliothek, Vienna, Cod. 1784. See Vignau-Wilberg 2017, no. A2. In fol. 40r, for instance, he copied almost verbatim the brown long-headed grasshopper (Plate 45), the dragonfly (order Odonata, Plate 55), and the striped crane fly (possibly *Nephrotoma flavescens*, Plate 57). In contrast, he pictured the identical dragonfly in fol. 379v along with ventral and dorsal views of the painted lady butterfly (Plate 8) in miniature with simplified coloring to that in *Ignis*, though only measuring a few millimeters. Black and white images of the volume are available at https://digital.onb.ac.at/RepViewer/viewer.faces?doc=DTL_5091323&order=1&view=SINGLE.

69 The hummingbird hawkmoth (*Macroglossum stellaturum*) in *Insects and the Head of a Wind God* (see above, note 57) lacks the numerous hairs of the body and the subtle modeling of the round form evident in Plate 23 of *Ignis*. At times, the treatment of insect faces and poses in his cabinet miniatures tends toward the amusing and whimsical, characteristics rarely seen in *Ignis*. See, for instance, *Arrangement of Flowers in a Vase with Insects* and *Still Life with Flowers, a Snail, and Insects* (see above, note 57).

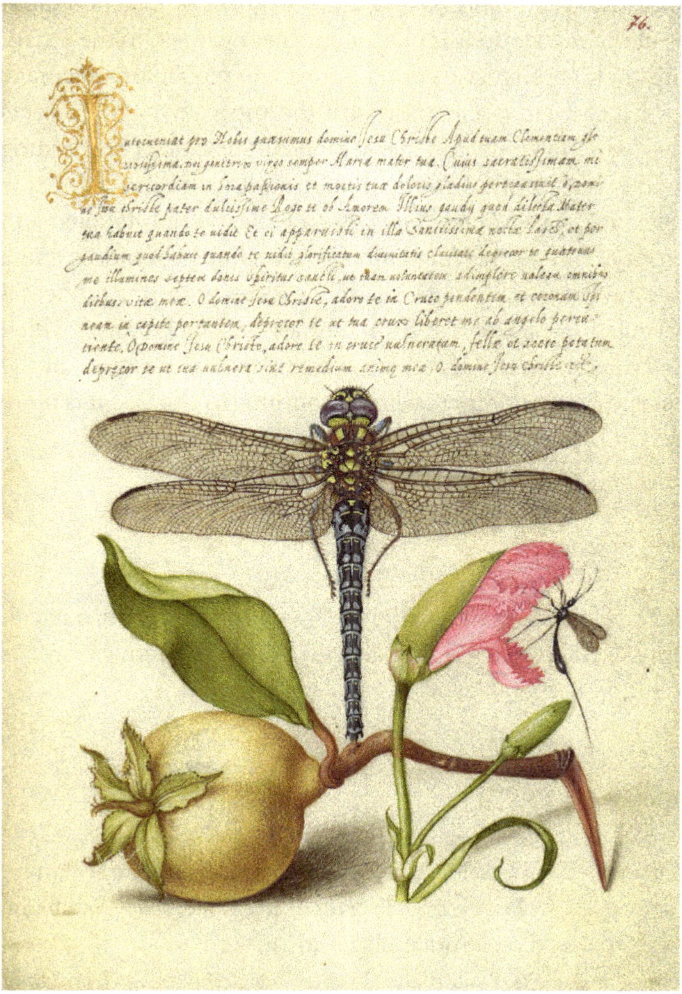

FIGURE 2.14 Joris Hoefnagel and George Bocskay, *Dragonfly, Pear, Carnation, and Insect*, in: *Mira Calligraphiae Monumenta*, 1561–1562, illumination added 1591–1596, watercolor, gold and silver paint, and ink, 166 × 124 mm (leaf), Los Angeles, CA, J. Paul Getty Museum, Ms. 20 (86.MV.527), fol. 76

© GETTY MUSEUM, LOS ANGELES. HTTPS://WWW.GETTY.EDU/ART/COLLECTION/OBJECT/103S03

Hoefnagel's sensitive use of artistic materials and his attention to detail resulted in some of the most closely observed depictions of insect color in early modern Europe. His understanding of the properties of his pigments and his skill in selecting them for their abilities to communicate not only color but also the effects of light and texture, combined with his manual control and powers of observation, set *Ignis* apart from the works of his contemporaries. While some of the inscriptions in the volume bear witness to the artist's preoccupation with their emblematic connotations, the meticulous accuracy of the watercolors betrays his fascination with the intricacy of the insects themselves, lending visual form to the many biblical quotations celebrating the beauty and endless variety of God's creation. In addition to its undeniable personal and emblematic significance, *Ignis* was a collection of detailed and true-to-life renditions of insects, serving both as a storehouse of motifs and a record of the artist's intensive study. This essay constitutes a starting point and a preliminary investigation of Hoefnagel's materials and techniques, which have received little attention in the past. Future technical examinations of the works of Hoefnagel and his contemporaries will continue to situate *Ignis* more firmly in the context of the art and natural history of its time.

Acknowledgements

The authors wish to acknowledge contributions and support from our colleagues at the National Gallery of Art, in particular Kathryn Morales, Brooks Rich, Michelle Stein, and Barbara Berrie.

Bibliography

Alberghina, Maria Francesca, Emanuela Massa, Anna Pelagotti and Salvatore Schiavone. 2014. "I segreti dei colori." In *Jacopo Ligozzi, "Pittore Universalissimo"* Exhibition Catalogue. Palazzo Pitti, Florence. Livorno: Sillabe Ed., 312–315.

Barnard, Peter C., 2018. "Bat-Fowlers, Pooters and Cyanide Jars: A Historical Overview of Insect Collecting and Preservation." In *Naturalists in the Field: Collecting, Recording and Preserving the Natural World from the Fifteenth to the Twenty-First Century*, Arthur MacGregor (ed.). [Emergence of Natural History 2]. Leiden: Brill, 646–685. https://doi.org/10.1163/9789004323841_023.

Bass, Marisa Anne. 2019. *Insect Artifice: Nature and Art in the Dutch Revolt.* Princeton: Princeton University Press.

Bass, Marisa Anne. 2020. "Joris Hoefnagel and the Kunstkammer: Review of Thea Vignau-Wilberg, *Joris and Jacob Hoefnagel: Art and science around 1600.*" *Zeitschrift für Kunstgeschichte* 83: 122–134. https://doi.org/10.1515/ZKG-2020-1006.

Bescoby, Jenny, Judith Rayner, Janet Ambers and Duncan Hook. 2007. "New Visions of a New World: The Conservation and Analysis of the John White Watercolours," *The British Museum Technical Research Bulletin*, 1: 9–22.

Briquet, Charles-Moïse. 1985. *Les filigranes: dictionnaire historique des marques du papier des leur apparition vers 1282 jusqu'en 1600*. [2nd ed.]. New York: Hacker Art Books.

Burmester, Andreas, Ursula Haller and Christoph Krekel. 2005. "Munich *Taxae* Project: The Kolberg Inventory List of 1589." In *Art of the Past: Sources and Reconstructions*, Mark Clark, Joyce H. Townsend and Ad Stijnman (eds.). London: Archetype Publications, 44–48.

Cambefort, Yves. 2006. "A Sacred Insect on the Margins: Emblematic Beetles in the Renaissance." In *Insect Poetics*, Eric C. Brown (ed.). Minneapolis: University of Minnesota Press: 200–223.

Carroll, Margaret D. 2020. "*Insect Artifice: Nature and Art in the Dutch Revolt*: Marisa Anne Bass." *Renaissance Quarterly*, 73 (3): 1016–1017.

Chastel, André. 1984. *Musca Depicta*. Milan: F.M. Ricci.

Conigliello, Lucilla. 1991. "Pesci, crostacei e un'iguana per l'imperatore Rodolfo II." *Paragone* 42 (493–495): 22–29.

DaCosta Kaufmann, Thomas, and Virginia Roehrig Kaufmann. 1991. "The Sanctification of Nature: Observations on the Origins of Trompe l'Oeil in Netherlandish Book Painting of the Fifteenth and Sixteenth Centuries." *The J. Paul Getty Museum Journal* 19: 43–64.

Dreyer, Peter. 1986. "Zeichnungen von Hans Verhagen dem Stummen von Antwerpen: Ein Beitrag zu den Vorlagen der Tierminiaturen Hans Bols und Georg Hoefnagels." *Jahrbuch der Kunsthistorischen Sammlungen in Wien* 82–83 (1986–87): 115–144.

Egmond, Florike. 2017. *Eye for Detail: Images of Plants and Animals in Art and Science, 1500–1630*. London: Reaktion Books Ltd.

Egmond, Florike, ed. 2018. *Conrad Gessners "Thierbuch": die Originalzeichnungen*. Translated by Gisella Vorderobermeier. Darmstadt: wbg Edition.

Fiorillo, Flavia, Lucia Burgio, Christine Slottved Kimbriel and Paola Ricciardi. 2021. "Non-Invasive Technical Investigation of English Portrait Miniatures Attributed to Nicholas Hilliard and Isaac Oliver." *Heritage*, 4 (3): 1165–1181. https://doi.org/10.3390/heritage4030064.

Fitzhugh, Elizabeth West. 1997. "Orpiment and Realgar." In *Artists' Pigments: A Handbook of Their History and Characteristics*. Vol 3. Elizabeth West Fitzhugh (ed.). Washington, D.C.: National Gallery of Art, in association with Archetype Publications Ltd.: 47–80.

Haag, Sabine. 2015. *Echt tierisch! Die Menagerie des Fürsten*. Vienna: KHM-Museumsverband.

Hendrix, Lee. 1984. "Joris Hoefnagel and the Four Elements: A Study in Sixteenth-Century Nature Painting." PhD diss., Princeton University.

Hendrix, Lee. 1995. "Of Hirsutes and Insects: Joris Hoefnagel and the Art of the Wondrous." *Word & Image* 11 (October): 373–390.

Hendrix, Lee, and Thea Vignau-Wilberg. 1992. *Mira Calligraphiae Monumenta: A Sixteenth-Century Calligraphic Manuscript Inscribed by Georg Bocskay and Illuminated by Joris Hoefnagel.* Malibu: J. Paul Getty Museum.

Hooke, Robert. 1665. *Micrographia, or Some Physical Descriptions of Minute Bodies Made by Magnifying Glasses* … London: Printed by J. Martyn and J. Allestry.

Howard, Helen. 2006. "Shells as Palettes and Paint Containers in England." In *Medieval Painting in Northern Europe: Techniques, Analysis and Art History*, Jilleen Naldolny (ed.). London: Archetype, 202–214.

Jarvis, Charles E. 2018. "'Take With You a Small Spudd or Trowell': James Petiver's Directions for Collecting Natural Curiosities." In *Naturalists in the Field. Collecting, Recording and Preserving the Natural World from the Fifteenth to the Twenty-First Century*, Arthur MacGregor (ed.). [Emergence of Natural History 2]. Leiden: Brill, 212–237. https://doi.org/10.1163/9789004323841_009.

Jorink, Eric. 2007. "Between Emblematics and the 'Argument from Design': The Representation of Insects in the Dutch Republic." In *Early Modern Zoology: The Construction of Animals in Science, Literature and the Visual Arts*, Karl A.E. Enenkel and Paul J. Smith (eds.). Leiden: Brill, 147–175.

Koning, Jan de, Gerda van Uffelen, Alicja Zemanek and Bogdan Zemanek. 2006. *Drawn after Nature: The Complete Botanical Watercolours of the 16th-Century Libri Picturati.* Zeist: KNNV Publishing.

Koreny, Fritz. 1988. *Albrecht Dürer and the Animal and Plant Studies of the Renaissance.* United Kingdom: Little, Brown.

Kusukawa, Sachiko. 2014. "Conrad Gessner on an '*Ad Vivum*' Image." In *Ways of Making and Knowing: The Material Culture of Empirical Knowledge*, Pamela H. Smith, Amy R.W. Meyers, and Harold J. Cook (eds.). Ann Arbor: The University of Michigan Press, 330–356.

Kusukawa, Sachiko. 2019. "*Ad vivum* Images and Knowledge of Nature in Early Modern Europe". In *Ad Vivum?* Thomas Balfe, Joanna Woodall, and Claus Zittel (eds.). Leiden: Brill, 89–121.

Mander, Karel van. 1994. *Karel van Mander, the Lives of the Illustrious Netherlandish and German Painters* … Hessel Miedema (ed. and trans.). Doornspijk: Davaco.

Moffet, Thomas. 1658. *The Theatre of Insects: Or, Lesser Living Creatures.* In *The History of Four-Footed Beasts and Serpents* … Ed. and trans. by Edward Topsell. London: Printed by E. Cotes.

Moog, Gerhard. 2021. "Skins and Hides for Making Parchment: An Overview of Parchment's Histologic Features that Are Helpful when Identifying Raw Source Material," Caroline Danforth (trans.). *Art in Translation* 13 (4): 419–431. https://doi.org/10.1080/17561310.2021.1951968.

Morales, Kathryn. 2022a. *DC107632_Hoefnagel_1987.20.5.9*, Analysis Report, National Gallery of Art, Scientific Research Department.

Morales, Kathryn. 2022b. *DC107639_Hoefnagel_1987.20.5.14*, Analysis Report, National Gallery of Art, Scientific Research Department.

Morales, Kathryn. 2022c. *DC1076333_Hoefnagel_1987.20.5.19*, Analysis Report, National Gallery of Art, Scientific Research Department.

Morales, Kathryn. 2022d *DC107635_Hoefnagel_1987.20.5.43*, Analysis Reports National Gallery of Art, Scientific Research Department.

Morales, Kathryn. 2022e. *DC1077636_Hoefnagel_1987.20.5.44*, Analysis Report, National Gallery of Art, Scientific Research Department.

Morales, Kathryn. 2022f. *DC108237_Hoefnagel_1987.20.5.61*, Analysis Report, National Gallery of Art, Scientific Research Department.

Morgan, Nigel. 2016. "Painting with Gold and Silver." In *Colour: The Art & Science of Illuminated Manuscripts*. Stella Panayotova (ed.). London: Harvey Miller Publishers, 192–219.

Mysels, Karol J. 1981. "Visual Art: The Role of Capillarity and Rheological Properties in Painting," *Leonardo* 14 (1): 22–27.

Neri, Janice. 2011. *The Insect and the Image: Visualizing Nature in Early Modern Europe, 1500–1700*. Minneapolis; London: University of Minnesota Press.

Power, Henry. 1664. *Experimental Philosophy in Three Books*. London: printed by T. Roycroft, for John Martin, and James Allestry.

Reeds, Karen. 2006. "Leonardo da Vinci and Botanical Illustration: Nature Prints, Drawings and Woodcuts ca. 1500." In *Visualizing Medieval Medicine and Natural History, 1200–1550*. Jean A. Givens, Karen M. Reeds, Alain Touwaide (eds.), Aldershot: Ashgate, 205–237.

Rikken, Marrigje E. 2016. "Dieren verbeeld. Diervoorstellingen in tekeningen, prenten en schilderijen door kunstenaars uit de Zuidelijke Nederlanden tussen 1550 en 1630." PhD diss., Leiden University Centre for the Arts in Society (LUCAS), Humanities, Leiden University.

Russo, Alessandra. 2015. "A Contemporary Art from New Spain." In *Images Take Flight: Feather Art in Mexico and Europe, 1400–1700*, Alessandra Russo, Gerhard Wolf, and Diana Fane (eds.). Munich: Hirmer, 22–63.

Scheller, Robert W. 1995. *Exemplum: Model-Book Drawings and the Practice of Artistic Transmission in the Middle Ages (ca. 900–ca. 1470)*. Amsterdam: Amsterdam University Press.

Schenck, Kimberly, and Stacey Sell. 2022. "Insect Transfer Prints in Joris Hoefnagel's *Ignis* from the *Four Elements*." *Master Drawings* 60 (2): 161–170.

Schmückle, Miron. 2017. "Nature Description and Nature Allegory in Joris Hoefnagel's Cabinet Miniatures." *Brukenthal. Acta Musei* 12: 263–283.

Seehafer, Michèle. 2021. "Shimmering Virtue: Joris Hoefnagel and the Uses of Shell Gold in the Early Modern Period." In *Materialized Identities in Early Modern Culture,*

1450–1750: Objects, Affects, Effects, Susanna Burghartz, Lucas Burkart, Christine Göttler, and Ulinka Rublack (eds.). Amsterdam: Amsterdam University Press, 281–322.

Shelley, Marjorie. 2020. "Joris Hoefnagel's Insects." *Metropolitan Museum Journal* 55 (1): 26–42. https://www.journals.uchicago.edu/doi/10.1086/712768.

Smith, Pamela H, and Tonny Beentjes. 2010. "Nature and Art, Making and Knowing: Reconstructing Sixteenth-Century Life-Casting Techniques." *Renaissance Quarterly* 63 (1): 128–179. https://doi.org/10.1086/652535.

Swammerdam, Jan. 1758. *The Book of Nature; or the History of Insects ... With the Life of the Author, by Herman Boerhaave*. translated by Thomas Flloyd. revised and improved by Notes from Reaumur and others by John Hill, M.D. London: Printed for C.G. Seyffert.

Swan, Claudia. 2005. *Art, Science, and Witchcraft in Early Modern Holland: Jacques de Gheyn II (1565–1629)*. New York: Cambridge University Press.

Thomson, George. 2012. *Insectorum sive Minimorum Animalium Theatrum: The Butterflies and Moths*. Waterbeck, Scotland: George Thomson.

Turner, Nancy K. 2017–2018. "Reflecting a Heavenly Light: Gold and Other Metals in Medieval and Renaissance Manuscript Illumination." In *Manuscripts in the Making: Art & Science*, Volume Two. Stella Panayotova and Paola Ricciardi (eds.). London and Turnhout: Harvey Miller, 81–96.

Turner, Nancy K. 2023 "The Materiality of Joris Hoefnagel's Insect Artifice" *Facture*. 6: 58–89.

Vermeylen, Filip. 2010. "The Colour of Money: Dealing in Pigments in Sixteenth Century Antwerp" in *Trade in Artists' Materials: Markets and Commerce in Europe to 1700*. Jo Kirby, Susie Nash and Joanna Cannon (eds.). London: Archetype Publications, 356–365.

Vignau-Wilberg, Thea. 1985. "Joris Hoefnagels Tätigkeit in München" *Jahrbuch der Kunsthistorischen Sammlungen in Wien*. 81: 103–167.

Vignau-Wilberg, Thea. 1994. *Archetypa Studiaque Patris Georgii Hoefnagelii, 1592: Natur, Dichtung und Wissenschaft in der Kunst um 1600 = Nature, Poetry and Science in Art around 1600*. Exhibition Catalogue. Staatliche Graphische Sammlung, Munich. Munich: Staatliche Graphische Sammlung.

Vignau-Wilberg, Thea. 2007a. "Flowers for His Mother: An Unknown Cabinet Miniature by Joris Hoefnagel." *Master Drawings* 45 (4): 522–526.

Vignau-Wilberg, Thea. 2007b. "In Minimis Maxime Conspicua: Insektendarstellungen um 1600 und die Anfänge der Entomologie". In *Early Modern Zoology: The Construction of Animals in Science, Literature and the Visual Arts*, Karl A.E. Enenkel and Paul J. Smith (eds.). Leiden: Brill, 2007, 217–243.

Vignau-Wilberg, Thea. 2017. *Joris and Jacob Hoefnagel: Art and Science around 1600*. Berlin: Hatje Cantz.

Weiler, Christina. 2011. *Von Fischen, Vögeln und Reptilien: Meisterwerke aus den kaiserlichen Sammlungen*. Vienna: Kremayr & Scheriau.

CHAPTER 3

"More True to Nature Than Paintings": Lepidochromy and the Color of Butterflies

V.E. Mandrij

1 Introduction

During the eighteenth and nineteenth centuries, a specific technique to depict and preserve the colors of the wings of butterflies and moths came into use among entomologists and amateurs of natural history: Lepidochromy.[1] The process, which involved the transfer of the colorful scales covering lepidopterans' wings onto a support such as paper, porcelain, or glass with the help of pressure and an adhesive, produced imprints that I will subsequently refer to as lepidochromes. Mid-nineteenth-century examples now conserved in the Staatliches Museum für Naturkunde Stuttgart demonstrate the practicality of the technique: Their maker, Friedrich Bauer, displayed and organized hundreds of lepidopterans according to their family by gluing the corresponding lepidochromes on large sheets of paper (Fig. 3.1).[2]

Besides historical lepidochromes, textual sources describing the technique have also come down to us: Some of its users (henceforth referred to as lepidochromists) published instructions in treatises on insects and books of secrets,[3] with many claiming that it was eminently superior to painting when it came to representing the colors of lepidopterans' wings. In his book *Der Schmetterlingsabdruck* (*The Butterfly Imprint*, 1845), Friedrich Mühlecker maintained that lepidochromes were "the most faithful copy of nature, and no

1 The term "lepidochromy," first employed in the late nineteenth century in contradistinction from decalcomania, a tracing technique, is a neologism that combines the Greek words *lepís* ("scale") and *chroma* ("color"). See Orousset 2008, 47.

2 The folder containing the sheets bears the date "1856" and gives Friedrich Bauer's profession as "veterinary." The Staatliches Museum für Naturkunde, or in short: Naturkundemuseum, also conserves a large collection of lepidochromes on cardboards; organized in boxes according to their family and furnished with information about species, sex, and place and year of the specimens' collection, they, too, were created by Bauer, but in the 1860s and 1870s.

3 See examples of these sources in Steensma 1999, 65–66; Orousset 2008; Wingreen-Mason 2014; Péru 2016. Books of secrets were compilations of recipes describing the 'how-to' of various techniques.

 | DOI:10.1163/9789004684553_004

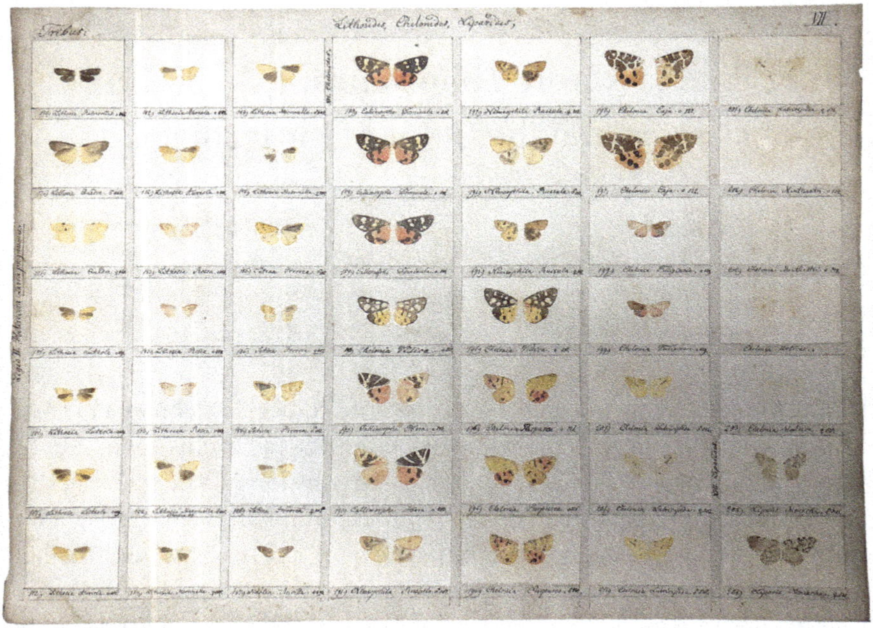

FIGURE 3.1 Friedrich Bauer, sheet VII with lepidochromes on paper from the families "Lithosides, Chlorides, Liparides", 66 × 44 cm (leaf), in: Friedrich Bauer, *Sammlung auf Papier übergetragener württemberg. Schmetterlinge*, 1856, lepidochromes on paper glued on a sheet, Stuttgart, Staatliches Museum für Naturkunde Stuttgart © V.E. MANDRIJ

painter would be able to reproduce so deceptively the brilliance, the shimmer, and the freshness of the colors and every single trait of the drawing, no matter how small."[4] Heinrich Martin Gottfried Köster (1734–1802), a history professor at the University of Frankfurt am Main, was convinced that the process yielded "a butterfly on paper" which was "more true to nature than any other form of painting." Furthermore, lepidochromes could be stored in albums, which protected the delicate organic scales from pests, sunlight, and adverse atmospheric conditions, and required significantly less care than notoriously fragile dry specimens.[5]

4 "Es ist die treueste Copie der Natur, und kein Maler wäre im Stande, den Glanz, den Schimmer und die Frische des Kolorits und jeden einzelnen noch so kleinen Zug der Zeichnung so täuschend wieder zu geben." Mühlecker 1845, 42. All translations are mine unless otherwise stated.

5 Köster 1793, 17: 554; Latreille 1804, 1: 164; Poulin 1876, 6.

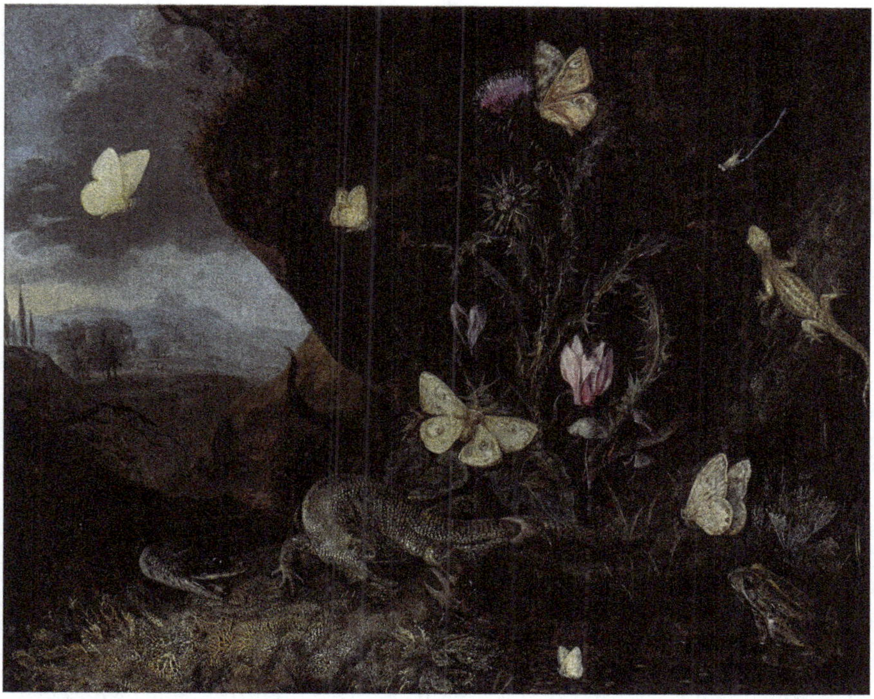

FIGURE 3.2 Otto Marseus van Schrieck, *Landscape with Insects, Cyclamen and Thistle*, no date, oil on canvas, 48.7 × 37.3 cm, Rotterdam, Museum Boijmans van Beuningen, 2206 (OK)

The use of fragments of butterfly and moth wings in paintings had begun much earlier: In the sixteenth century, for example, the Flemish miniaturist Joris Hoefnagel (1542–1600) transferred scales of lepidopterans on vellum (Fig. 2.11),[6] while almost a century later, the Dutch painter Otto Marseus van Schrieck (*c.*1620–1678) employed the technique of lepidochromy in still-life and *sottobosco* paintings on canvas and panel (Fig. 3.2).[7] There are several possible explanations for the fact that none of the latter's lepidochromes have survived intact: Environmental factors such as light (ultraviolet light in particular), oxygen, air temperature, and/or chemical reactions with the materials (lead white,

6 See Schenck and Sell 2022, as well as the same authors' contribution to this volume. See also Shelley 2020.

7 The presence of butterfly materials in paintings by Marseus and Johann Falch (1687–1727) was documented for the first time in Beier 1987.

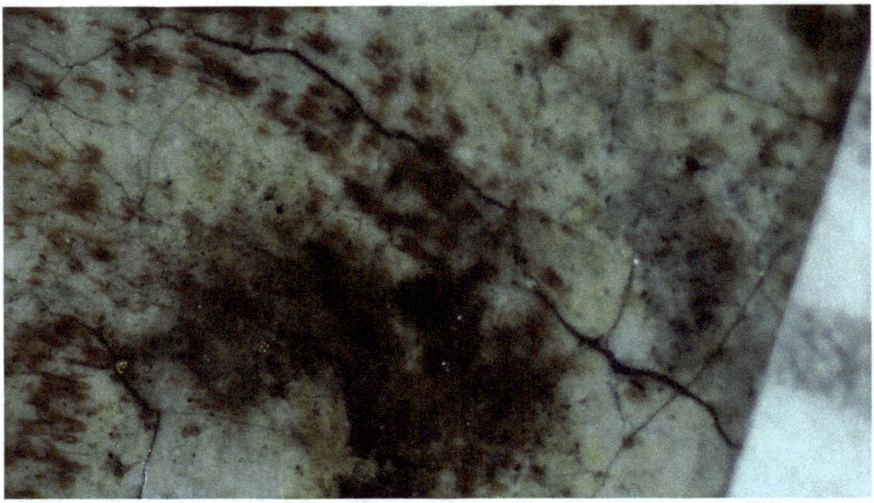

FIGURE 3.3 Microscopic view showing butterfly scales in the lepidochrome of a small emperor moth (*Saturnia pavonia*), detail of Fig. 3.2

oil, varnish, etc.) used in the painting have discolored the scales, and cleaning processes may have removed many of them altogether.[8] While Hoefnagel's lepidochromes were comparatively well protected between the pages of an album, many lepidopterans in Marseus's *sottoboschi* now show the lead white substrate, a local layer placed on the imprimatura especially for making the imprints, or have been overpainted. Microscopic examination is the only way to detect remaining butterfly scales, which look like tiny brown or transparent seeds due to the loss of their original colors, as can be seen from this microscopic photograph of the small emperor moth (*Saturnia pavonia*) in the *sottobosco* now in the museum Boijmans van Beuningen in Rotterdam (Fig. 3.3).

Marseus, who is the main focus of this chapter, worked in a context where painting was theorized as a mirror of nature, with pictorial mimesis aimed at

8 I am thankful to Florent Figon and Marianne Elias for discussing the discoloration of the scales with me. Further scientific research is needed to identify its causes, but Crane 1954, 109–110, Imafuku et al. 2002, 177, Brehm et al. 2008, 45, and Ayre and Bevan 2016 point out light as a key factor in the discoloration of dry specimens. Mike Fitton and Pamela Gilbert (1994, 119) emphasize that light, insects, and damp are serious threats for dry specimens kept in cabinets. Steensma (1999, 63), Berthier et al. (2006, 183), Menu et al. (2006, 599), Breazeale (2007, 527), and Wedemeyer (2007, 27) have argue that restoration removed the scales from Marseus's paintings.

making present that which was absent.[9] It is thus fair to wonder why he used actual butterfly scales in his paintings. Was his goal to represent the colors of lepidopterans, their brilliance and luminous sheen as praised by the lepidochromists, in a more true-to-nature manner than would have been possible with paint? Frequently discussed in seventeenth-century color theories, the imitation of shimmering colors and iridescence certainly presented a formidable artistic challenge.[10] Could lepidochromes have helped the artist to achieve specific optical effects, such as the shimmering blue colors of certain species, more efficiently?

The present chapter aims to answer these questions by discussing two key characteristics of lepidochromy on paper and canvas: First, the transfer of butterfly scales onto another support reversed their original orientation, a process that was liable to alter the resulting colors; and second, microscopic examinations have demonstrated that Marseus often covered parts of his lepidochromes with paint, which raises questions as to the limitations of the visual effects that the technique was capable of producing.[11]

Despite the increasing interest in Marseus's artworks and painting techniques, his use of lepidopterous colors has to date received little scholarly attention.[12] As the three known early modern texts that mention lepidochromy in connection with Dutch oil painters do not provide much information regarding the technique itself,[13] later lepidochromy recipes will be analyzed instead. Although posterior in date and devised for use with a paper support, the recipes in question nonetheless shed light on the above-mentioned issues, while allowing us to reflect on Marseus's motivation for employing the technique. Together with data gathered from the microscopic analysis of the artist's Rotterdam *sottobosco*, these written sources will be compared with my own experiments with lepidochromy recipes. The latter method, also

---

9 This conception of painting in Dutch art theories was championed by the painter and art theorist Samuel van Hoogstraten (1627–1678), who was a friend of Marseus's. On the former's art theory, see Weststeijn 2008, Blanc 2008, and the introduction by Celeste Brusati in Hoogstraten et al. 2021.

10 Leonhard 2014, 55.

11 A third aspect, namely the use of varnish in seventeenth-century oil paintings and its impact on the perception of colors underneath, cannot be discussed here due to its considerable complexity. I address this question in detail in my PhD thesis.

12 On the use of lepidochromy in the *sottobosco* paintings, see Beier 1987, 35–8; Weber 1989, 36; 1993; 1995; Steensma 1999, 62–64; Hartwieg and Goltz 2000, 44–45; Hildebrecht 2004, 137n84; Breazeale 2007; Wedemeyer 2007, 9–14; Berthier et al. 2008; Jorink 2014, 222; Kremkau 2020, 13–23; Leonhard 2020, 166; Mandrij 2021b; Lamb 2022; Hermens 2022.

13 Swammerdam 1669, 129 and table of content; Hoogstraten 1678, 218 and 241; Weyerman 1729, 4: 211.

referred to as 're-enactment' or 'replication,' should not be interpreted as an attempt at precise historical reconstruction[14] – rather, my goal was to form an idea of what the intended colors and visual effects may have been, as Marseus's originals are now irretrievably lost. Given that the iridescent blue colors visible on the wings of peacock butterflies (*Aglais io*) and some species of gossamer-winged butterflies (Lycaenidae), especially members of the subfamily Polyommatinae, are a recurring feature of Marseus's *sottoboschi*,[15] it was their wings that I used in my research.[16]

In the following sections, I will first outline some basic facts regarding lepidopterans' anatomy and coloration, before introducing Marseus and the overarching context in which he employed lepidochromes in oil paintings. I will then examine a number of lepidochromy recipes in detail, which will lead me to the above-mentioned issue of the color changes brought about by the transfer of the insects' scales and the addition of paint. In a final step, I will discuss my own experiments against the backdrop of the lepidochromes created in the nineteenth century by Friedrich Bauer.

2 The Colors of Moth and Butterfly Wings

Moths and butterflies have two pairs of colored and patterned wings.[17] While the motifs are symmetrical across the insects' longitudinal axis, the pattern on the pair of forewings (anterior) usually varies from that on the pair of hindwings (posterior). A wing is made up of a double membrane with a network of tubular veins. Since the dorsal (seen from above) and ventral (seen from below) sides of the wing develop independently, their patterns and coloration are often different.[18] The richly variegated wing patterns are generated by the juxtaposition of tiny scales.[19] Scales are mobile microstructures attached to the membrane with a stalk implanted in a socket. Their structure consists of

14 See Dupré et al. 2020.

15 The many different species of *Polyommatinae* are often challenging to identify. See "polyommatinae" in Savela n.d.: https://www.nic.funet.fi/pub/sci/bio/life/insecta/lepidoptera/ditrysia/papilionoidea/lycaenidae/polyommatinae/.

16 The necessary materials were generously provided by ETH Zurich, the Naturmuseum Augsburg, the Koninklijk Belgisch Instituut voor Natuurwetenschappen in Brussels, and the Naturkundemuseum in Stuttgart. Only existing specimens from the respective collections were used, and no living insects were harmed.

17 In this chapter, I use butterflies for speaking about lepidopterans in general.

18 Berthier 2007, 22.

19 Hence the term "lepidopterans" (see above, note 1), *ptera* meaning "wing".

a thin blade with two sides: Whereas the lower surface (or lamina, to use the appropriate scientific term) faces the membrane and has a smooth and fairly uniform surface, the upper lamina (i.e., the one that is visible to the world) features longitudinal ridges intersected by cross-ribs, which lend the surface a grid-like appearance.[20] This network of ridges and ribs forms openings that reveal the hollow interior of the scale,[21] which could thus be described as being composed of alternating air gaps and chitinous cuticle.[22] This anatomical feature can be found in all scales, but slight structural changes occur depending on their color. Essentially, colors can be produced in two different ways. On the one hand, scales may contain pigments, which absorb some wavelengths and reflect others, the latter being the ones that are visible;[23] colors produced in this manner (for instance, yellow, red, and brown) are called pigmentary colors. Second, scales can exhibit more complex structures with a thin film that generates iridescent colors through interference. Here, the refraction of the light on the surfaces of the scales produces structural (or physical) colors such as blue, purple, and green, the hues of which vary depending on the viewing angle.[24]

In general, the wings' surfaces are covered with two layers of scales: The ground scales, which are touching the membrane, and the cover scales situated above.[25] In most species, the scales are organized in regular rows that overlap each other like roof tiles.[26] A historical hand-colored plate published by German jurist and naturalist Martin Frobenius Ledermüller (1719–1769) documents in words and image the structure of butterfly wings. The individual scales, which he called "feathers," and the venation, which he observed under the microscope, are represented in painstaking detail (Fig. 3.4).[27] It is precisely those "feathers" that are transferred onto another surface through the technique of lepidochromy: The layers of colorful scales are the only parts of the insect's wings that remain in a lepidochrome.

20 Ghiradella 1984, 637.

21 Ghiradella 1984, 637.

22 This is a slightly simplified description, as scale architecture can differ from species to species. See Stavenga 2014, 110.

23 Zhang et al. 2015, 68.

24 Ghiradella 1991, 3492. In many species, colors are generated through a combination of pigmentary and structural coloration. On iridescence, see also the chapter by Kay Etheridge in this volume as well as Berthier 2007, Kinoshita 2008, and Debat et al. 2018.

25 Zhang et al. 2015, 70.

26 Vukusic, Sambles, and Ghiradella 2000, 61.

27 "Federn". Ledermüller 1763, 2: 19. I am grateful to Giulia Simonini for drawing my attention to this source.

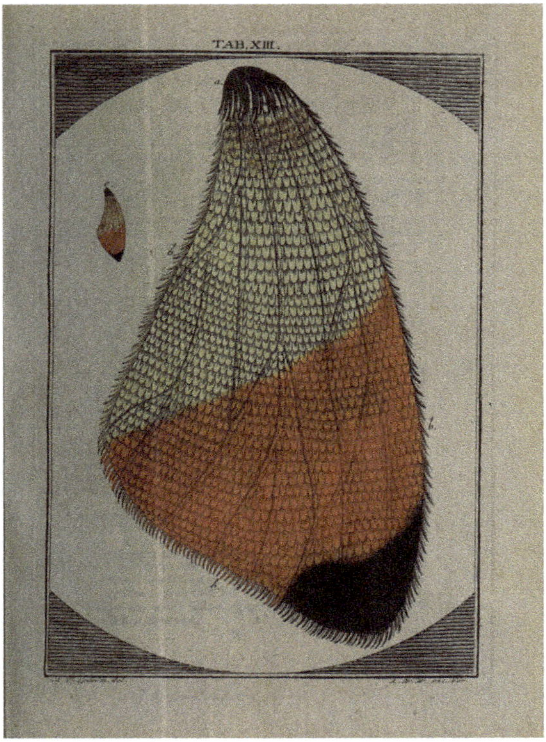

FIGURE 3.4
Martin Frobenius Ledermüller (painter), Adam Wolfgang Winterschmidt (etcher/engraver), *Der ganze Flügel eines Sommer-Vogels*, in: Martin Frobenius Ledermüller, *Mikroskopische Gemüths- und Augen-Ergötzung*, 1763, vol. 1, p. 25, Plate 13, engraving on paper, watercolor, London, Wellcome Collection, EPB/C/32690
© PUBLIC DOMAIN. HTTPS://WELLCOMECOLLECTION.ORG/WORKS/R9AW36XN/ITEMS?CANVAS=69

3 Marseus's *Sottoboschi* and the Lepidochromists

Before entering into the technical minutiae of lepidochromy, I would like to introduce various users of the technique. Trained as a still-life painter in the Netherlands, Otto Marseus van Schrieck travelled to Italy in the 1650s and developed the *sottobosco* paintings.[28] This subcategory of still lifes represents small-scale animals that are unusual subjects for oil painting, such as reptiles, amphibians, and insects, together with different kinds of flora in dark forest surroundings. The case study chosen for this chapter is emblematic of this iconography. The Rotterdam painting, made at an early stage of Marseus's career in Italy, displays in a horizontal format a rocky background with an Italianate landscape on the left.[29] In the center, a lizard is engaged in a fight against a

28 The term *sottobosco*, which recalls the Italian origins of this iconography, is anachronical but commonly used in reference to the paintings in question. On this issue, see Steensma 1999, 131–133 and Mandrij 2021a, 193–195.

29 The painting is not dated, but Steensma (1999, cat. nr B1.25) has identified compositional features typical of Marseus's Italian phase.

crab and a snake at the foot of a thistle surrounded by butterflies, moths, mushrooms, and cyclamens. A frog is watching the scene, while a gecko is scaling the precipice to catch a dragonfly. My own examination of this artwork under the microscope has yielded the result that all lepidopterans bear evidence of the use of lepidochromy.

The large variety of species present in Marseus's *sottoboschi* is evidence of the artist's interest in accurately depicting the flora and fauna that he observed around him. In Rome, the members of the society of Northern painters to which he belonged, the so-called *Bentvueghels* (Dutch for "Birds of a feather"), called him *Snuffelaer* ("Sniffer") because of his habit of wandering around the forest in search of animal specimens including "unusually coloured and speckled snakes, lizards, caterpillars, spiders, butterflies, and strange plants and herbs."[30] According to some anecdotes, Marseus gathered in his studio living animals to paint after life; the above-cited quote certainly suggests that he had a particular interest in the color of specimens.[31]

Marseus worked for the free market, but he was also active in natural history circles. Prominent scholars, collectors, and naturalists who commissioned drawings from him include Cassiano dal Pozzo (1588–1657) and Agnes Block (1629–1704), the former during Marseus's stay in Rome and the latter after the artist's return to Amsterdam around 1660.[32] Art historians and historians of science have convincingly argued that Marseus's *sottoboschi* were linked to contemporary discourses on natural history and empirical research on insects.[33] In the early 1660s, the life and reproduction cycles of insects were the objects of investigations carried out by many naturalists, including the Dutch alchemist and oil painter Johannes Goedaert (1617–1668), the Italian scholar Francesco Redi (1626–1697), and the Dutch naturalist Jan Swammerdam (1637–1680),[34] with the latter thanking Marseus for his significant observations regarding wasp parasitism that interrupted the metamorphosis of butterflies.[35]

30 "vremd gekleurde of gespikkelde slangen, hagedissen, rupsen, spinnen, flintertjes, en vremde gewassen en kruiden". Houbraken 1718, 358. English translation in Houbraken 2021, 358.

31 For further quotations regarding Marseus's relationship to animals, see Mandrij 2021b, 284–286, and, in reference to snakes, Hildebrecht 2004, ch. 3 and 4.

32 On Dal Pozzo, see Rutgers 2012. On Block, see Seelig 2017, 156–168 and Powell 2023.

33 See especially Hildebrecht 2004; Leonhard 2009, 2011, 2013, 2020; Jorink 2014; essays in Seelig 2017.

34 Goedaert 1660; Redi 1668; Swammerdam 1669. For additional information regarding these scholars, see Cobb 2006, ch. 3 and 5; Ogilvie 2008 and Ogilvie 2014; Jorink 2011 and Jorink 2014; Leonhard 2020, passim; essays in Hond, Jorink, and Mulder 2022.

35 Swammerdam 1737, 709. See Hildebrecht 2004, 222; Lenteren and Godfray 2005, 18.

The exact connection between Marseus's artworks and later lepidochromists has yet to be explored. It is clear, however, that Marseus had a number of followers, including several artists who worked in Germany in the eighteenth century who could have passed on his knowledge about lepidochromy.[36] Mentioned in books of secrets and entomological treatises, the technique was used by French and British missionaries in India as well as by French and German naturalists.[37] In this section, I focus on a selection of the latter: Johann Friedrich Naumann (1780–1857), Christian Ludwig Landbeck (1807–1890), Friedrich Berge (1816–1883), Friedrich Mühlecker, Conrad W. Zirkler, Jean-Baptiste François Rozier (1734–1793), H. Poulin, and Pierre-André Latreille (1762–1833). Naumann was a German ornithologist and taxidermist whose work in turned served as a source of inspiration for later lepidochromists such as Landbeck, who referred to his predecessor in one of his own publications.[38] Berge, a German naturalist, included a lepidochromy recipe in his book on butterflies that went through several reprints.[39] No biographical details are known about Mühlecker and Zirkler, but each of them dedicated an entire booklet to the lepidochromy technique.[40] In 1771, Rozier, a French botanist, published the first known lepidochromy recipe in the French language,[41] which specifies materials similar to those used later by his compatriots Poulin and Latreille.[42] As mentioned above, besides its eminent utility with regard to the accurate depiction of lepidopterans, lepidochromy made it possible to compile a large collection of species that was not only easy to store, but also to transport.[43] Hence, given the fact that eighteenth- and nineteenth-century approaches to the taxonomy of lepidopterans were mostly based on the color of their wings, lepidochromes could serve as reliable tools for their identification and classification.[44]

36 Marseus is widely regarded as a pioneer of lepidochromy in oil painting. See Beier 1987. On Marseus's network of painters, see Steensma 1999, 92–98; Leonhard 2020, 83–86.

37 One example of a recipe found in a book of secrets is Anonymous 1731, 19–20. On French and British missionaries, see A.M.C. 1879, 3 and Cave 2010, 154–155, respectively.

38 Naumann 1848, 143–155; Landbeck 1843, 263.

39 Berge 1863, 40–43.

40 Mühlecker 1845; Zirkler 1847.

41 Rozier 1777, 1: 52–53. I quote from this edition because I could not locate a copy of the first edition published in 1771.

42 Poulin 1876 and Poulin 1899; Latreille 1804, 1: 309–115. Little is known about Poulin. Judith Elisabeth Weiss (2022) gives his first name as 'Henri' without providing further biographical details. According to Jean Orousset (2008, 47 and 57), Poulin worked in a bank in Paris.

43 Rozier 1777, 1: 52; Köster 1793, 17: 554; Latreille 1804, 1: 315; Landbeck 1843, 259; Mühlecker 1845, 58.

44 Naumann 1848, 180. Today, lepidochromes are considered useless for scientific purposes, among other things because they do not preserve the membranes and other body parts of the insects which are important features for their identification and classification.

4 Latreille's Lepidochromy Technique

What the authors listed above all have in common is that their respective books and treatises on natural history and/or entomology contain step-by-step instructions on how to make lepidochromes. While the materials involved vary from source to source, all recipes are based on the same fundamental elements: The wings of dead butterflies or moths, leaves of paper, glue, and pressure. In this section, I discuss reconstructions made with Latreille's recipe, which I selected on the grounds that it incorporates improvements over previous approaches such as Rozier's.[45] A brief overview of Latreille's method will allow me to explain how lepidochromes were created on paper by printing both sides of the lepidopteran's wings simultaneously.

The first step proposed by Latreille is the fabrication of a suitable glue by saturating water with gum arabic and a third of tragacanth.[46] A thin layer of this adhesive must then be brushed on a piece of thick and uniform paper folded in two.[47] With regard to the arrangement of the butterfly's severed wings on the glue-covered surface, Latreille stipulates a specific sequence of operations: first the forewings, then the hindwings, leaving between each pair enough space for the body.[48] The dorsal side of the butterfly touches the coated paper, while the ventral side is facing towards the lepidochromist. The actual transfer of the scales is achieved by folding the paper on the wings and pressing them softly with the hand or with a roller of wood.[49] While Latreille specifies that pressure is to be applied for no longer than a minute, my own experiments have demonstrated that leaving the paper with the wings under press until the glue has dried yields better results than opening the paper while the glue is still wet.[50] Finally, Latreille suggests to open the paper, to remove

---

45 Latreille 1804, 1: 309. The experiments in question were carried out at the Cologne Institute for Conservation Science (CICS) with the support of Doris Oltrogge in 2019.

46 Latreille 1804, 1: 309. I did not add alum, an optional ingredient mentioned by Latreille. According to Latreille, preparing the glue with salt eliminates the shiny aspect of the gum, but I did not notice a discernable difference. The addition of tragacanth is optional, but I added it in the quantity proposed by Latreille as it also features in other recipes.

47 Latreille 1804, 1: 309–311. Latreille (1804, 1: 312–313) further advised to print only one side of the wings at a time, but I did not find the process more difficult when transferring both sides simultaneously. Pre-folding the paper does not form part of Latreille's method. Proposed by other lepidochromists, I found this trick to be particularly useful.

48 Latreille 1804, 1: 310–311. As the forewings always overlap the hindwings they must be transferred first.

49 Latreille 1804, 1: 311.

50 Latreille (1804, 1: 311–312) found that the dried glue complicated the opening of the folded paper, which is why he also proposed to press the lepidochromes between two pieces of humid flannel.

the membranous parts of the wings, and to paint the body taking the remains of the specimen as a model.[51]

In what follows, I will address Poulin's and Landbeck's opinion regarding a specific aspect of lepidochromes, namely their color.

5 Pigmentary and Structural Colors and How to Maintain Them

Latreille described lepidochromy as a technique to "fix on paper the scales of the wings of the butterflies without altering their color."[52] However, his method produces an inversion of the scales, with the lower lamina (i.e., the scale surface that is normally touching the membrane) now facing upwards. This issue was noticed by Poulin, who pointed out in his recipe that lepidochromes produced in such a manner merely show the "underside of the scales, and not their actual front," which in turn has an effect on the visible colors.[53] Landbeck, having examined under the microscope the scales of an unspecified butterfly, judged the lower lamina's color to be "worse" than the upper's,[54] a difference in coloration that presented a serious disadvantage for taxonomical purposes, as one species could easily be mistaken for another.[55] While this difference is not noticeable in many species, Landbeck argued that it adversely affected the colors of the "most beautiful" ones; moreover, a "false shimmer" could be produced on the lepidochrome that was absent on the real specimen, a phenomenon I will discuss below.[56]

What Landbeck, Poulin, and other lepidochromists also noticed was that they were unable to reproduce iridescent blue and purple in their lepidochromes.[57] Poulin attempted to explain this failure by arguing that "the blue color does not, so to speak, exist, or at least, it is only the result of a reflection."[58] Although the physical phenomenon of iridescence in moths and butterflies was not fully understood in the late nineteenth century, Poulin was essentially correct: As

51 Latreille 1804, 1: 312. 12 useful colors for this task are listed at the beginning of the recipe.

52 "fixer sur du papier les écailles des ailes des papillons sans altérer leur couleur." Latreille 1804, 1: 309.

53 "que le dessous des écailles et non leur véritable face." Poulin 1876, 15.

54 "die schlechter gefärbte Wurzelhälfte" (Landbeck 1843, 259). Naumann (1848, 175–176), too, noticed this change of colors.

55 Landbeck 1843, 260.

56 "bei den schönsten Arten"; "ein falscher Schiller." Landbeck 1843, 260.

57 Landbeck 1843, 263; Zirkler 1847, 4; Naumann 1848, 177–178; Poulin 1876, 16; Poulin 1899, 22.

58 "La couleur bleue n'existe pour ainsi dire pas, ou du moins, elle n'est que le résultat d'un reflet." Poulin 1899, 22.

explained above, it occurs through reflection of the light on the structured surfaces of their countless scales. Yet, Poulin believed that the effect is absent when the scales' orientation is inverted.

To counter this problem, Landbeck and Poulin added an extra step to their techniques that would allow the specimen's true colors to be maintained: A second transfer of the scales.[59] According to Poulin, this could be accomplished by applying "with a small brush ... a light coat of varnish on the whole surface, right on the scales."[60] When the varnish was neither liquid nor yet completely dry, Poulin recommended to "place carefully the varnished print onto the chosen paper, pressing a little with your finger, so that the scales can adhere to it thanks to the varnish, which forms a glue while drying, and you put the whole under press."[61] The last step is to separate the two pieces of paper to obtain the final imprint. As glue residue can tone down or cancel the iridescent effect of structural colors due to its effect on the refraction of light on the scales, Poulin advised his readers to soak the imprint for a couple of hours to dissolve the glue completely and achieve vivid blue colors.[62]

However, contrary to the arguments presented so far in favor of a second transfer, some practitioners did not notice any adverse effects on the colors of their specimens after applying the single-transfer method.[63] Latreille, for instance, objected to the necessity of a second transfer, claiming that "these feathers have the same color and the same vivacity on both sides, of which one can convince oneself by the counter-impression"[64] – if he provided his readers

59 Poulin (1876, 16) called the second transfer a "contre-épreuve" ("counterproof") or "impression au vernis" ("impression with varnish"). He used the same terminology in his later publication, where he also spoke of a "turnaround" ("retournement"). See Poulin 1899, 18. In the world of printing, the term "counterproof" denotes a print produced by transferring the inked motif a second time, which restores its original orientation. See Seigneur 2004. Landbeck (1843, 263) employed the term "Wiederabdruck" ("reprint").

60 "à l'aide d'un petit pinceau ... une légère couche de vernis sur la surface entière, à même les écailles." Poulin 1876, 17.

61 "Vous placez avec soin, en appuyant un peu avec le doigt, l'impression ainsi recouverte de vernis sur le papier choisi, de manière à ce que les écailles puissent y adhérer par la force du vernis qui forme colle en séchant, et vous mettez le tout sous presse." Poulin 1876, 17.

62 Poulin 1876, 23–24 and 1899, 23.

63 Rozier 1777, 1: 54. Latreille (1804, 1: 314) argued that the second transfer was not easily executed and could damage the butterfly.

64 "ces plumes ont la même couleur et la même vivacité de deux côtés, ce dont on peut se convaincre par la contre-impression". Latreille 1804, 1: 313. Rozier (1777, 1: 54) made similar observations. It is not known if Latreille and Rozier examined the coloration of scale surfaces with a microscope or with the naked eye. While Berge (1863, 43) acknowledged that both sides of the scales have different colors, he did not support the utility of a second transfer.

with instructions for making a second transfer with varnish, he did so only to demonstrate that there was, in fact, no appreciable difference.[65]

Other lepidochromists proposed to solve the problem of color discrepancies by adding paint. As in Latreille's recipe, some authors advised to paint the body to complete the representation of a lepidopteran, but also to mend and fill out the areas where the scales had failed to transfer.[66] Zirkler suggested to use watercolor for imprints made with gummed water, and oil colors for those transferred with varnish;[67] Landbeck remarked that matte-looking lepidochromes should be retouched with colors following the model of a real specimen;[68] and Naumann proposed to paint scales whose pigmentary colorsappeared faded or dull.[69] In particular, paint was recommended to restore the structural colorsthat no longer reflected the light and had therefore been completely altered, in which case Naumann's advice was to scrape off the blue and green scales and to repaint the colors after a model.[70] Discussing the *Apatura* genus, which displays structural purple colors, Landbeck criticized Naumann's suggestion to paint the removed scales, claiming that the value of the lepidochrome would decrease.[71] Berge, meanwhile, proposed to imitate bluish-purple iridescence by dusting dry pigments of ultramarine and carmine on the lepidochromy with the help of a small, porous bag,[72] adding that that gold, silver, or bronze particles could increase the shimmering effect.[73]

65 As Latreille (1804, 1: 314) pointed out, the varnish must be as clear as possible so as not to impart a yellow tinge to the scales. However, Latreille did not clarify whether he was referring to iridescent or pigmentary colors.

66 Landbeck 1843, 263; Mühlecker 1845, 51; Zirkler 1847, 14 and 21; Naumann 1848, 174.

67 Zirkler 1847, 34.

68 Landbeck 1843, 263. The colorants suggested by Landbeck include a pigment lake made with tobacco (leaves?), gamboge, copper acetate, and black ink.

69 Naumann 1848, 174–177. Naumann recommended the use of water-soluble organic colorants and watercolors, such as brown obtained from tobacco, gamboge, verdigris, and copper green.

70 Naumann 1848, 177.

71 Landbeck 1843, 263. Instead, Landbeck suggested to make a second transfer of the scales to render the effect visible.

72 Berge 1863, 42–43. Berge proposed to fix the dry pigments with the same glue used for the lepidochrome (diluted with spirit). Pigmentary colors were to be retouched using (water-soluble) fish gall as binding agent for pigments and dyes. In terms of colors, Berge listed Paris green and verdigris, lead chromate, red lead, cinnabar, carmine lake, indigo, cobalt blue, ultramarine, gamboge, Cassel brown, and bronze.

73 Berge 1863, 43. Naumann (1848, 178) proposed a similar method to represent the shimmering colors of the purple emperor.

6 Practical Experiments and Comparison with Bauer's Lepidochromes

Recent research on the colors of lepidopterans' scales has partly answered the question of whether the laminae have different colors. As explained above, the surface of the lower lamina is featureless, whereas the upper exhibits complex, grid-like structures. In this section, I will explain that this difference impacts how we perceive scales featuring pigmented and structural colors. I will then present the results of my own reconstructions of lepidochromy recipes, which will allow us to evaluate the contradictory positions with regard to the coloration of lepidochromes espoused by Poulin and Landbeck on the one hand, and Latreille on the other.

As a 2014 study by Doekele Stavenga et al. demonstrates, the lower laminae of isolated red and black scales of a red admiral (*Vanessa atalanta*) exhibit a metallic blue reflection. What their research also shows is that the orientation of an isolated scale producing the structural color of a peacock butterfly has no impact on its iridescence – in other words, it does not matter whether the lower or the upper lamina faces upwards and refracts the light.[74] In theory, this implies that while a butterfly's pigmentary colors should appear noticeably different in a single-transfer lepidochrome, its metallic scales should maintain their structural iridescence even when inverted.

Stavenga et al. further explain that the coloration of scales changes when they are stacked on a specimen's membrane – for instance, the two layers of red-pigmented scales on the forewings of peacock butterflies increase color saturation of the areas in question.[75] Regarding the structural colors, the study shows that the metallic blue coloration of this butterfly species is much more intense and visible when the scales are situated over a layer of dark ground scales – placed over entirely unpigmented scales, they would appear white.[76] Regarding some species of Lycaenidae exhibiting iridescent blue colors, the reasons are different: Their scales contain a brown pigmentation that changes the scale's reflectance and therefore enhances the refractive index of the chitinous material.[77]

74 Stavenga, Leertouwer, and Wilts 2014, 2174–2175.

75 Stavenga, Leertouwer, and Wilts 2014, 2178.

76 Stavenga, Leertouwer, and Wilts 2014, 2176.

77 Serge Berthier confirmed this to me in a personal conversation. Ren et al. 2020 identified brown pigments in the scales of three species of Lycaenidae that have metallic (but not blue) colors. On the iridescent blue colors of blue morpho butterflies (genus Morpho), see Berthier 2007, 26; Debat et al. 2018.

With these considerations in mind, I will now present the results of my own experimentations with lepidochromy on paper following recipes in order to assess whether the inverted position of the scales impacts the pigmentary colors, and whether the structural colors can be preserved. Since I was unable to successfully execute a second transfer, I will compare the colors of my single-transfer lepidochromes with a dry specimen of a similar species.[78] For this purpose, I have selected a peacock butterfly, a species often present in Marseus's paintings (Fig. 3.5a). The dorsal forewings are rusty red with black and brown motifs on the edge. On the apex, the eyespots are black, yellow, and iridescent blue. The dorsal hindwings display the same colors, with grey variations. The eyespots contain much structural blue colorsurrounded by black. The areas of the wings close to the body seem to be sprinkled with gold, which recalls the shimmering effects praised by the lepidochromists.

One of the lepidochromes prepared for the purposes of this study shows that pigmentary colors such as red are indeed much duller once the scales have been transferred onto the paper (Fig. 3.5b).[79] Many scales remained on the wing membranes, which was bound to have an impact on coloration given that intense pigmentary colors are the product of compact clusters of stacked scales, as explained in the above-mentioned paper by Stavenga et al. Another attempt on thicker paper yielded a better result (Fig. 3.6a).[80] With the exception of an area on the right forewing, almost all scales were successfully transferred, which renders the butterfly's characteristic patterns clearly visible. The rusty red seems slightly lighter and matter than on the specimen.[81] Furthermore, moving my second reconstruction under direct light produces a shimmering effect in the darkest parts: Brown and black are tinted with gold, while the iridescence of the eyespots and areas on the edge of the forewings recalls the sheen of an oil slick, bringing to mind the "false shimmer" that irritated

78 The second transfer probably failed because of the type of varnish used, which was too liquid. Instead of producing a sticky film on the surface of the lepidochromes, the varnish was soaked up by the paper. Other possible explanations are that the paper was too porous or should have been sized, a subject on which the lepidochromists remain silent.

79 For this attempt, 1.4 g of gum arabic and 0.6 g of tragacanth were dissolved in 18 ml of water, without salt, following the recipe of Latreille 1804, 1: 309–315. I applied a thin layer of glue on a thick paper, pressed the wings with my fingers, and left the lepidochrome to dry under a weight.

80 For this attempt, 1.4 g of gum arabic and 0.6 g of tragacanth were dissolved in 18 ml of water, with the addition of 3 gr of marine salt, again following the recipe in Latreille 1804, 1: 309–315.

81 Naumann (1848, 176) and Landbeck (1843, 263) observed that the lepidochromes sometimes appeared matte. Zirkler (1847, 4) noticed the same with regard to structural colors that had lost their iridescence, an issue I discuss further below.

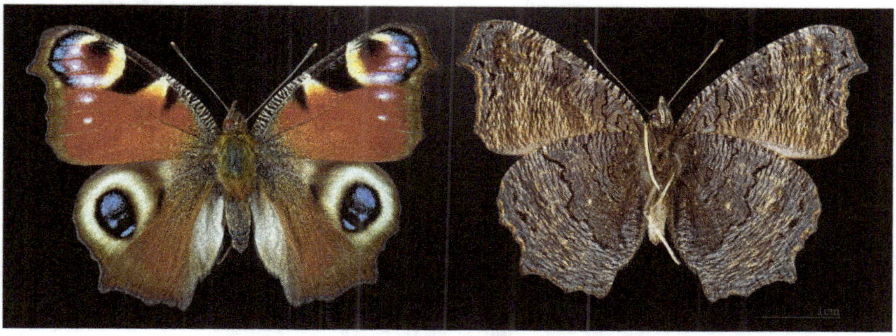

FIGURE 3.5A A dry specimen of the peacock butterfly (*Aglais io*). Public domain
PHOTO CREDITS: DIDIER ESCOUENS

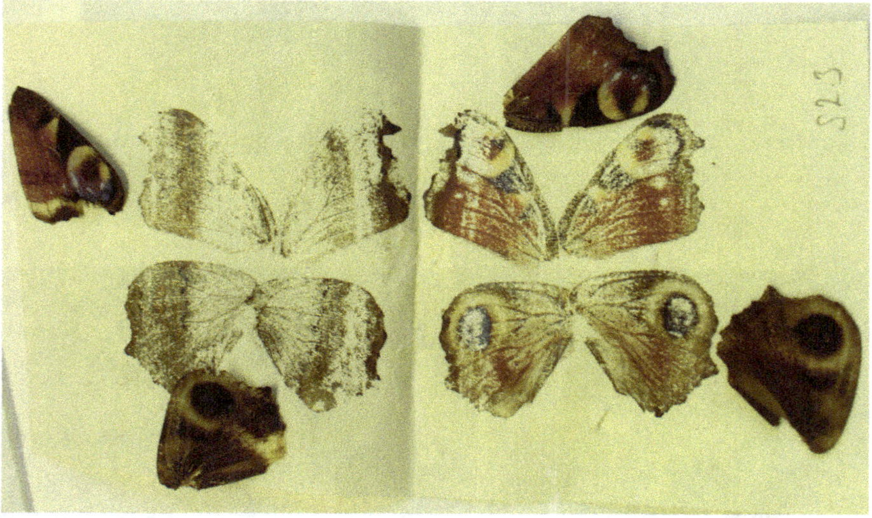

FIGURE 3.5B Peacock butterfly (*Aglais io*), lepidochrome on paper after Pierre-André Latreille's recipe, 2019
© V.E. MANDRIJ

Landbeck. This effect could either be produced by the inverted scales themselves in the manner described by Stavenga et al., or by the layer of dried gum arabic underneath.[82] According to my experiments, Poulin's and Landbeck's assessment that lepidochromes differ from real specimens with regard to

82 The glue used for this reconstruction contained salt, an element which, according to Latreille (1804: 1, 309), was supposed to prevent this very effect.

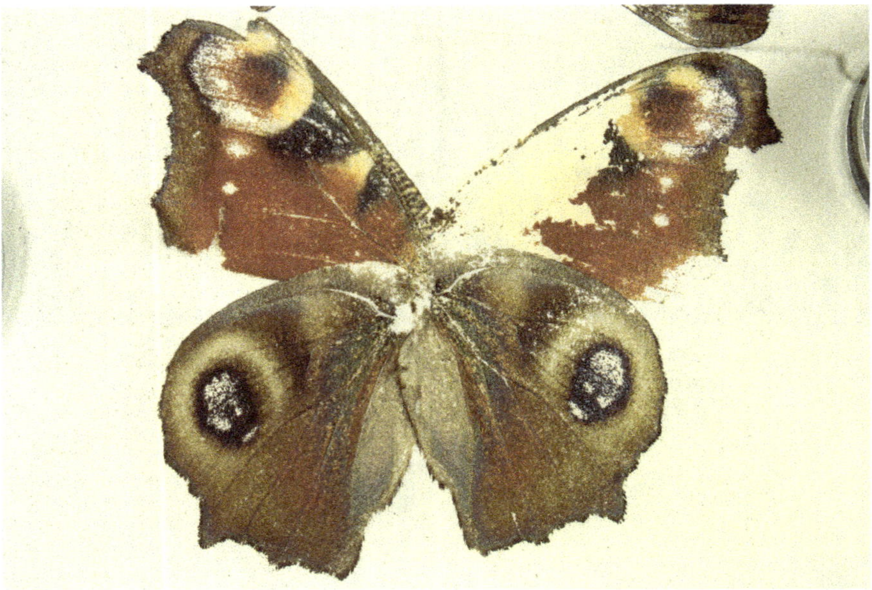

FIGURE 3.6A Peacock butterfly (*Aglais io*), lepidochrome on paper after Pierre-André Latreille's recipe, 2019
© V.E. MANDRIJ

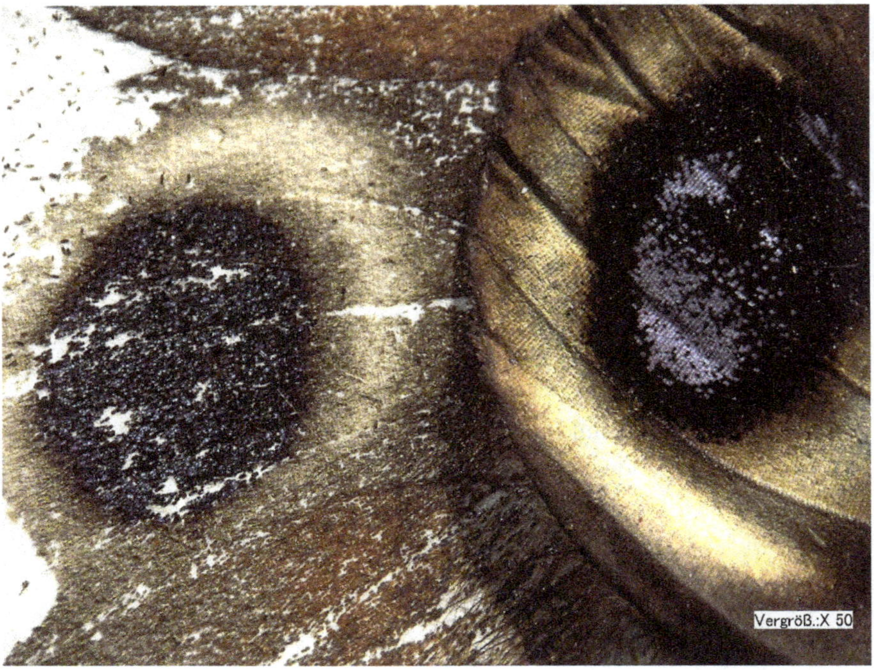

FIGURE 3.6B Microscopic view of a peacock butterfly (*Aglais io*): lepidochrome on paper after Pierre-André Latreille's recipe and the hindwing of a dry specimen, 2019
© V.E. MANDRIJ

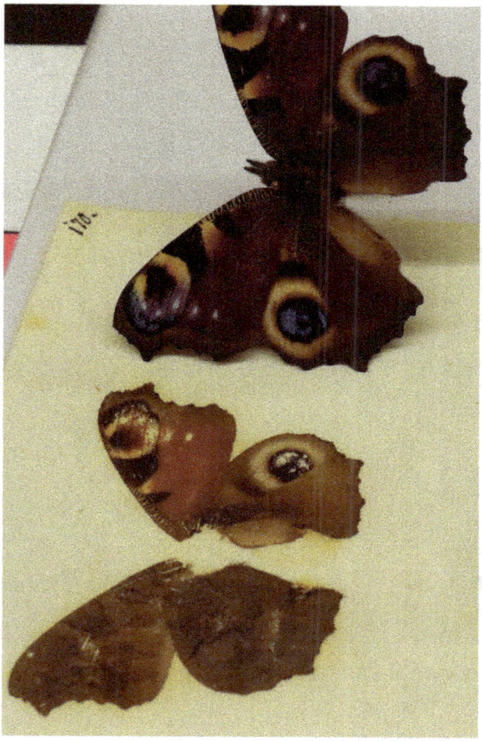

FIGURE 3.7
Comparison between a dry specimen of the peacock butterfly (*Aglais io*) and a lepidochrome of the same species on paper, from the collection of lepidochromes on cardboards by Friedrich Bauer, box 5, *c.*1840, 80 × 120 mm, Staatliches Museum für Naturkunde Stuttgart
© V.E. MANDRIJ

pigmentary colors is correct. As has also become clear, however, the resulting discrepancies are comparatively minor: While visible with the naked eye, they do not have a dramatic impact on the visibility of the patterns and colors.

This brings us to the question of whether it is possible to maintain the structural colors with a single transfer. In the second lepidochrome, the scales with structural colors have been transferred, but they appear white. Under the microscope, iridescent blue particles are visible, although the effect is much less intense than on the specimen (Fig. 3.6b). An examination of a lepidochrome by Friedrich Bauer shows similar lacunae in the structural colors when compared to a dry specimen (Fig. 3.7).

As peacock butterflies have a comparatively small number of structural scales, it is instructive to examine species such as the blue gossamer-winged butterflies (Lycaenidae) that are likewise present in Marseus's *sottoboschi*. The dorsal wings of the specimen depicted here are completely covered with iridescent blue, while small eyespots decorate the ventral wings in various colors such as white, brown, and black (Fig. 3.8a). My attempt to transfer the scales of an iridescent blue butterfly failed: The blue colors turned brown (Fig. 3.8b). As explained above, it is likely that the scales of this family contain a certain

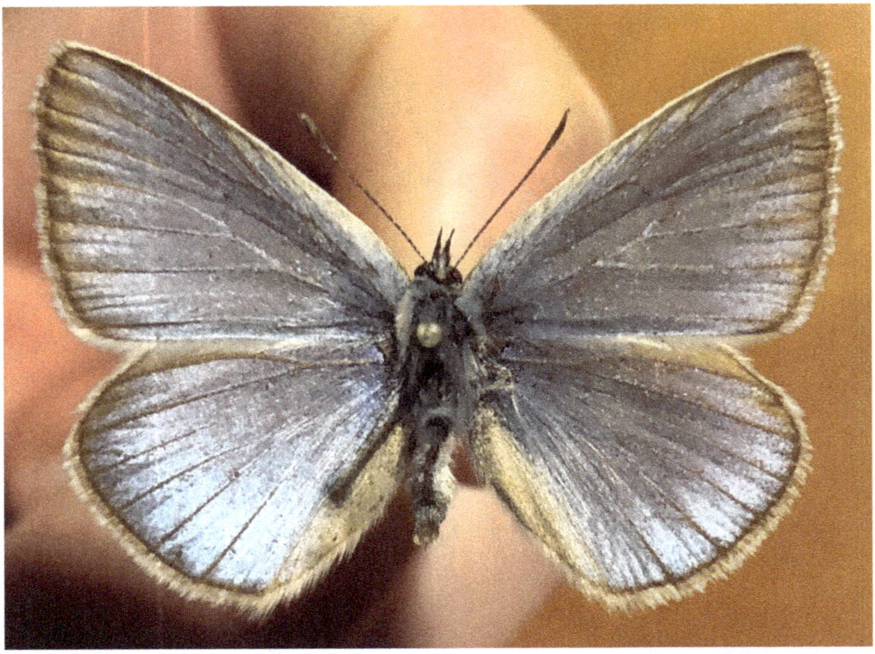

FIGURE 3.8A Dry exemplar of a male specimen of the Lycaenidae family, dorsal side © V.E. MANDRI

quantity of brown pigment, probably melanin, that helps to increase the iridescent effect. The brown color of the lepidochromes is thus probably the color of the scales when iridescence does not take place, in our case as the result of the destruction of their structure during transfer. We can assume that the outcome of my experiment resembles the results obtained by Poulin and Landbeck, which explains why they argued for a second transfer.

However, as Friedrich Bauer's lepidochromes demonstrate, it is possible to maintain the iridescent colors of butterflies even with a single transfer.[83] His collection of lepidochromes contains several examples of species of Lycaenidae which show the ventral and dorsal faces of two of their wings, with a visible

83 Contemporary artists Maximilian Prüfer and Lisa Albrecht have both managed to maintain structural colorswhen employing the technique of lepidochromy in their works, Prüfer using glue on paper and Albrecht oil on paper; both agree that shimmering colors are more visible on a black support. On Prüfer's technique, see Mandrij 2021b. I am grateful to the artists for discussing their techniques with me. Landbeck (1843, 260), too, acknowledged the efficacy of using black paper for printing blue scales.

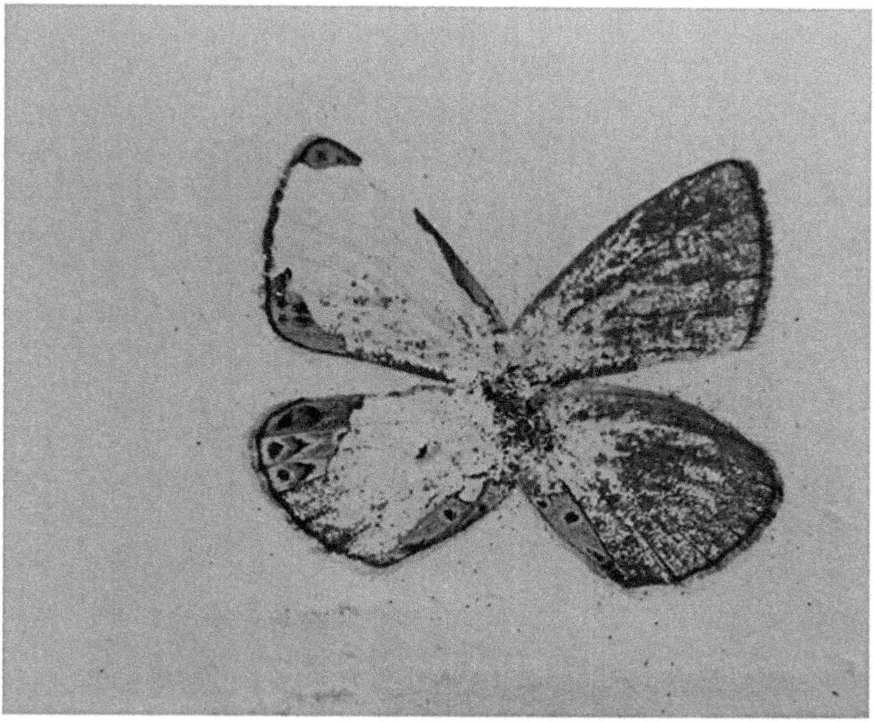

FIGURE 3.8B Butterfly from the Lycaenidae family, lepidochrome on paper after Pierre-André Latreille's recipe, 2019
© V.E. MANDRIJ

fold in the paper clearly attesting that he produced them in one go. Many lepidochromes in Bauer's collection show brown colors like my own example, a few exhibit little iridescence, but some display a striking structural blue as in this example of a "*Lycaena adonis*" (from the Lycaenidae family) (Fig. 3.9).[84] Not all scales were successfully transferred, but the hues of blue change to great effect according to the light angle. By demonstrating that the inverted orientation of the scale has no impact on its capacity to generate iridescence, this lepidochrome refutes the arguments advanced by Poulin and Landbeck, who believed that two transfers were required to preserve structural colors.

84 The example is taken from Bauer's second collection of lepidochromes on cards. On the large sheets, some lepidochromes also exhibit iridescent colors, but they are not as striking as the one shown here.

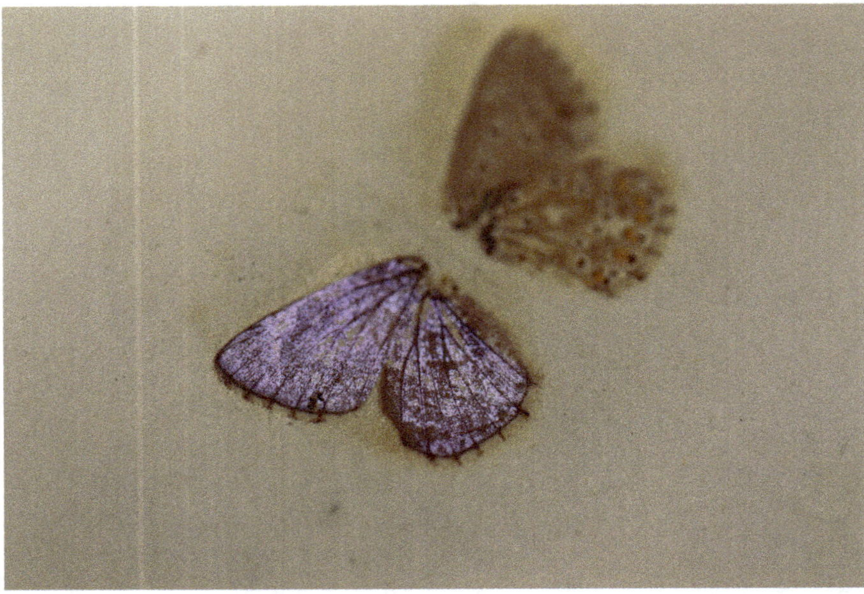

FIGURE 3.9 Friedrich Bauer, "*Lycaena adonis*," a butterfly from the Lycaenidae family, in Friedrich Bauer, lepidochrome on cardboards, box 4 "Lycaenides, Erycinides," *c.*1850, 8 × 12 cm, Stuttgart, Staatliches Museum für Naturkunde Stuttgart © V.E. MANDRIJ

7 Marseus's Lepidochromes in the *Sottoboschi*

After this brief digression into the eighteenth and nineteenth centuries, we will now return to Otto Marseus van Schrieck and the question of colors in the *sottoboschi*. Consequently, this section will illustrate the artist's modus operandi by combining the results of technical examinations of his Rotterdam painting with an analysis of lepidochromy recipes and a description of the outcomes of my own reconstructions. As the *sottobosco* in question is painted in oil on canvas, I will then explore the extent to which the colors of lepidochromes change when they are created with different materials than those described in the previously discussed lepidochromy recipes.

In the Rotterdam *sottobosco*, Marseus pressed the severed wings of lepidopterans on the surface of the painting before removing them again, thereby producing lepidochromes with a single transfer. The adhesive was probably the almost dried paint layer, which has been identified as lead white.[85] Known

85 As of yet, no pigment analysis has been carried out for Marseus's Rotterdam *sottobosco*. Bodo Beier (1987, 35), Susanna Steensma (1999, 62), Babette Hartwieg and Michael van

to dry quickly, this pigment is sticky enough to remove the scales from a butterfly's wings when mixed with oil. Using a layer of lead white was probably the most practical and efficient way to make lepidochromes. This choice, as I will show below, had an impact on the colors of the transferred scales, both pigmentary and structural.

My examinations under the microscope have shown that while scales are present in some of the lepidochromes of the Rotterdam *sottobosco* , they have been severely discolored by various factors (Fig. 3.3).[86] Many have turned brown or exhibit different degrees of transparency, which makes it very difficult to detect them. While it is thus entirely impossible to determine the number of transfers performed by Marseus on the basis of color discrepancies, the thick veins that left indentations during the pressing process provide an important clue. The characteristic imprints in the paint layer are clearly visible in this example of an unidentified species of butterfly flying on the left of the composition in the Marseus's Rotterdam *sottobosco* (Fig. 3.10). Here, the fine patterns of the venation on the ventral side of this lepidochrome's right forewing are very well defined, and must be distinguished from the craquelure that forms on the surface of paintings as a result of the aging process.[87] At least in this particular case, the imprint of the venations leaves no doubt that Marseus executed a single transfer of the lepidopteran's scales, since the veined membranes that produce such traces are destroyed in the initial transfer and cannot therefore be visible in a subsequent one.

Some scholars, such as Serge Berthier and a team of conservators, have come to a different conclusion.[88] Based on their technical examinations of a *sottobosco* attributed to Marseus from the Musée des Beaux-Arts in Grenoble,

---

der Goltz (2000, 44), Jakob Wedemeyer (2007, 13), Rebekka Kremkau (2020, 17), Sophie Lamb (2022, 37), and Erma Hermens (2022, 143) have deduced from their technical examinations of paintings by Marseus, Johann Falch, and Rachel Ruysch (1664–1750) that the artists used lead white pigment under the lepidochromes, yet not all of them agree that the paint layer was used to make the transfer. Jakob Wedemeyer informed me that a layer of oil could have been used as adhesive and that he had restored paintings by Marseus where, on rare occasions, lepidochromes appeared to have been produced without a lead white underlayer. Weber (1989, 36; 1993, 27; 1995, 214), too, has suggested that Ruysch and Elias van den Broeck (1649–1708) may have added a layer of glue and used a light ochre substrate instead of lead white.

86 As stated in note 8, more scientific research is needed to establish the causes of this discoloration.

87 The brownish colors stem from varnish that filled the hollow lines left by the veins. See Kremkau 2020, 18.

88 Berthier, Menu, and Mottin 2006; Menu 2006; Berthier et al. 2008. Hermens (2022, 143) explores the idea of a second transfer in her analysis of Marseus's technique based on Berthier's findings.

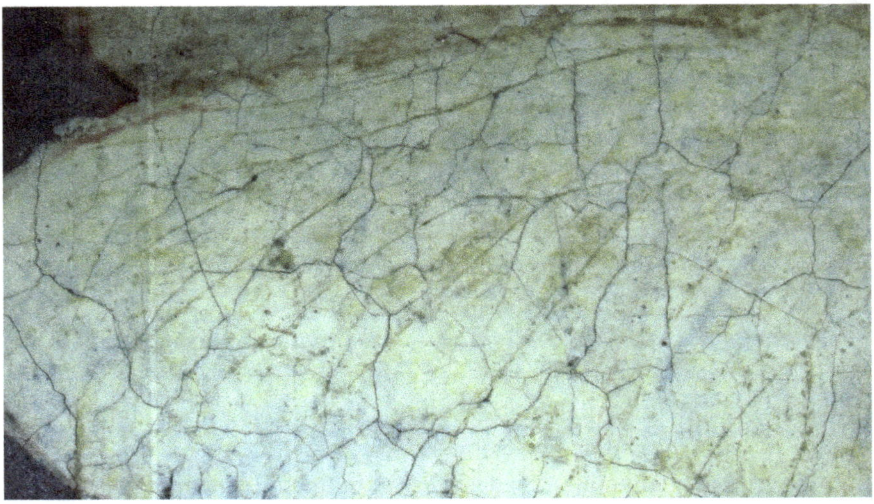

FIGURE 3.10 Microscopic view showing the imprint of veins in the lepidochrome of an unidentified species of butterfly, detail of Fig. 3.2

they argued that the imprint of a peacock butterflyshowed "the right side of the scales," which would prove that the painter made two transfers.[89] The execution of a second transfer cannot be completely ruled out given that the technique applied may vary from painting to painting. Yet, my own examinations yielded different results and I am reticent to generalize this hypothesis to other *sottoboschi*. First of all, the semi-transparency of the scales as seen in Berthier et al.'s microscopic photograph makes it quite difficult to discern how they are oriented.[90] Furthermore, many of the lepidochromes I have examined under magnification show clear evidence of contact between the wing and the painting surface: The veins have been imprinted into the paint material. Without carrying out my own microscopic examinations, I cannot assess if such indentations are present in the Grenoble *sottobosco* as well. Berthier and his team did not comment on this feature one way or the other, but the digital

89 Quoted in Menu 2006, 599. See the microscopic image in Berthier et al. (2008, Fig. 3). The painting in question was probably executed by an anonymous follower of Marseus. See Steensma 1999, cat. nr. B 3.55.

90 This tallies with my own examinations of other paintings; Kremkau (2020, 17–18) agrees with this assessment.

microscopic image of the peacock butterfly in this painting seems to show that the veins were also printed.[91]

Marseus's damaged lepidochromes have long since lost their original colors, but my reconstructions of lepidochromy on canvas nonetheless allow us to consider the impact of this particular support material on the representation of colors and iridescence. In general, I was able to produce lepidochromes on canvas easily without adding any extra adhesive. Applying a thin layer of modern tube oil paint on a prepared canvas and letting it dry for three days made it possible to transfer the scales of the butterfly wings by tapping the wings directly with the fingertips and then rubbing them energetically with the nail, adding a piece of paper in between.[92] In my lepidochrome of a peacock butterfly, almost all scales were transferred, and the patterns on the wings are clearly visible (Fig. 3.11).

As explained above, the inversion of scales that occurs in single-transfer lepidochromes does not preclude the accurate, true-to-nature depiction of lepidopterans, small color variations notwithstanding. It is therefore legitimate to wonder whether Marseus used the lepidochromy technique for the optical effect achieved by the natural scales' coloration. My own lepidochrome of a peacock butterfly demonstrates that the pigmentary colors red and brown appear darker on canvas than on the original dry specimen; it also lacks the shiny effect exhibited by its counterpart on paper.[93] That said, employing the natural colors of lepidopterans nevertheless increased the lifelikeness of their depictions compared to the use of paint.

This brings us to the issue of why Marseus's lepidochromes have been painted over, and by whom. The author of the added colors is sometimes challenging to identify: Marseus himself painted many lepidochromes, but other hands also intervened to counter the deterioration of their original colors. For instance, the butterfly, probably from the Lycaenidae family, that is flying from left to right in front of the brown rock in the Rotterdam *sottobosco*, has been overpainted with orange (Fig. 3.12). This paint layer clearly covers the

91 This previously unpublished image was taken by Bruno Mottin and is referred to as 'dc34945' in the documentation of the Centre de Recherche et Restoration des Musées de France (C2RMF).

92 For this experiment, I employed Norma-brand zinc titanium white (made by Schmincke); the canvas was prepared with a size made of animal glue and white chalk under the assistance of Peter Vogel (ABK Stuttgart) for which I am thankful. The lead white pigments used by Marseus in the seventeenth century are now very difficult to obtain due to their high toxicity. Compared to modern materials, historical lead white pigment is much more malleable and was probably at least as efficient as the paint I used.

93 It is unlikely that Marseus used gum arabic on oil because oil and water are immiscible.

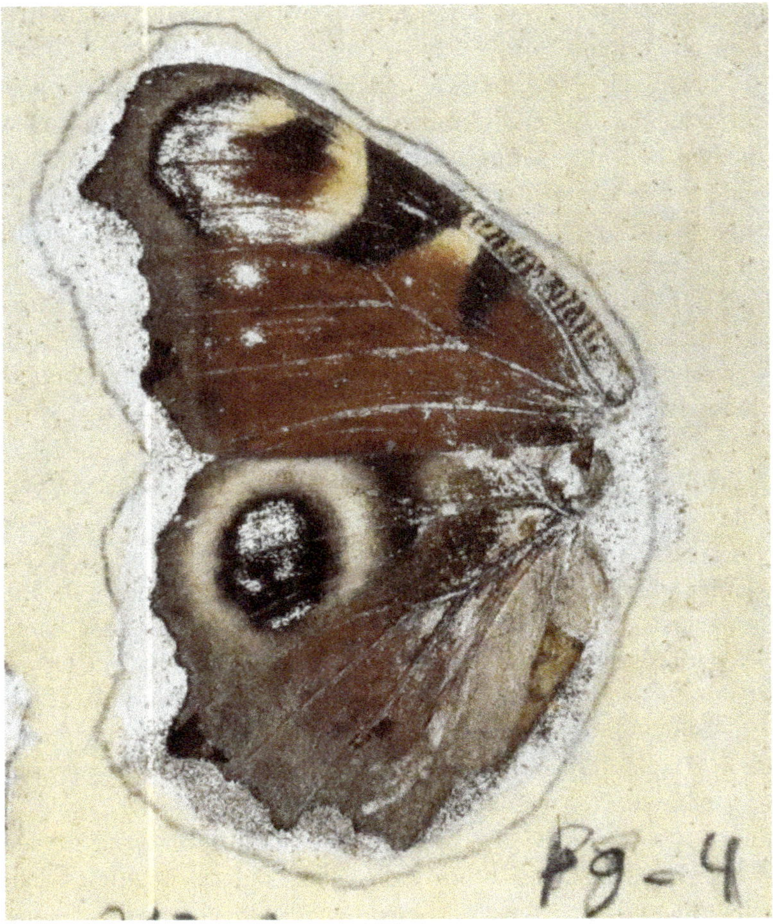

FIGURE 3.11 Peacock butterfly (*Aglais io*), lepidochrome on canvas with a layer of zinc titanium white oil paint of the Norma brand (Schmincke), 2022
© V.E. MANDRIJ

craquelures, which is a sign of an addition made a long time after the original painting was executed. In some cases, however, crack patterns in the paint layers strongly suggest that the colors were added in Marseus's time, opening up the possibility that he made these additions himself.[94] In the case of the brown paint on the wings of the small emperor moth from the Rotterdam *sottobosco*,

94 Marseus could have added paint during the initial painting process or years later. Likewise, another early modern painter could have made the additions at a later point in time, which would then also exhibit craquelure.

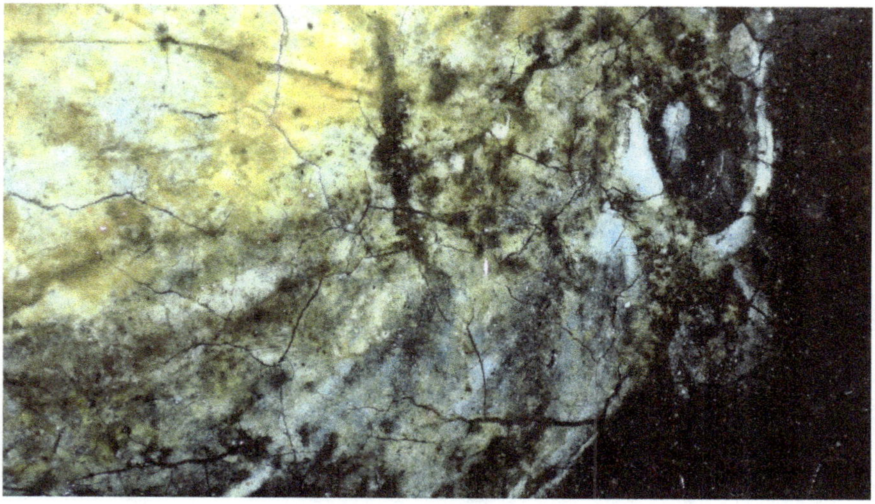

FIGURE 3.12 Microscopic view showing orange glazes on the lepidochrome of a member of the Lycaenidae family, detail of Fig. 3.2

for example, this is a likely scenario, since the color is very similar to that used for depicting its body (Fig. 3.13).[95]

There is a number of practical reasons that could have induced Marseus to paint over his lepidochromes. As pointed out by later lepidochromists, paint can mitigate changes in coloration and fills the gaps where scales have failed to transfer.[96] Furthermore, knowing that the pigmentary colors of his butterflies and moths would lose their vividness over time, the painter may have attempted to delay their discoloration.[97] Another factor of particular relevance in the context of oil painting is the application of a final varnish that saturates the colors underneath and modifies their perception – an effect that Marseus may well have anticipated and tried to preempt.[98]

---

95 Made during my examination of the painting under the microscope, this observation requires further verification.

96 Lamb (2022, 40) is of the same opinion.

97 There are limits to this explanation, however. For one, Marseus's approach to retouching his butterflies and moths was actually quite systematic: For instance, he always added red, black, and blue paint on the forewings of red admirals (*Vanessa atalanta*) but left the real scales of the ventral hindwings visible, which suggests that he made a deliberate choice regarding which original patterns to display and which to modify.

98 On the visual effect of varnish, see, for instance, Rie 1987 and 1989, Stols-Witlox 2001, Phenix and Townsend 2020. I propose an interpretation in my PhD thesis. See Mandrij 2023.

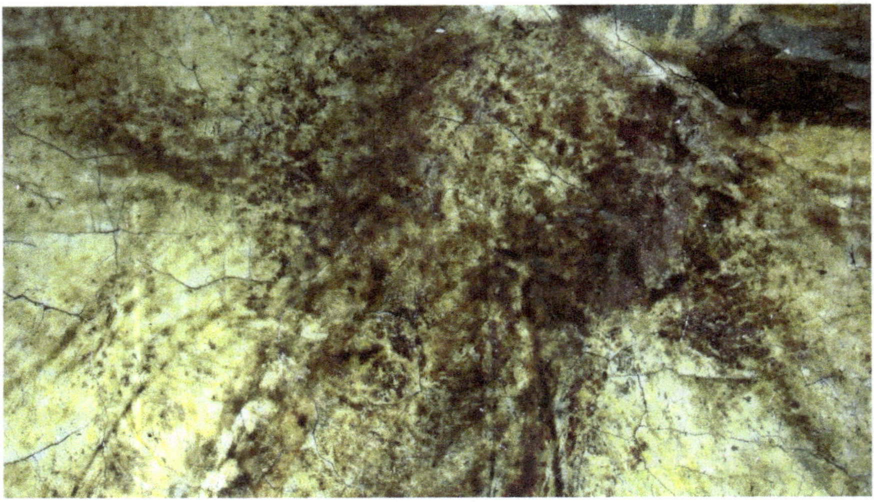

FIGURE 3.13 Microscopic photograph showing brown paint added onto a lepidochrome of a small emperor moth (*Saturnia pavonia*), detail of Fig. 3.2

Another question is whether Marseus was able to reproduce iridescent colors with lepidochromes, an optical effect that is particularly challenging to imitate with regular paint. Many of his *sottoboschi* depict species with structural colors, such as peacock butterflies and some species of Lycaenidae that are difficult to identify given the lepidochromes' damaged state. All blue areas of his lepidochromes have been painted over, and there is evidence that Marseus himself was the author. For example, the lepidochrome of a member of the *Phengaris* genus at the bottom of the Rotterdam painting is seen in profile with both faces of its wings visible. My microscopic examinations have confirmed that a large number of transparent scales are still present on the dorsal side of the right forewing that is normally fully iridescent, which means that Marseus transferred the scales of this insect's wings (Fig. 3.14). However, this evidence does not allow us to conclude whether or not the iridescent effect was originally visible. Small areas of cracked blue paint are visible here and there, which suggests that Marseus retouched the areas where the scales had not transferred. Another possible interpretation is that the wing was originally completely overpainted with blue, and that most of the paint layer has flaked off in the meantime.[99] Existing scholarship on Marseus's lepidochromes takes

99 I am thankful to Sophie Lamb for this suggestion.

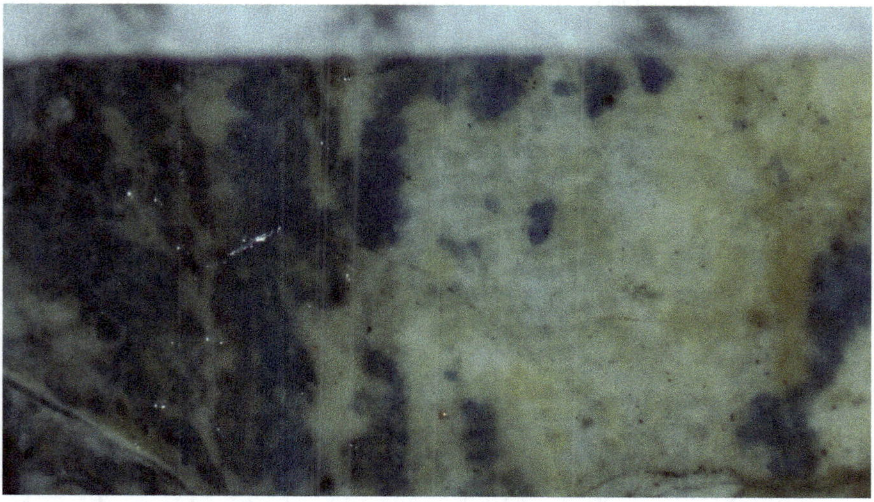

FIGURE 3.14 Microscopic view showing ranges of scales with blue paint on the lepidochrome of a member of the *Phengaris* genus, detail of Fig. 3.2 © COURTESY OF THE MUSEUM BOIJMANS VAN BEUNINGEN. PHOTO CREDITS: EVA VAN ZUIEN AND THE AUTHOR

it for granted that he was unable to maintain the iridescence of the depicted specimens, to which he responded by overpainting their scales with blue.[100] Berthier and his colleagues provide a physical explanation of why such an approach might have been necessary: Iridescence in lepidopterans is produced by the air compartments contained in their scales, whose refractive index differs from that of the surrounding chitin; bringing the scales into contact with a liquid such as oil or vanish fills these air gaps with a material that has a similar refractive index, and thus annihilates the effect.[101]

I would interject that the addition of paint does not amount to conclusive proof of the impossibility of maintaining the iridescence of lepidochromes executed in oil on canvas. Here, too, significant insights are yielded by the historical reconstruction of the technique. My initial attempt to produce a canvas-based lepidochrome of a species of gossamer-winged butterfly using a modern tube oil paint layer as adhesive ended in failure, like my previous experiment with a paper base.[102] My first assumption was that pressing and rubbing the wings with my fingernails (with a protective paper in between)

100 Steensma 1999, 62–63; Berthier 2008, 56; Kremkau 2020, 18–19; Lamb 2022, 41; Hermens 2022, 143.
101 Berthier et al. 2008, 53–56.
102 Either the membrane did not detach from the paint, or the colors turned brown.

had crushed the scales to the point where they could no longer produce an iridescent effect. However, applying similar pressure to a wing of the same species on a layer of modern black tube paint led to a different result:[103] As Figure 3.15 shows, the structural colors are clearly visible and change their hue according to the incidence of light, which confirms that a dark underlayer enhances iridescence. The same holds true for the blue eyespots of the peacock butterfly (Fig. 3.16). Notably, the pigmentary colors are less vivid than on a white substrate.

As can be seen from my reconstruction, the use of oil paint as an adhesive does not, in and of itself, render blue iridescence impossible. One important finding is that the color of the underlayer plays a major role with regard to the visibility and intensity of both structural and pigmentary colors. It is conceivable that future attempts to maintain the visibility of structural colors with white oil paint on canvas will be as successful as Bauer's lepidochromes on paper. For the time being, we cannot say with certainty whether the scales in Marseus's paintings were originally shiny and iridescent on a lead white substrate, and if so, for how long – one possibility is that he overpainted the blue parts after some time had elapsed because the shimmering colors had disappeared; another is that varnishing the lepidochromes had cancelled their iridescence as explained by Berthier et al. We can conclude, however, that Marseus was not primarily interested in utilizing the natural iridescent effect of butterfly scales in his paintings: Instead of employing a black underlayer to increase the luminosity of the iridescent colors, he favored white to emphasize the visibility of the pigmentary colors. He may, however, have explored other painting techniques to imitate the iridescence of blue species of butterflies – it is known, for instance, that he used the technique of glazing a silver leaf with blue pigments to achieve this effect.[104]

103 For this experiment, I used Norma-brand ivory black by Schmincke.

104 The peacock butterfly in Marseus's *sottobosco* conserved in Dessau was made with glazed silver leaf (see Steensma 1999, 58 and Wedemeyer 2007, 28). A species of Lycaenidae in *Plants and Insects*, 1665, oil on canvas, 102.3 × 75.8 cm, The Hague, Mauritshuis, inv. no. 532 (Steensma 1999, cat. nr. B1.52, https://www.mauritshuis.nl/en/our-collection/artworks/532-plants-and-insects/) and the blue morpho in *Large Thistle*, oil on canvas, 132 × 90 cm, Munich, Alte Pinakothek, inv. no. 1966 (see Steensma 1999, cat. no. B1.80, https://www.sammlung.pinakothek.de/de/artwork/y7GE6oM4PV/otto-marseus-van-schrieck/die-grosse-distel) were probably painted with shell silver and/or gold under blue glazes. There is a high probability that Marseus used shell silver to depict a species of fritillary in a watercolor, but no further analyses have been carried out so far to confirm this hypothesis. See Otto Marseus van Schrieck, *Anemone and Queen of Spain fritillary* (*Issoria lathonia*)gouache on paper, 278 × 192 mm, Knowsley Hall, The Derby Collection, Album W292, f.44r, reproduction published in Seelig 2017, 158 (Fig. 127).

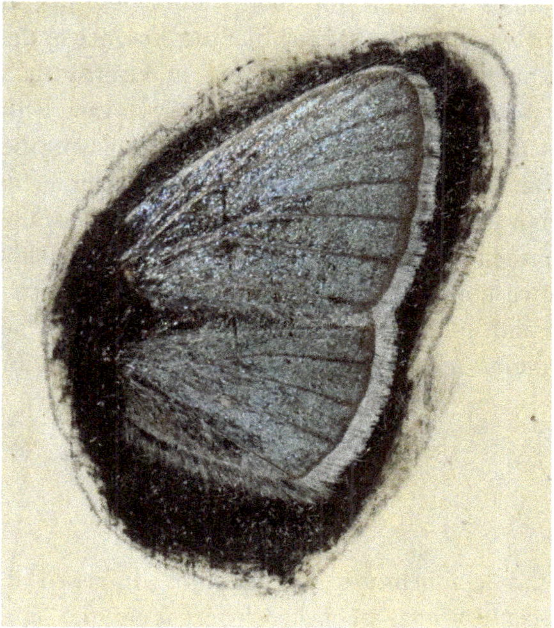

FIGURE 3.15
Species of Lycaenidae, lepidochrome on prepared canvas with a layer of ivory black oil paint of the Norma brand (Schmincke), 2022

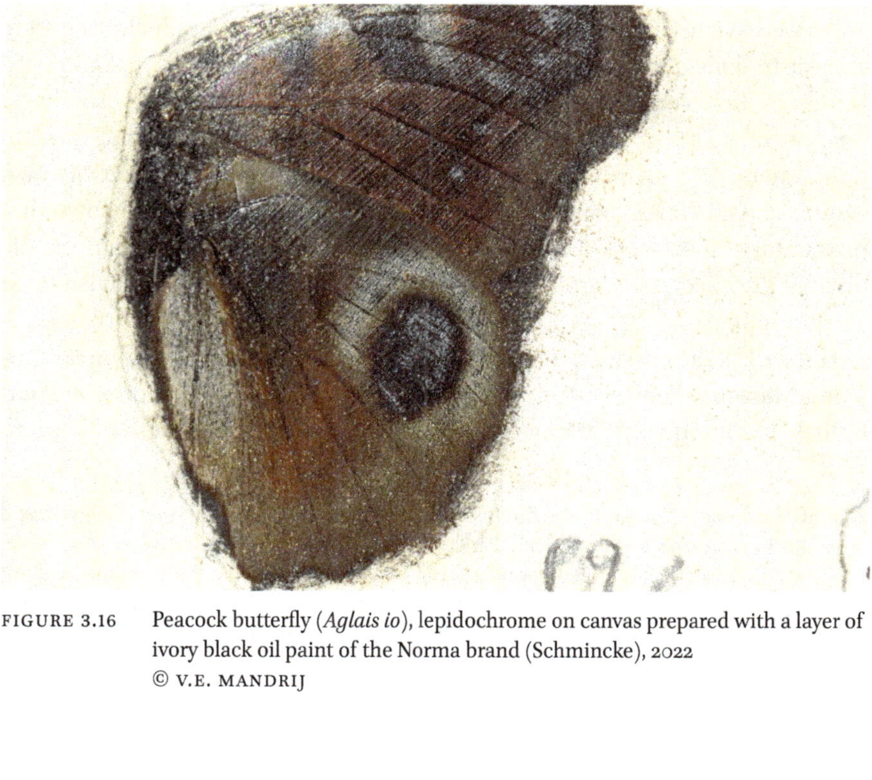

FIGURE 3.16 Peacock butterfly (*Aglais io*), lepidochrome on canvas prepared with a layer of ivory black oil paint of the Norma brand (Schmincke), 2022

In light of these observations regarding the moths and butterflies in Marseus's *sottoboschi*, it seems reasonable to suggest that the presence of the scales of lepidopterans in his paintings was more important to the artist than their original colors. Marseus's use of lepidochromes was fully in tune with the contemporary European culture of collecting, studying, and displaying the "wonders of Nature" as well as *artificialia* in curiosity cabinets in order to celebrate the awe-inspiring power of their Creator.[105] By making use of the scales of lepidopterans to represent them in his paintings, Marseus implicitly acknowledged nature's creative potential; and by selectively adding costly pigments such as vermilion and ultramarine to their delicate wings,[106] he effectively blurred the boundaries between reality and illusion, between art and nature.[107]

8 Conclusion

This chapter investigated the lepidochromy technique as discussed in eighteenth- and nineteenth-century texts published by its users, the lepidochromists, and as applied in the work of the Dutch painter Otto Marseus van Schrieck.

As we have seen, approaches to and experiences with lepidochromy were subject to considerable variation: Some lepidochromists such as Poulin and Landbeck believed that the scales' laminae had different colors, so that a second transfer was necessary to return the scales to their original orientation; others, such as Latreille, observed that there was no difference. My own experiments with lepidochromy recipes on paper and canvas have shown that pigmentary colors sometimes become darker and less vibrant than on the original dry specimens, and that the structural colors are not easily visible on a white support.

Unlike Friedrich Bauer, who depicted lepidopterans for taxonomical purposes, Marseus's goal was to create vivid, lifelike impressions of moths and butterflies that interacted with an overarching narrative. The addition of paint to

105 On curiosity cabinets, see, for instance, Impey and MacGregor 1985, Bergvelt and Kistemaker 1992, and Daston and Park 1998.

106 Lamb (2022, 41) identified vermilion and ultramarine (that is, the pigment made of lapis lazuli) in the paint covering the lepidochromes in *Flowers, insects and reptiles*, 1673, 69.5 × 52.7 cm, Cambridge, Fitzwilliam Museum, inv. no. 0303 (Steensma 1999, cat. no. B1.87).

107 On Marseus's ambition to blur the boundaries between art and nature, see Weber 1993, Leonhard 2020, 120 and Mandrij 2021b.

the scales helped to integrate the butterfly materials into the respective composition, while also opening up a competitive discourse about nature's artistic potential and his own. By making lepidochromes, the painter in effect created a *paragone* between art and nature within an image that was truer to nature than painting.

Acknowledgements

This research was carried out under the auspices of the graduate research program *Changing Frames. Art History and Art Technology in Exchange* (Universität Konstanz/Staatliche Akademie für Bildenden Künste Stuttgart, 2018–2023) financed by the Volkswagen Stiftung. I am very grateful to the following scholars for their support: Karin Leonhard (Universität Konstanz); Doris Oltrogge (CICS); Grace Touzel (Natural History Museum London); Ruben Suykerbuyk (Museum Boijmans van Beuningen Rotterdam) and Eva van Zuien; Marianne Elias (Museum National d'Histoire Naturelle Paris); Florent Figon (Université Grenoble Alpes); Serge Berthier (Paris, Sorbonne Université); Katharina Schmidt-Loske (Biohistoricum Bonn); and the conservators Rebekka Kremkau, Sophie Lamb, and Jakob Wedemeyer. I would also like to thank Regina Jäckel (Naturmuseum Augsburg), Stefan Kerkhof (Koninklijk Belgisch Instituut voor Natuurwetenschappen in Brussel), Hossein Rajaei (Staatliches Museum für Naturkunde Stuttgart), and Melissa Whitaker (ETH Zurich) for generously providing specimens for my experiments.

Bibliography

A.M.C. 1879. *A Guide to Nature-Printing: Butterflies and Moths*. London: Harrison.

Anonymous. 1731. *The Art of Drawing and Painting in Watercolours*. London: Printed for J. Peele.

Ayre, Evelyn, and George Bevan. 2016. "Calibrated UV Reflectance Photography of Hebomoia Glaucippe Sulphurea." *Collection Forum* 30 (1): 34–50. https://doi.org/10.14351/0831-4985-30.1.34.

Beier, Bodo. 1987. "'Contre-Epreuves' in der barocken Stillebenmalerei." *Maltechnik/Restauro* 1 (93): 35–39.

Berge, Friedrich. 1863. *Schmetterlingsbuch: oder Allgemeine Naturgeschichte der Schmetterlinge und besondere der Europäischen Gattungen*. Third edition. Stuttgart: R. Thienemann's Verlag.

Bergvelt, Ellinnoor, and Renée Kistemaker. 1992. *De wereld binnen handbereik. Nederlandse kunst- en rariteitenverzamelingen, 1585–1735*. Exhibition Catalogue. Amsterdam: Amsterdams Historisch Museum.

Berthier, Serge. 2007. *Iridescences: The Physical Colours of Insects*. New York: Springer.

Berthier, Serge, J. Boulenguez, Michel Menu, and Bruno Mottin. 2008. "Butterfly Inclusions in Van Schrieck Masterpieces. Techniques and Optical Properties." *Applied Physics* 92: 51–57. https://doi.org/10.1007/s00339-008-4480-8.

Berthier, Serge, Michel Menu, and Bruno Mottin. 2006. "Otto Marseus Van Schrieck, chardon, reptile et papillons: La nature dans l'œuvre." In *Couleur & Temps. La couleur en conservation et restauration* 12ème journée d'études de la SFIIC (Section française de l'insitut international de conservation). Paris: Institut national du patrimoine, 177–186.

Breazeale, William. 2007. "Nature and a New Drawing by Otto Marseus van Schrieck." *Master Drawings* 45: 527–533.

Brehm, Gunnar, Konrad Fiedler, Christoph L. Häuser, and Helmut Dalitz. 2008. "Methodological Challenges of a Megadiverse Ecosystem." In *Gradients in a Tropical Mountain Ecosystem of Ecuador*, Erwin Beck, Jörg Bendix, Ingrid Kottke, Franz Makeschin, and Reinhard Mosandl (eds.). [Ecological Studies 198]. Berlin, Heidelberg: Springer Berlin Heidelberg, 41–47.

Cave, Roderick. 2010. *Impressions of Nature: A History of Nature Printing*. London, New York: British Library, Mark Batty Publisher.

Cobb, Matthew. 2006. *The Egg & Sperm Race: The Seventeenth-Century Scientists Who Unravelled the Secrets of Sex, Life and Growth*. London: The Free Press.

Daston, Lorraine, and Katharine Park. 1998. *Wonders and the Order of Nature, 1150–1750*. New York; Cambridge (Mass.): Zone Books: Distributed by MIT Press.

Debat, Vincent, Serge Berthier, Patrick Blandin, Nicolas Chazot, Marianne Elias, Doris Gomez, and Violaine Llaurens. 2018. "Why Are Morpho Blue?" In *Biodiversity and Evolution*, Philippe Grandcolas and Marie-Christine Maurel (eds.). Amsterdam: Elsevier, 139–74. https://doi.org/10.1016/B978-1-78548-277-9.50009-7.

Downey, John C., and Arthur C. Allyn. 1975. "Wing-Scale Morphology and Nomenclature." *Bulletin of the Allyn Museum* 31: 1–32.

Dupré, Sven, Anna Harris, Julia Kursell, Patricia Lulof, and Maartje Stols-Witlox (eds.). 2020. *Reconstruction, Replication and Re-Enactment in the Humanities and Social Sciences*. Amsterdam: Amsterdam University Press. https://doi.org/10.2307/j.ctv1bofvx7.

Fitton, Mike, and Pamela Gilbert. 1994. "Insect Collections." In *Sir Hans Sloane: Collector, Scientist, Antiquary – Founding Father of the British Museum*, Arthur MacGregor (ed.). London: British Museum Press, 112–23.

Ghiradella, Helen. 1984. "Structure of Iridescent Lepidopteran Scales: Variations on Several Themes1." *Annals of the Entomological Society of America* 77 (6): 637–645. https://doi.org/10.1093/aesa/77.6.637.

Ghiradella, Helen. 1991. "Light and Color on the Wing: Structural Colors in Butterflies and Moths." *Applied Optics* 30 (24): 3492–3500. https://doi.org/10.1364/AO.30.003492.

Goedaert, Johannes. 1660. *Metamorphosis naturalis ofte historische beschryvinghe van den oirspronck, aerd, eygenschappen ende vreemde veranderinghen der wormen, rupsen, maeden, vliegen, witjens, byen, motten en dierghelijcke dierkens meer; niet uyt eenighe boecken, maer alleenelijck door eygen ervarentheydt uytgevonden, beschreven, ende na de konst afgeteyckent.* Middleburg: Apud Jacobum Fierensium.

Hartwieg, Babette, and Michael van der Goltz (eds.). 2000. *Bildproduktion. Über die Herstellung niederländischer Gemälde im 17. Jahrhundert.* Exhibition Catalogue. Hannover: Niedersächsisches Landesmuseum.

Hermens, Erma. 2022. "Crawly Creatures, Forest Floors and Butterfly Wings Otto Marseus van Schrieck and Sottobosco." In *Crawly Creatures. Little Animals in Art and Science,* Jan de Hond, Eric Jorink, and Hans Mulder (eds.). Exhibition Catalogue. Amsterdam, Rijksmuseum. 30 September 2022–15 January 2023. Amsterdam: Rijksmuseum, 139–149.

Hildebrecht, Douglas R. 2004. "Otto Marseus van Schrieck (1619/20–1678) and the Nature Piece. Art, Science, Religion and the Seventeenth-Century Pursuit of Natural Knowledge." Unpublished doctoral dissertation, Ann Arbor: University of Michigan.

Hond, Jan de, Eric Jorink, and Hans Mulder (eds.). 2022. *Crawly Creatures. Little Animals in Art and Science.* Exhibition Catalogue. Amsterdam, Rijksmuseum. 30 September 2022–15 January 2023. Amsterdam: Rijksmuseum.

Hoogstraten, Samuel van. 1678. *Inleydinge tot de hooge schoole der schilderkonst: anders de zichtbaere werelt.* Rotterdam: François van Hoogstraeten.

Hoogstraten, Samuel van. 2021. *Samuel van Hoogstraten's Introduction to the Academy of Painting, or, The Visible World.* Edited by Celeste Brusati. Translated by Jaap Jacobs. Texts & Documents. Los Angeles: Getty Research Institute.

Horn, Hendrik J. 2000. *The Golden Age Revisited: Arnold Houbraken's Great Theatre of Netherlandish Painters and Paintresses.* Doornspijk: Davaco Publisher.

Houbraken, Arnold. 1718. *De Groote Schouburgh Der Nederlantsche Konstschilders En Schilderessen.* Amsterdam: Gedrukt voor den autheur, daar de zelve ook te bekomen zyn.

Impey, Oliver, and Arthur MacGregor (eds.). 1985. *The Origins of Museums: The Cabinet of Curiosities in Sixteenth and Seventeenth Century Europe.* Oxford: Clarendon Press.

Jorink, Eric. 2011. "Beyond the Lines of Appelles. Johannes Swammerdam, Dutch Scientific Culture and the Representation of Insect Anatomy." In *Art and Science in the Early Modern Netherlands/Kunst en wetenschap in de vroegmoderne Nederlanden,* Eric Jornik and Bart Ramaker (eds.). [Netherlands Yearbook for History of Art/Nederlands Kunsthistorisch Jaarboek 61]. Zwolle: Wbook, 148–83.

Jorink, Eric. 2014. "Snakes, Fungi and Insects. Otto Marseus van Schrieck, Johannes Swammerdam and the Theory of Spontaneous Generation." In *Zoology in Early Modern Culture. Intersection of Science, Theology, Philology, and Political and*

Religious Education, Karl A.E. Enenkel and Paul J. Smith (Eds.). [Intersections 7]. Leiden: Brill, 197–234.

Kinoshita, Shuichi. 2008. *Structural Colors in the Realm of Nature*. Singapore: World Scientific Publishing Company.

Köster, Heinrich Martin Gottfried. 1793. *Deutsche Encyclopädie oder Allgemeines Real-Wörterbuch aller Künste und Wissenschaften*. Vol. 17. Frankfurt am Main: Barrentrapp und Weuner.

Kremkau, Rebekka. 2020. "Schmetterlingsflügel, Moos und Eidechsenhaut. Eine vergleichende Studie zu künstlerischen Techniken in Sottobosco-Gemälden. Vol. 3: Theoretischer Teil." Unpublished master thesis, Dresden: Hochschule für Bildende Künste Dresden.

Lamb, Sophie. 2022. "Material and Technique as a Presentation of Knowledge in Otto Marseus van Schrieck's Flowers, Insects and Reptiles." *Hamilton Kerr Institute Bulletin* 9, 32–44.

Landbeck, Christian Ludwig. 1843. "Das Abdrucken der Schmetterlinge auf Papier." In *Das Buch der Welt*, Carl Hoffmann (ed.). Stuttgart: Hoffmann'sche Verlags Buchhandlung, 259–63.

Latreille, Pierre-André. 1804. *Histoire naturelle, générale et particulière des crustacés et des insectes*. Vol. 1. Paris: F. Dufart.

Ledermüller, Martin Frobenius. 1763. *Mikroskopische Gemüths- und Augen-Ergötzung: Bestehend in Ein Hundert nach der Natur gezeichneten und mit Farben erleuchteten Kupfertafeln, sammt deren Erklärung*. Vol. 2. Nuremberg: Verlegt von Adam Wolffgang Winterschmidt, kupferstecher in Nürnberg, gedruckt von Christian de Launoy.

Lenteren, Joop C. van, and Charles J. Godfray. 2005. "European Science in the Enlightenment and the Discovery of the Insect Parasitoid Life Cycle in the Netherlands and Great Britain." *Biological Control* 32 (1): 12–24. https://doi.org/10.1016/j.biocontrol.2004.08.009.

Leonhard, Karin. 2009. "Pictura's Fertile Field: Otto Marseus van Schrieck and the Genre of Sottobosco Painting." *Simiolus: Netherlands Quarterly for the History of Art* 34 (2): 95–118.

Leonhard, Karin. 2011. "Painted Poison: Venomous Beasts, Herbs, Gems, and Baroque Colour Theory." In *Art and Science in the Early Modern Netherlands/Kunst en wetenschap in de vroegmoderne Nederlanden*, Eric Jorink and Bart Ramaker (eds.). [Netherlands Yearbook for History of Art/Nederlands Kunsthistorisch Jaarboek 61]. Zwolle: WBooks, 116–47.

Leonhard, Karin. 2013. *Bildfelder: Stilleben und Naturstücke des 17. Jahrhunderts*. Berlin: Akademie Verlag.

Leonhard, Karin. 2014. "Larve und Mimikry: Etymologisches und Entomologisches im 17. Jahrhundert." In *Maske, Maskerade und die Kunst der Verstellung: Vom Barock bis zur Moderne*, Christiane Kruse (ed.). [Wolfenbütteler Arbeiten zur Barockforschung 52]. Wiesbaden: Harrossowitz, 173–92.

Leonhard, Karin. 2020. *The Fertile Ground of Painting. 17th-Century Still Lifes and Nature Pieces*. London, Turnhout: Harvey Muller Publishers.

Mandrij, V.E. 2021a. "Otto Marseus van Schrieck and the Sottobosco Paintings: A Hybrid Subject as a Form of Automimesis." In *Questioning Pictorial Genres in Dutch Seventeenth-Century Art*, Marije Osnabrugge (ed.). [Gouden Eeuw. New Perspectives on Dutch Seventeenth-Century Art 2]. Turnhout: Brepols, 193–214.

Mandrij, V.E. 2021b. "Painted by Nature, Printed by Artists: Butterfly Materials in the Work of Otto Marseus van Schrieck and Maximilian Prüfer." *Humans and Other Animals/Mensen en andere diere*, Erik Jorink, Joanna Woodall, and Edward H. Wouk [Netherlands Yearbook for History of Art/Nederlands Kunsthistorisch Jaarboek 71]. Leiden: Brill, 276–312. https://doi.org/10.1163/22145966-07101011.

Mandrij, V.E. 2023. "Ornamenting a Painting with a Butterfly. Otto Marseus van Schrieck (c. 1620–1678) and the Lepidochromy Technique." Unpublished doctoral dissertation. Konstanz: University of Konstanz.

Menu, Michel. 2006. "Analysis of Works of Art Down to the Nanometric Scale." *Microelectronic Engineering* 83 (4–9): 597–603. https://doi.org/10.1016/j.mee.2005.12.020.

Mühlecker, Friedrich. 1845. *Der Schmetterlingsabdruck. Ein Beitrag zu Verbreitung reinerer Naturkenntnis unter der Jungend*. Stuttgart: J. Mackendorf.

Naumann, Johann Friedrich. 1848. *Taxidermie, oder, die Lehre Thiere aller Klassen: Am einfachsten und zweckmässigsten für Naturaliensammlungen auszustopfen und aufzubewahren*. Halle: C.A. Schwetschke und Sohn.

Ogilvie, Brian W. 2008. "Nature's Bible: Insects in Seventeenth-Century European Art and Science." *Tidsskrijft for Kulturforkning* 7 (3): 5–21.

Ogilvie, Brian W. 2014. "Order of Insects. Insect Species and Metamorphosis between Renaissance and Enlightenment." In *The Life Sciences in Early Modern Philosophy*, Ohad Nachtomy and Justin E.H. Smith (Eds.). Oxford; New York, NY: Oxford University Press, 222–245.

Orousset, Jean. 2008. "Un art oublié : La lépidochromie." *L'Entomologiste* 64 (1): 47–58.

Péru, Laurent. 2016. "Papillons Imprimés." *Insects* 183 (4): 27–29.

Phenix, Alan, and Joyce Townsend. 2020. "A Brief Survey of Historical Varnishes." In *Conservation of Easel Paintings*, Joyce Hill Stoner and Rebecca Rushfield (eds.), second edition. London: Routledge, 262–73. https://doi.org/10.4324/9780429399916.

Poulin, H. 1876. *Lepidoptères. Procédé pour fixer sur le papier les couleurs des ailes du papillon et principalement pour obtenir les couleurs bleues*. Paris: Deyrolle.

Poulin, H. 1899. *La lépidochromie. L'Art de décalquer et de fixer les couleurs des ailes du papillon*. Paris: Henri Laurens.

Powell-Warren, Catherine. 2023. *Gender and Self-Fashioning at the Intersection of Art and Science: Agnes Block, Botany, and Networks in the Dutch Seventeenth Century.* [Studies of Early Modernity in the Netherlands]. Amsterdam: Amsterdam University Press.

Redi, Francesco. 1668. *Esperienze intorno alla generazione degl'insetti.* Florence: All'insegna della Stella.

Rie, E. René de la. 1987. "The Influence of Varnishes on the Appearance of Paintings." *Studies in Conservation* 32 (1): 1–13. https://doi.org/10.1179/sic.1987.32.1.1.

Rie, E. René de la. 1989. "Old Master Paintings: A Study of the Varnish Problem." *Analytical Chemistry* 61 (21): 1228–1240. https://doi.org/10.1021/ac00196a003.

Rozier, François. 1777. *Introduction aux observations sur la Physique, sur l'Histoire Naturelle et sur les Arts.* Vol. 1. Paris: chez le Jay, Libraire, rue Saint-Jacques, au Grand Corneille. Barrois, l'Aîné, Libraire, Quai des Augustins.

Rutgers, Jaco. 2012. "Otto Marseus van Schrieck in Rome: een opdracht van Cassiano dal Pozzo." *Oud Holland* 125: 59–64. https://doi.org/10.1163/18750176-90000023.

Savela, Markku. n.d. "Lepidoptera and Some Other Life Forms". Accessed June 04, 2024. https://www.nic.funet.fi/pub/sci/bio/life/intro.html.

Schenck, Kimberly, and Stacey Sell. 2022. "Insect Transfer Prints in Joris Hoefnagel's Ignis from the Four Elements." *Master Drawings* 60 (2): 161–170.

Seelig, Gero, ed. 2017. *Medusa's Menagerie. Otto Marseus van Schrieck and the Scholars.* Exhibition Catalogue Medusa's Menagerie, Staatliches Museum Schwerin 7 July–15 October 2017, Rijksmuseum Twenthe, Enschede 5 November 2017–11 March 2018. Munich: Hirmer.

Shelley, Marjorie. 2020. "Joris Hoefnagel's Insects." *Metropolitan Museum Journal* 55: 27–43.

Stavenga, Doekele G. 2014. "Thin Film and Multilayer Optics Cause Structural Colors of Many Insects and Birds." *Materials Today: Proceedings*, Living Light: Uniting biology and photonics – A memorial meeting in honour of Prof Jean-Pol Vigneron, 1 (January): 109–21. https://doi.org/10.1016/j.matpr.2014.09.007.

Stavenga, Doekele G., Hein L. Leertouwer, and Bodo D. Wilts. 2014. "Coloration Principles of Nymphaline Butterflies – Thin Films, Melanin, Ommochromes and Wing Scale Stacking." *Journal of Experimental Biology* 217 (12): 2171–2180. https://doi.org/10.1242/jeb.098673.

Steensma, Susanna. 1999. *Otto Marseus van Schrieck: Leben und Werk.* Hildesheim: Georg Olms.

Stols-Witlox, Maartje. 2001. "Final Varnishes for Oil Paintings in Holland, 1600–1900. Evidence in Written Sources." *Zeitschrift für Kunsttechnologie und Konservierung* 2: 241–84.

Swammerdam, Jan. 1669. *Historia insectorum generalis, ofte agemene verhandeling der bloeelose dierkens.* Utrecht: Merinardus van Dreunen.

Swammerdam, Jan. 1737. *Bybel der natuure.* Leiden: I. Severinus, B. van der Aa, P. van der Aa.

Vukusic, P., J.R. Sambles, and H. Ghiradella. 2000. "Optical Classification of Microstructure in Butterfly Wing-Scales." *Photonics Science News* 6: 61–66.

Weber, Gregor J.M. 1989. *Stilleben alter Meister in der Kasseler Gemäldegalerie.* Gutemberg: Meslungen.

Weber, Gregor J.M. 1993. "Stilles Leben am Erdboden." *Kunst & Antiquitäten* 1 (2): 24–29.

Weber, Gregor J.M. 1995. "Elias van den Broeck." In *The Lure of Still-Life,* Jacopo Lorenzelli and Eckart Lingeauber (ed.). Exhibition Catalogue. Düsseldorf: Galerie Lingenauber, 212–19.

Wedemeyer, Jakob. 2007. "Insekten- und Pflanzenabdrücke in der holländischen Stilllebenmalerei des 17. Jahrhunderts." Unpublished master thesis, Dresden: Hochschule für Bildende Künste Dresden Studiengang Kunsttechnologie, Konservierung und Restaurierung von Kunst- und Kulturgut.

Weiss, Judith Elisabeth. 2022. "Faunal Nature Printing: Traces of Life in Sherman Denton's Moths and Butterflies (1900)." Accessed June 04, 2024. https://blog.khi.fi.it/2022/01/faunal-nature-printing-traces-of-life-in-sherman-dentons-moths-and-butterflies-1900/.

Weststeijn, Thijs. 2008. *The Visible World: Samuel van Hoogstraten's Art Theory and the Legitimation of Painting in the Dutch Golden Age.* Amsterdam: Amsterdam University Press.

Weyerman, Jacob Campo. 1729. *De levens-beschryvingen der Nederlandsche konst-schilders en konst-schilderessen.* Vol. 4. Dordrecht: Blussé en Zoon.

Wingreen-Mason, Daria. 2014. "Lepidochromy: Butterfly Transfer Prints." Accessed June 04, 2024. https://blog.library.si.edu/blog/2014/10/31/lepidochromy-butterfly-transfer-prints/.

Zhang, Di, Wang Zhang, Jiajun Gu, Tongxiang Fan, Qinglei Liu, Huilan Su, and Shenmin Zhu. 2015. "Inspiration from Butterfly and Moth Wing Scales: Characterization, Modeling, and Fabrication." *Progress in Materials Science* 68: 67–96. https://doi.org/10.1016/j.pmatsci.2014.10.003.

Zirkler, Conrad Wilhelm. 1847. *Genaue Beschreibung einer neuen Methode des Abdruckens der Schmetterlinge.* Tübingen: Guttenberg.

CHAPTER 4

The Biology of Color in Insects

Kay Etheridge

1 Introduction

For centuries, naturalists have used patterns of color to aid in the identification and classification of animals such as insects. Early classification schemes were based on differences and similarities in adult forms of animals, and it follows that the careful rendition of color in images of insects was often critical for this purpose. Representation of accurate, lifelike color in images of organisms made for study or in published works was therefore the focus of much attention by early modern artist-naturalists. However, insect color in the visible spectrum signifies a variety of things and can tell us much more about these animals than where they belong in a classification scheme.[1] Color plays a substantial role in insect ecology and behavior, serving to attract mates, warn predators, and camouflage the animals. Often, the color patterns in an insect such as the Teucer owl butterfly (*Caligo teucer*) as depicted by Maria Sibylla Merian (1647–1717) in a painting on vellum[2] (Fig. 4.1) have evolved to serve multiple roles, and examples will be discussed herein.

Additionally, color can vary within a species and even change over time in individual insects. Color variations can confound the process of identification, but the plasticity of this trait reveals additional information about the biology of insects. In concert with aspects of their behavior and ecology, color can be a part of an insect's physiology, reflecting developmental stages, changes in diet, thermoregulatory adaptations, or an immunological response. Representation of color in insects extends to the description and depiction of the various life stages, which in the case of moths and butterflies, the focus of this chapter, typically bear no similarity in color among the larvae, pupae, and adults. Classification schemes and systematics are based largely on adult animals,

1 Unlike humans, many animals including some insects also detect ultraviolet light as well as visible light. For examples of ultraviolet marking patterns in butterflies, see Rutowski et al. 2007, 280–290.

2 Merian created luxury sets of paintings on vellum of the images for her 1705 *Metamorphosis Insectorum Surinamensium*, which she sold to wealthy collectors to help finance publication of the book. The image of the Teucer owl butterfly, whiptail lizard, and butterfly appear in Plate 23 of *Metamorphosis*. For more on this topic, see Heard 2023, 132–142.

© KAY ETHERIDGE, 2025 | DOI:10.1163/9789004684553_005

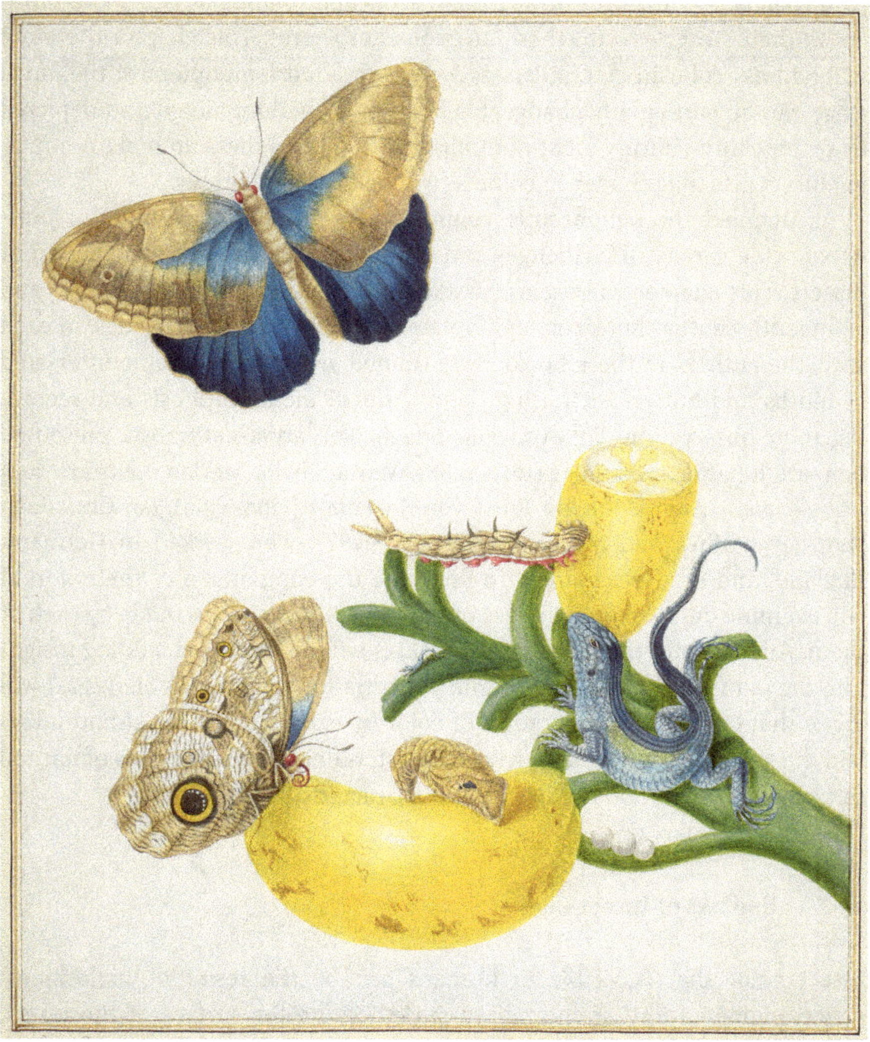

FIGURE 4.1 Maria Sibylla Merian, Teucer owl butterfly (*Caligo teucer*) on banana (*Musa* sp.) with whiptail lizard (*Cnemidophorus lemniscatus*), in: Maria Sibylla Merian, *Metamorphosis Insectorum Surinamensium*, 1702/3, Plate 23, counterproof on vellum, watercolor and bodycolor, 357 × 274 mm (leaf), London, Royal Collection Trust, RCIN 921178

but important clues to selection pressures and evolution are found in all life stages.[3] For strictly practical purposes, knowing the identity of a larval stage

3 See the chapter co-authored by Stefanie Jovanovic-Kruspel and Harald Bruckner in this volume, as well as Beth Tobin's contribution.

(caterpillar) is critical for raising the insects and to aid gardeners and farmers in their struggles with these ravenous herbivores. This chapter therefore will address color in caterpillars as well as in adult lepidopterans; the larval stage can be particularly changeable in color, something noted and depicted by seventeenth-century artist-naturalists who studied these animals. A pupating insect also may change in color, as will be discussed briefly.

As detailed throughout this volume, the representation of color presents a wide variety of challenges and can be a weak link in the depiction of insects. This chapter will use as case studies a selection of seventeenth- and eighteenth-century books on lepidopterans with hand-colored plates. In each case, the authors of these books were trained artists who became interested in moths and butterflies, raising them through metamorphosis and recording their observations in word and image. The artist-naturalists presented here are Johannes Goedaert (1617–1668), Maria Sibylla Merian, Eleazar Albin (1690–*c.*1742), August Johann Rösel von Rosenhof (1705–1759), Jan Christiaan Sepp (1739–1811), and Moses Harris (1730–1788), who worked in Germany, England, and the Netherlands. To fine-tune the comparison of their work, I will examine differences in images of the same insect species made by each of them. Additionally, the accompanying texts will be discussed, adding insight into the naturalists' thoughts regarding this trait. Based on this analysis, I will argue that careful documentation of color by early modern artist-naturalists laid the groundwork for our current understanding of the processes of natural selection and evolution of this physical characteristic.

2 Sources of Insect Colors

Insect color that is visible to humans can be the result of bioluminescence, pigmentation, or specialized structures on the surface of the insect. Bioluminescence, generated by chemical reactions in specialized organs such as those in fireflies, is the rarest source of color and will not be discussed further here. In most insects, color comes from pigments in the cuticle, which is the outermost layer of the exoskeleton. The exoskeleton consists of other layers below the cuticle and serves as protection for the insect as well as the structure to which muscles are attached – analogous in some functions to our endoskeleton but growing instead on the outside of the animal. In order to grow, larvae must shed this relatively inflexible exoskeleton, and this can be part of the process of color change in some species. The cuticle of the exoskeleton can be modified to produce structures like pupae, lepidopteran wings and their scales, as well as structural color as discussed below. Pigmentation of the cuticle is a complex process and still not completely understood; as such, it is beyond the scope of this chapter.[4] However, the means by which pigments imbedded in the cuticular structures of insects generate our perception of color is well established, and these compounds have been studied extensively.[5] A pigment resulting in the perception of red color reflects red light wavelengths back to our retinas while absorbing other wavelengths of light. Our brains process only what is reflected back to our retinas and not what is absorbed by the pigments; black pigments absorb almost all visible light, reflecting no colored wavelengths.

Naturalists as early as the entomologist and artist Maria Sibylla Merian understood that the scales on the wings of moths and butterflies, which we now know are modified cuticle, were responsible for their color and pattern. Merian used magnification to observe the scales of butterfly wings, in one instance describing them as being arranged in an "orderly and regular" manner "like roof tiles."[6] She knew that if these delicate scales were brushed off, the remaining wing was semi-transparent. In her first book on European insects, she wrote:

> At this point the attentive reader or observer may note (just incidentally) that the colors of all moths or moth-birds (be they of one color, or however they might appear) are not durable or, we might say, permanent colors, but merely like the purest fine-cut wool which is then artfully strewn on them, as if Nature had dusted it on or (so to speak) applied it in fine detail, to give such an insect a far lovelier appearance. For if one were simply to brush off the wings, there would be nothing left but a very thin creature of isinglass and not at all pretty. Of summer-birds it will be noted that their color is also easily brushed off, but it more resembles very finely ground flour than the wool just described.[7]

---

4 Sugumaran 2009, 523–525.

5 For a review, see Badejo et al 2020, 2.

6 "als de pannen op de daaken, ... ordentlyk en regulier." Merian 1705, text with Plate 9. Translation from Merian et al 2016.

7 "Hierbey nun kan der günstige Leser oder Beschauer (als im vorübergehen) so viel merken, daß die Farbe aller Motten, oder Motten-vögelein (sie mögen gleich eine Farbe haben, oder aussehen wie sie immer wollen) keine selbständige, oder so zu sagen, fest-anklebende Farbe, sondern nur wie eine von der allerreinsten, gehackten, darnach künstlich darauf gestreueten Wolle sey, so die Natur gleichsam darauf gestäubt, oder Miniatur-weise (so zu reden) darauf gelehnt; solchem Vögelein ein weit schönerers Ansehen zu machen: Denn so man nur die

Merian used powdered pigments to make her paints, and brushing the colored scales off of the wings may have seemed like brushing powdery flour from a surface.

The metallic and iridescent sheen of some insects is a more complex physical phenomenon than the reflection of light from simple pigmented surfaces such as the wings described above by Merian. The shimmering effects and colors in iridescent insects result from physical structures on their surfaces, often in combination with pigmentation.[8] One example of structural color in insects is that of metallic gold or silver sheen, which appears in a number of beetles as well as the pupae of some butterflies.[9] The pigments in scales on the wings and body of the Teucer owl butterfly create what we see as the brown, tan, and cream colors, and the micro- and nanostructure of the wing scales provide a range of iridescent colors that can also appear metallic (Fig. 4.1). Various species of brilliant blue morpho butterflies from the neotropics have long been favorites of collectors because of their iridescent wings, but interestingly, biologists are still unsure of the exact roles of this trait in these and other iridescent insect species. However, the physical nature of the shimmering color is now understood thanks to the high magnification of electron microscopy, a twentieth-century technology.[10] Iridescence such as that seen in a morpho butterfly (genus *Morpho*) wing is an example of structural color, generated by small shifts of color that are perceived as our angle of viewing a surface changes. The shimmering effect of iridescence is created by minute specialized structures on the insect's surface (Fig. 4.2). In morpho butterflies, the nanostructures on the tiny ridges of the scales have a pyramidal, branched shape like an evergreen tree. Light striking the surface of scales is reflected back to the viewer at different angles by thousands of these branched structures; the change in angle of reflection alters the wavelength of the light, thus altering the perceived color. Any animal with photoreceptors such as those in our retinas will process these variations in reflected light as iridescence with the movement of the insect or the viewer.

Flügel abwischet, so were anderst nichts, als ein sehr dünnes Wesen von Heusenblasen übrig, und stunde nicht schön. An den Sommervögelein ist auch zu merke[n], daß zwar ihre Farb leicht abzuwischen, aber sie gleicht mehr einem allerreinst gemahltem Meel, als einer gedachten Wolle." Merian 1679, 66. Translation by Michael Ritterson; from Etheridge 2021, 279. Isinglass is a clear gelatin made from fish protein.

8 See for example Stavenga et al. 2015, 14.

9 Steinbrecht 1985, 745–762.

10 Debat et al. 2018, 139–174.

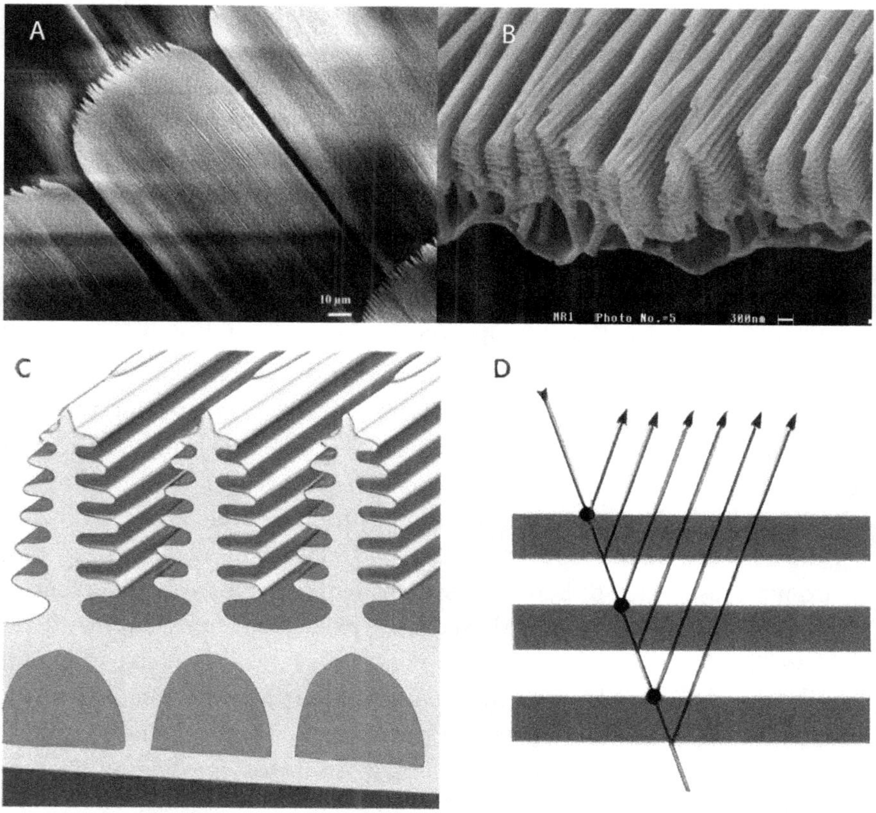

FIGURE 4.2 The micro- and nanostructures of the scales on the wings of *Morpho menelaus* are shown in images A and B respectively. The corrugated tree-like nanostructures (photo B and schematic C) create a complex phenomenon of physical color. Light striking the scales on the wings is reflected back through the layers of air and chitin in the branching lamellae (chitin layers), as represented by the arrows in schematic D. Depending upon the angle of viewing the surface of the wings, the color will shift from red to blue.
© ELSEVIER

3 Functions of Color in Insects

As in other animals, color in insects serves a variety of purposes, including that of physical protection against damaging radiation and aiding in thermoregulation.[11] Dark colors help an animal in a cool environment absorb

11 Poulton 1890 is the foundational work in this area. See Badejo et al. 2020 for a recent review of the functions of insect color.

more heat, and lighter colors reflect heat. Dark pigments partially block damaging ultraviolet radiation from getting past the insect's cuticle. Color can provide another kind of protection by camouflaging an insect so that it blends with its background, making detection by predators more difficult. This trait is evident in many species, and some animals even have the ability to change their color to match their background. As described below, many species of lepidopterans can alter their color in the larval or pupal stages.

Some insects have evolved a very different kind of protection and display bright colors as a warning to predators that they are toxic or distasteful. Such aposematic coloration is often red, orange, or yellow, colors that are avoided by predators innately or through learning by experience with a noxious prey species.[12] Another type of warning coloration involves mimicking the aposematic coloration of a dangerous or noxious species. Bearing the name of Batesian mimicry after the English naturalist Henry Walter Bates (1835–1892) who first described it in 1862, this evolutionary trait has evolved in a number of palatable lepidopterans – both caterpillars and adults across a range of species. An example discovered by Bates is shown in Figure 4.3; the *Dismorphia* species (family Pieridae) on the top and third row are palatable to birds, but they deter predators by closely mimicking the color patterns of the unpalatable Ithomiini (family Nymphalidae) shown in the second and fourth rows.[13] Needless to say, such mimics caused problems for early taxonomists, who might have assumed that these butterflies were closely related, when in fact they are in different taxonomic families.

Arrangement of color on the surfaces of insects can be used either to camouflage or to attract attention. Color patterns that simulate features such as large eyes are used by various caterpillars as well as adult moths and butterflies. Like the ventral side of the wings of the Teucer owl butterfly pictured by Merian (Fig. 4.1), the dorsal side of the wings of adult small emperor moths (*Saturnia pavonia*) feature prominent eyespots. Figures 4.4a and b show the emperor moths as depicted by Merian for her book on European insects (primarily lepidopterans), the 1679 *Der Raupen wunderbare Verwandlung und sonderbare Blumen-nahrung*.[14] The eyespots that appear on the wings of a number of

12 See for example Exnerová et al. 2007, 148–156.

13 Bates 1862, 495–566.

14 The complete title translates as *The wondrous transformation and particular food plants of caterpillars, wherein, by means of an entirely new invention the origin, food, and changes of caterpillars, worms, butterflies, moths, flies, and other such small creatures, together with their time, location, and characteristics, for the benefit of naturalists, artists, and garden lovers, are carefully investigated, briefly described, painted from life, engraved in copper, and personally published by Maria Sibylla Graff, daughter of the late Matthaeus Merian the*

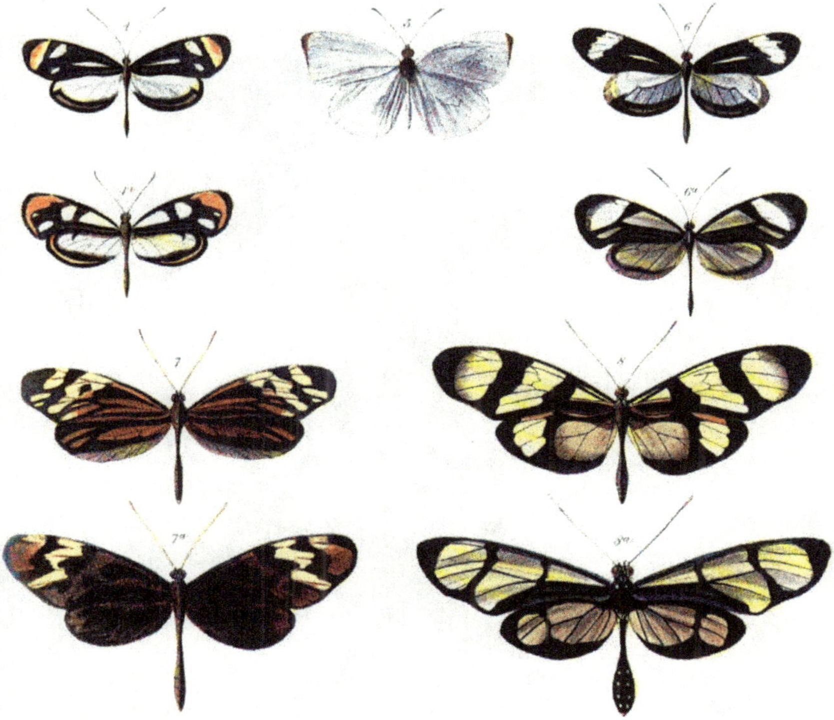

FIGURE 4.3 F.W. Robinson (illustrator and lithographer), Palatable *Dismorphia* species (family Pieridae, on the top and third row) and unpalatable *Ithomiini* (family Nymphalidae, in the second and fourth rows), in: Henry Walter Bates, "Contributions to an Insect Fauna of the Amazon Valley: Lepidoptera: Heliconidae," 1862, detail of Plate 56. Washington, Smithsonian Libraries, Cullman Library Rare Books, QL554 A1B3 quarto. Public Domain HTTPS://ARCHIVE.ORG/DETAILS/CONTRIBUTIONSTOI00BATE/PAGE/N77/MODE/2UP

lepidopterans have been shown to deter predatory birds, but like other insect color patterns, these may serve more than one role in a species.[15] A variety of caterpillars have coloration that takes the eyespots a step further, with additional colored markings that make their vulnerable larvae look like snakes (Fig. 4.5). Some caterpillars enhance this warning coloration with distinctive behaviors in response to a threat; these eyespots also appear to deter predators.[16] The

Elder. Translation by Michael Ritterson. Hereafter abbreviated as *Raupen* book. Regarding the emperor moth, see Etheridge 2021, 133.

15 Crees et al. 2021 329–337.

16 Hossie and Sherratt 2013, 383–389.

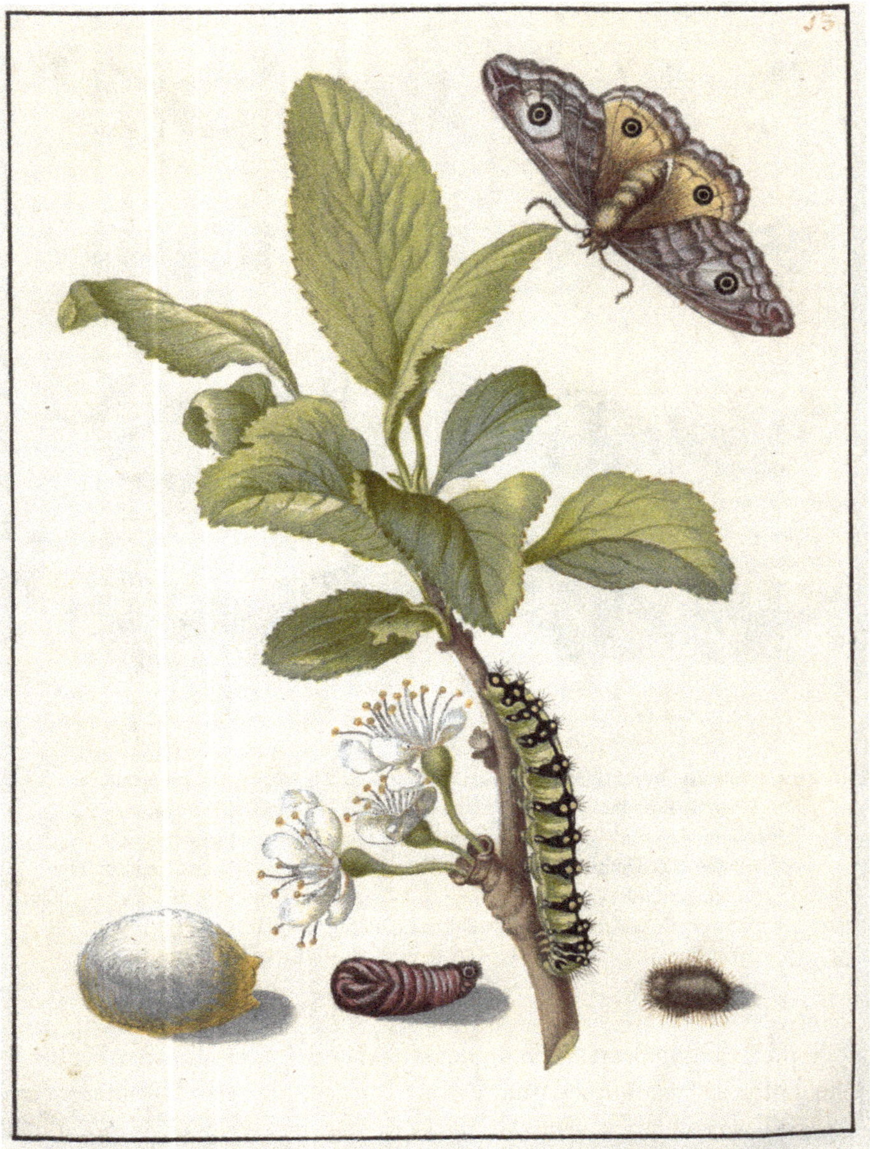

FIGURE 4.4A Maria Sibylla Merian, male small emperor moth (*Saturnia pavonia*) with larva, pupa, cocoon, and a shed larval exoskeleton (lower right), in: Maria Sibylla Merian, *Der Raupen wunderbare Verwandelung*, 1679, Plate 13, counterproof on paper, watercolor and body color, Leiden, Library of the Netherlands Entomological Society at Naturalis Biodiversity Center, NEV 70061

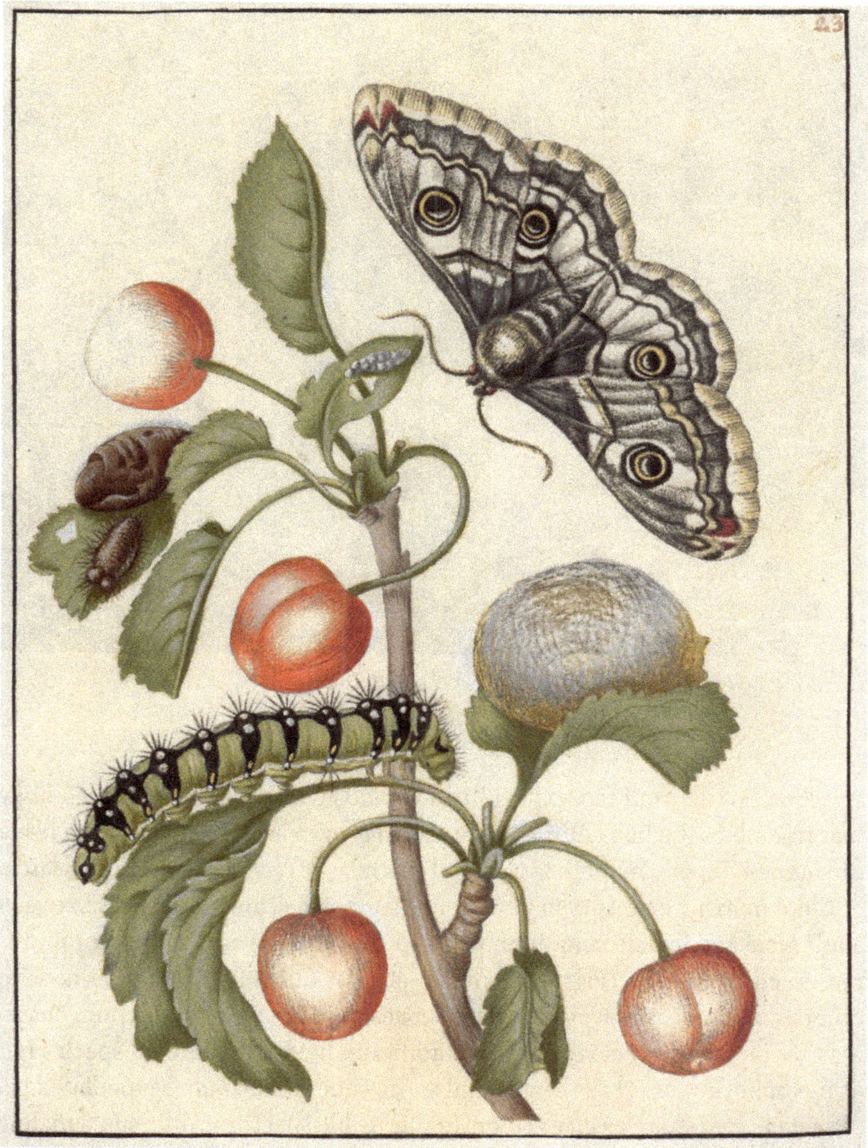

FIGURE 4.4B Maria Sibylla Merian, female small emperor moth (*Saturnia pavonia*) depicted with her eggs on the second leaf from the top, larva, pupa, and cocoon, in: Maria Sibylla Merian, *Der Raupen wunderbare Verwandelung*, 1679, Plate 23, counterproof on paper, watercolor and body color, Leiden, Library of the Netherlands Entomological Society at Naturalis Biodiversity Center, NEV 70061 © NEV LIBRARY/NATURALIS, LEIDEN

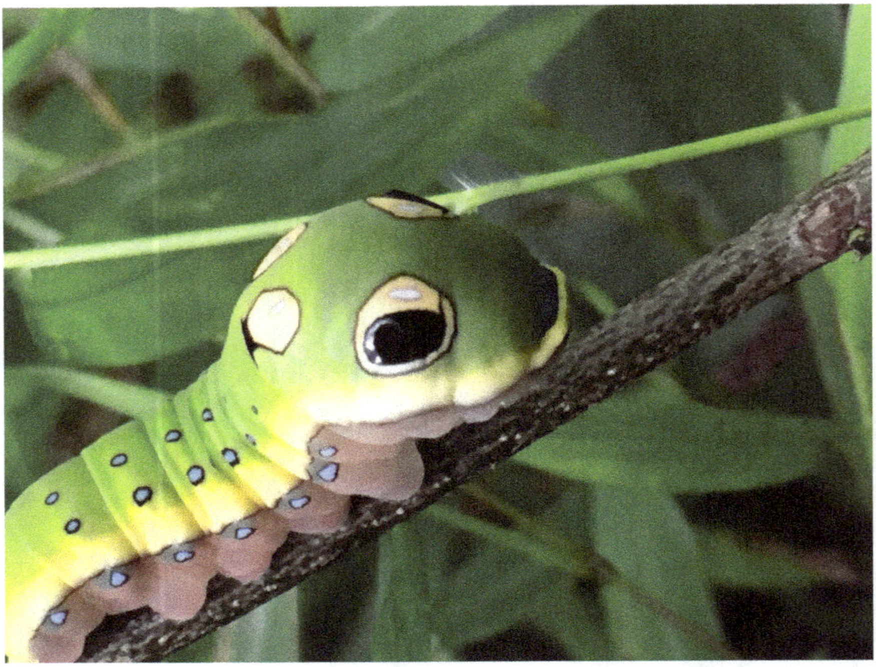

FIGURE 4.5 Larva of the spicebush swallowtail (*Papilio troilus*)

spicebush swallowtail caterpillar (*Papilio troilus*) puffs up the part of their body that resembles the head of the snake and may even sway back and forth like a threatened serpent; numerous videos of this may be found with a simple search.

Color in many insects is an important sexual signal, and differences between adult male and female coloration within a species such as the emperor moths presented by Merian (Figs. 4.4a and b) are not uncommon. In insects as in other animals, distinctive color often serves to facilitate reproduction, ensuring that an adult insect can discern another individual of its own species but of the opposite sex. These sexual and species identifiers may be mediated via ultraviolet patterns as well as colors in the visible spectrum, and this signaling saves the insects time and energy in finding mates. One of the earliest examples of color variation in insects noted by naturalists was sexual dimorphism in wing patterns and pigments, as will be discussed below.

4 Color Plasticity in Insects

Color changes in insects represent an important type of plasticity in response to differences in their environment. These changes may be long-term, resulting

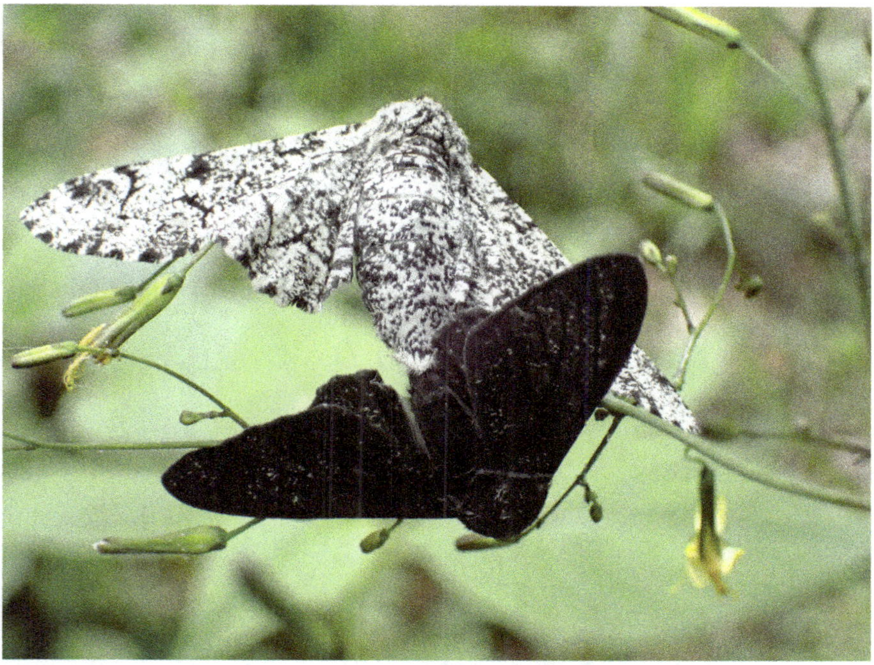

FIGURE 4.6 Two color morphs of the peppered moth (*Biston betularia*)

from evolution over generations through natural selection, or short term, occurring within the life of an individual. A well-known example of the former is the darkening in color of populations of adult peppered moths (*Biston betularia*) as the accumulation of soot generated in the industrial revolution darkened the bark of trees – light-colored individuals were more visible against the darkened bark and more likely to be preyed upon by birds.[17] Predation on the light-colored moths selected for the more camouflaged darker moths, which were more likely to survive to reproduce and pass on their genes for darker wing color. Over generations, darker individuals thus became more frequent in the population (Fig. 4.6). Evidence from recent research supports the protective nature of cryptic coloration in the peppered moth.[18] Geographic variations in insect color also may occur as populations of a species become reproductively isolated, or due to a number of other selection pressures.

Color changes within the lifetime of an individual insect are more common in the larval stage. Many caterpillars change color as they grow, developing new pigments and patterns as they shed their exoskeleton between the larval

17 Tutt 1896, 305–307.
18 Walton and Stevens 2018, 1–7.

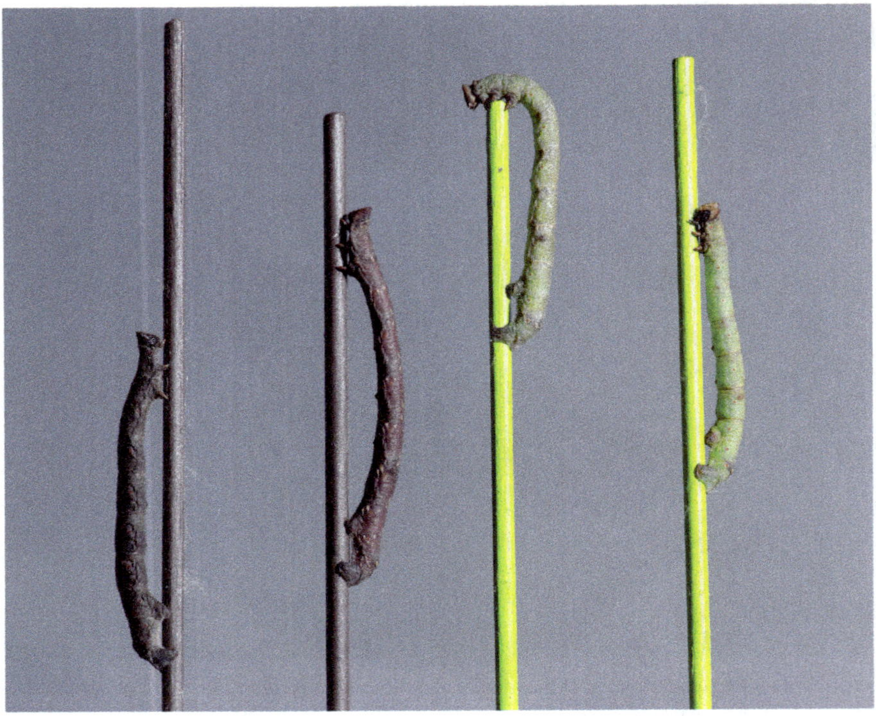

FIGURE 4.7 Larvae of peppered moths (*Biston betularia*) change color to mimic the colored dowels placed in their containers as artificial twigs. In Eacock et al., "Colour Change of Twig-Mimicking Peppered Moth Larvae Is a Continuous Reaction Norm That Increases Camouflage against Avian Predators," 2017, Fig. 2

stages, which are called instars. Such color plasticity may be triggered by a modification in environmental temperature, light quality, or diet.[19]

Reflecting the adaptive significance of being cryptic to predators is the fact that a number of caterpillars demonstrate remarkable abilities to adapt their color to their environment. These changeable larvae can generate pigments that closely mimic the colors of the plants on which they live and feed. Recent studies have shown that caterpillars of the peppered moth respond to visual cues that trigger color changes, resulting in a close match to their background (Fig. 4.7).[20] This adaptation in peppered moth larvae employs a very different mechanism from the long-term genetic change controlling the coloration of adults of the same species, but each adaptation results in increasing crypsis.

19 See for example Greene 1996, 277–285.
20 See for example Eacock et al. 2017, e3999.

Pupal color plasticity has also been documented in some species. For instance, the lemon emigrant butterfly (*Catopsilia pomona*) can approach a match to the background color of the leaves on which they pupate by varying their pigmentation from light brown to dark green.[21] Lepidopteran caterpillars and pupae are a highly nutritious and sought after food source for a wide range of predators, and because they have little to no ability to escape, the selection pressure for such camouflage abilities is profound.

5 Early Depictions of Color Variation

What we now understand about the roles of color in insects and the variability of their colors has its roots in early modern natural history. Johannes Goedaert and Maria Sibylla Merian were the earliest to picture the life stages of most of their insect subjects within one plate, making it clear which larva and pupa became which type of adult. Both Goedaert and Merian were trained as artists, but in raising and observing hundreds of types of insects through metamorphosis, they both made invaluable contributions to natural history.[22] Each of them etched their own plates as well as writing the text of their books. The time-lapse compositions of insect metamorphosis popularized by Goedaert and Merian showed stages that would not typically occur together in nature, but their plates had the advantage of representing temporal changes in their subjects. Prior to the first volume of Goedaert's *Metamorphosis naturalis* (1660), caterpillars and pupae, if depicted at all, were usually separated from the adult insects.[23] An example of this segregation of life stages may be seen in the animal encyclopedia of Johannes Jonston (1603–1675), who had his engravers copy figures from earlier works that presented them in similar fashion (Fig. 4.8).[24] Merian often included more than one larval stage in her plates, opening the door to show differences in color within the caterpillars of some species, and Goedaert described larval color change for at least one species, as will be described below. In her 1679 *Raupen* book and in all her subsequent publications, Merian also offered her readers careful descriptions of lepidopteran pupal color.[25] She appears to have been the earliest naturalist to depict sexual dimorphism in insects, showing distinctions in color and other

21 Yumnam et al. 2021, 331–341.

22 For additional background, see Etheridge 2021, 32–37.

23 See Goedaert and de May 1662, 9.

24 Jonston 1660, Plates 4 and 21. See also Etheridge 2021, 21–26.

25 Schmidt-Loske and Etheridge 2022, 100 and 106–107.

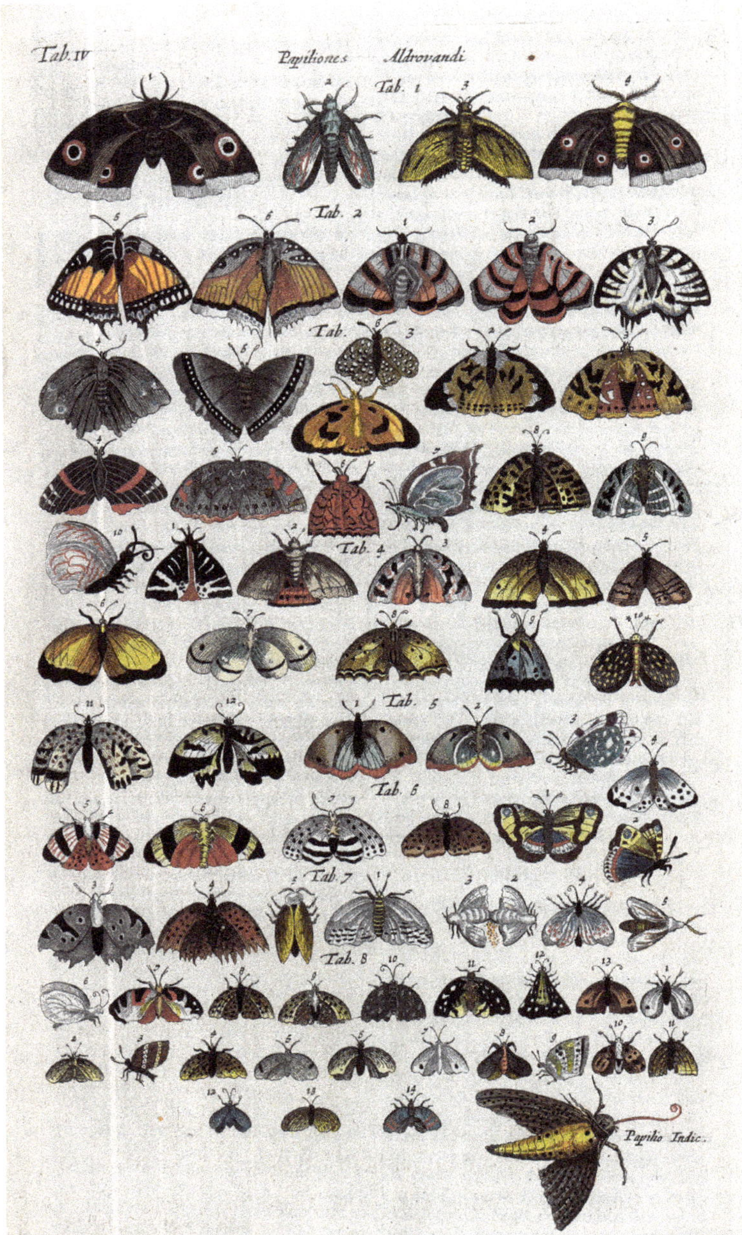

FIGURE 4.8 Artist unknown, adult and larval lepidopterans, in: Johannes Jonston, *Jonstons naeukeurige beschryving van de natuur der gekerfde of kronkel-tieren, slangen en draken*, 1660, Plates 4 and 21, intaglio prints on paper, watercolor and body color, Amsterdam, Universiteit van Amsterdam, Artis Bibliotheek van het Allard Pierson, AB 126:22,23 HTTPS://HDL.HANDLE.NET/11245/3.3422

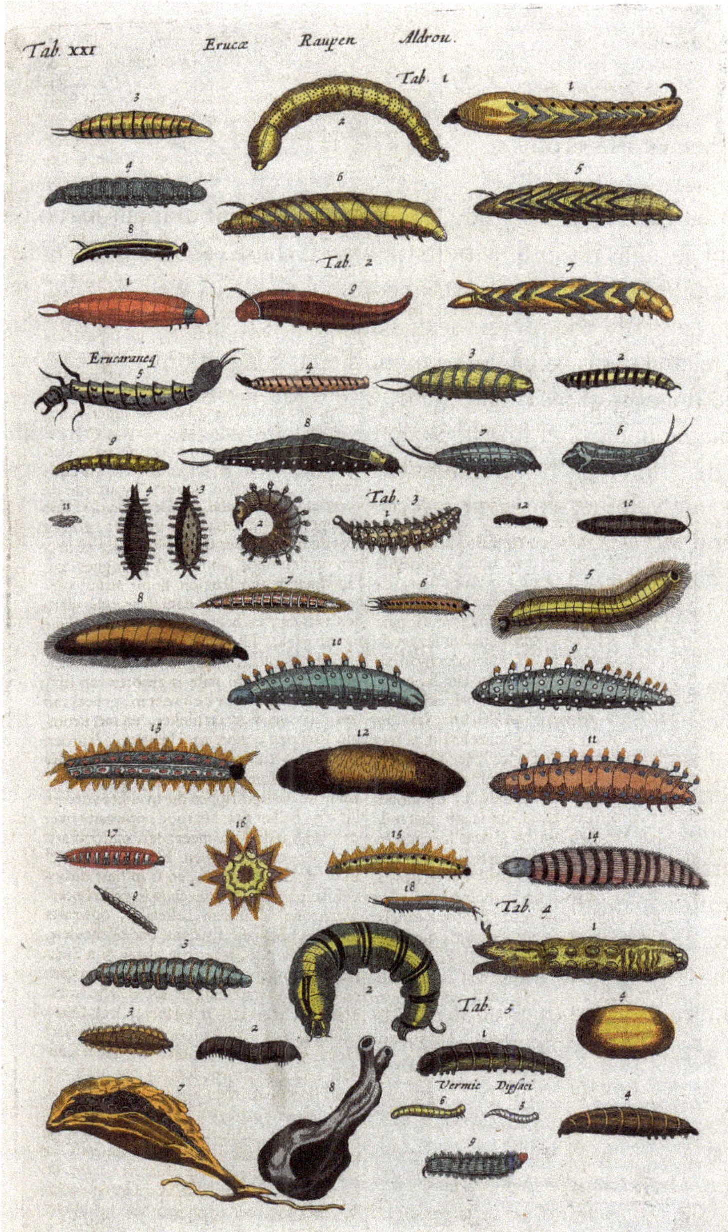
Tab. XXI
Erucæ
Raupen
Aldrou.
Tab. 1
Tab. 2
Erucaraneq
Tab. 3
Tab. 4
Tab. 5
Vermic Dipsaci

morphological characteristics between adult males and females of several species of lepidopterans.

6 Sexual Dimorphism

In addition to different reproductive structures, adult animals exhibit a wide variety of sexually dimorphic traits; in insects these can include overall physical size, antennae structure, and abdominal shape as well as color and pattern. Goedaert did not address sexual differences in insects, and this omission may be related to his belief in the spontaneous generation of insects. While he depicted the eggs of his subjects only rarely, Merian frequently included eggs in her plates, and in her first book on insects wrote that "... all caterpillars, as long as the adult insects have mated beforehand, emerge from their eggs ..."[26] In the same volume, she went on to accurately depict the differences in the males and females of several species, including the emperor moth (Figs. 4.4a and b), and she composed and etched separate plates for each sex.[27] Merian did not designate the sex of the insects in the text accompanying these plates, but she instead indicated her thoughts on their sex by adding eggs only to the plate showing the female of the species.[28] For example, she showed the differences in coloration evident in the male and female emperor moths and included eggs in the plate with the adult female (Fig. 4.4b).[29] Another clue as to the sex of the individual emperor moths is the larger body size of the female; throughout her work, Merian depicted all stages of insects as life-sized.

Naturalists who followed Merian were more direct in writing about the sexual differences in adult lepidopterans. One of the first to emulate her compositions and cite her work was Eleazar Albin. He appears to have been one of the earliest artist-naturalists to include both sexes of an insect species in one plate and to indicate the sex of each, as seen in the example from *A Natural History of English Insects* (1720) (Fig. 4.9).[30] In his text on the emperor moth,

26 "Daß insgemein alle Raupen aus ihrem Samen, so die Vögelein zuvor gepaart, hervorkommen." Merian 1679, preface iv [n.p.]. Translation from Etheridge 2021, 142.

27 Merian 1679, Plates 10, 17, 18, 21, 31, and 32. She created separate plates and text entries for the male female lappet moth (*Gastropacha quercifolia*), the gypsy or spongey moth (*Lymantria dispar*), the oak eggar (*Lasiocampa quercus*), and the emperor moth. The first plate in the book on the silkworm (*Bombyx mori*) is the only one in the book that depicts the male and female on same plate.

28 Etheridge 2021, 152, 192, 220.

29 Merian 1679, Plates 13 and 23.

30 Albin 1720. James Petiver (*c.*1665–1718) earlier published an illustrated pamphlet that showed adult butterflies of each sex in some sexually dimorphic species in Petiver 1717.

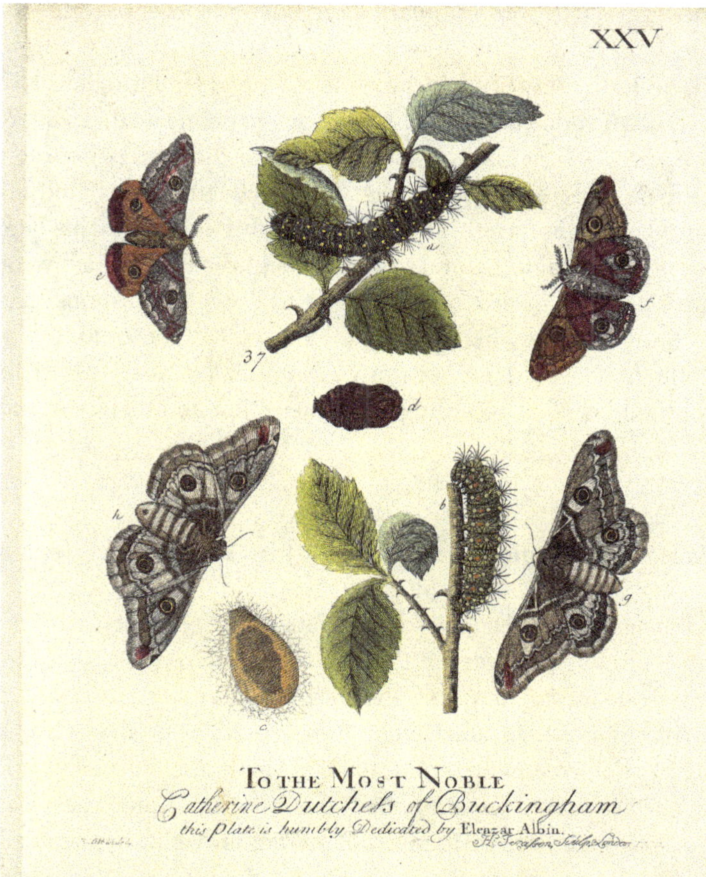

FIGURE 4.9 Eleazar Albin, male small emperor moth (*Saturnia pavonia*, dorsal view above left, ventral view above right), female small emperor moth (ventral view lower left, dorsal view lower right), pupa with and without cocoon, and two instars of larvae in: Eleazar Albin and William Derham, *A Natural History of English Insects*, 1720, Plate 25, etching on paper, watercolor and body color, Upperville, Virginia, Oak Spring Garden Foundation © OAK SPRING GARDEN FOUNDATION

Albin describes the color and size differences between male and female individuals and cites Merian's 1679 *Raupen* book. Albin is explicit about which sex he is describing, writing that "The male [is] much darker, and his under Wings

---

However, he did not know these were individuals of the same species, as evidenced by the fact that he gave each sex a different name. These pamphlets were primarily issued uncolored. See Vane-Wright 2020, 286–288.

Orange border'd with Red."[31] Albin similarly portrayed other sexually dimorphic species of lepidopterans in the same volume.

After Albin, it became common to include the male and female of sexually dimorphic insects within one plate and, in most cases, to delineate their differences within the text. Early naturalists addressing color differences between the male and female insects described what they saw, but they offered no explanations for what might have been a reason for this variation. This lack of speculation is not unusual in natural history texts of the time, which were largely descriptive and not prone to probing into the "why" questions that would be raised by evolutionary biologists of the future. The causes for many sexually dimorphic traits would not become clear until after Charles Darwin (1809–1882) proposed his ideas regarding sexual selection and its role in the evolution of traits.

7 Color Variation in Lepidopteran Larvae

Interestingly, Albin also pictured and described changes in the emperor moth's larval coloration,[32] although he was not the first naturalist to notice that caterpillars of one type could vary in color. Goedaert may have been the earliest to publish on larval color change, describing the puss moth caterpillar (*Cerura vinula*) as changing from green to red in October.[33] This would normally happen before this species enters pupation, but Goedaert's specimen died just before this stage of its metamorphosis. In lepidopterans for which he obtained the pupa and adult from a caterpillar, he included these in his plates. However, in this unusual plate, he etched two identical larvae. In at least some copies with hand-colored plates, these larvae are painted green and red respectively, as shown in the example housed at Artis Library in Amsterdam (Fig. 4.10). As was the case for many of his insects, Goedaert was unable to observe the complete life cycle for this moth, but his description of the alterations in the last larval stages of the puss moth are reasonably accurate; the caterpillars of this species do indeed turn from a bright green to a reddish-brown color before pupation.

Maria Sibylla Merian went on to make a broader study of color and its variation in lepidopterans. In the two self-published volumes of the *Raupen* books (1679 and 1683) and in *Metamorphosis Insectorum Surinamensium* (1705) – her

31 Albin 1720, text with Plate 25.

32 Albin 1720, text with Plate 25.

33 Goedaert and de Mey 1662, 1211–1212, text with Plate 65.

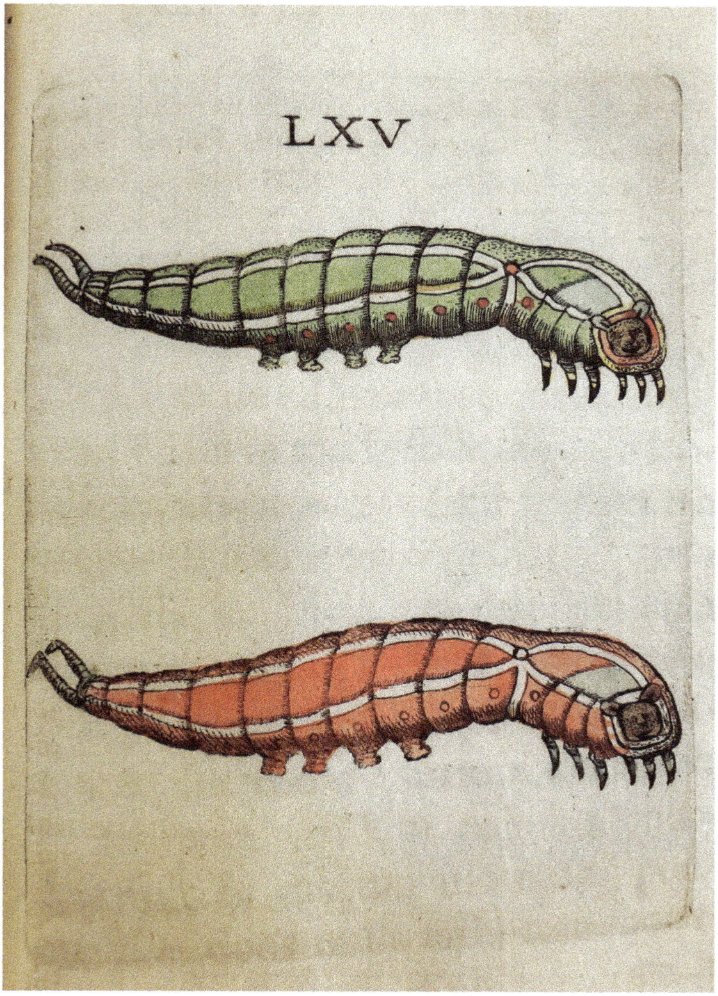

FIGURE 4.10 Johannes Goedaert, larvae of the puss moth (*Cerura vinula*), in: Johannes Goedaert, *Metamorphosis et historia naturalis insectorum*, 1662, vol. 1, Plate 65, etching on paper, watercolor and body color, Universiteit van Amsterdam, Artis Bibliotheek van het Allard Pierson, AB 181:27
© UVA/ARTIS BIBLIOTHEEK VAN HET ALLARD PIERSON

works on European and Surinamese lepidopterans, respectively – she depicted and described several examples of caterpillars of very different colors that metamorphosed into the same type of adult. In her 1679 *Raupen* book, she describes what appear to be her earliest observations of this phenomenon regarding lappet moth larvae (*Gastropacha quercifolia*) of various colors. In her text with Plate 17, she describes the upper caterpillar of the lappet moth as

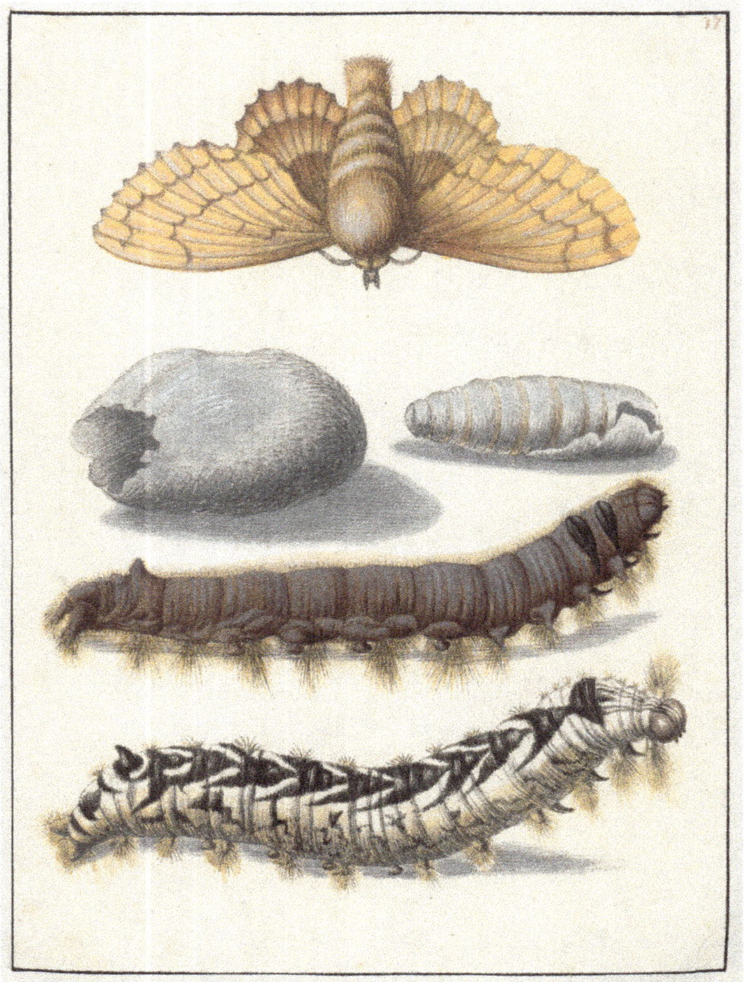

FIGURE 4.11 Maria Sibylla Merian, two larvae, pupa with cocoon, and adult of a male lappet moth (*Gastropacha quercifolia*), in: Maria Sibylla Merian, *Der Raupen wunderbare Verwandelung*, 1679, Plate 17, counterproof on paper, watercolor and body color, Leiden, Library of the Netherlands Entomological Society at Naturalis Biodiversity Center, NEV 70061
© NEV LIBRARY/NATURALIS, LEIDEN

white with black markings and the lower one as dark brown with two velvety black patches near its head (Fig. 4.11).[34] In Plate 32, under the adult female of the same species, she shows yet another color morph of the caterpillar, writing

34 Merian 1679, 35–36. Plate 17 that accompanies this entry is the only one in the book that does not include a host plant, because Merian did not know what these caterpillars ate. See Etheridge 2021, 220.

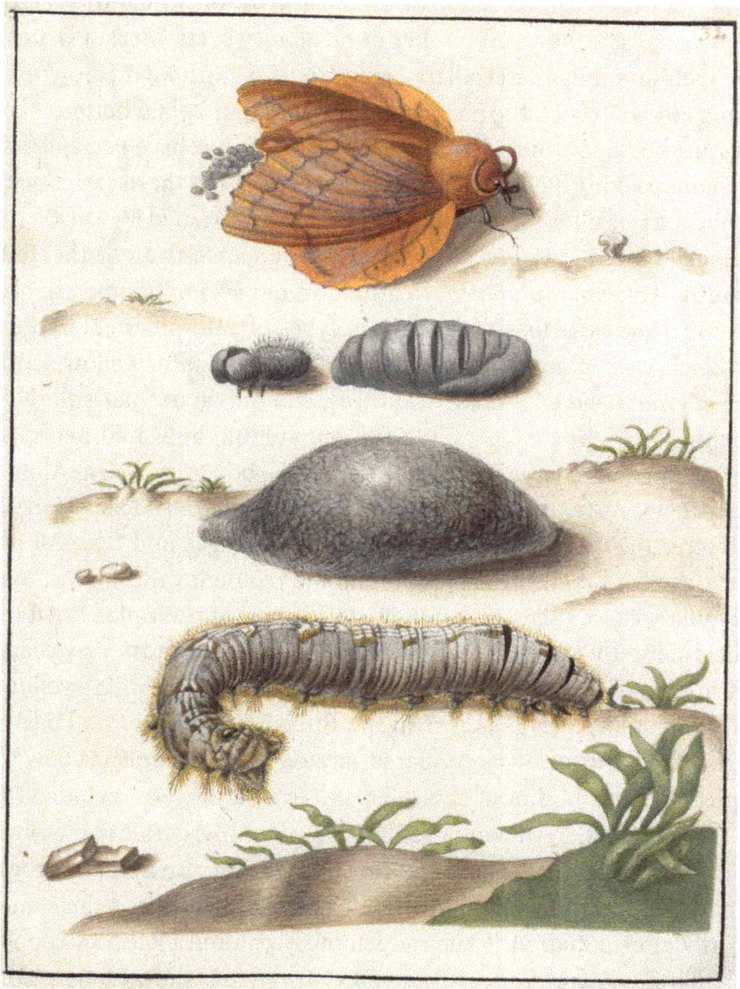

FIGURE 4.12 Maria Sibylla Merian, female adult of a lappet moth (*Gastropacha quercifolia*) laying eggs, with shed exoskeleton, pupa with cocoon, and larva, in: Maria Sibylla Merian, *Der Raupen wunderbare Verwandelung*, 1679, Plate 32, counterproof on paper, watercolor and body color, Library of the Netherlands Entomological Society at Naturalis Biodiversity Center in Leiden, NEV 70061
© NEV LIBRARY/NATURALIS, LEIDEN

that it is gray with black, velvety bands (Fig. 4.12).[35] Her amazement that the same type of adult moth resulted from such distinctive caterpillars led her to title the text with Plate 17 *Vermes miraculosi* ("miraculous worms"), a departure

35 Merian 1679, 65 and Plate 32.

from titling her entries with the name of the host plant on which the caterpillars were pictured as feeding. About her miraculous worms, Merian writes: "Now, where such dissimilar caterpillars come from, even though [they produce] the same kind of insect, the naturalists can perhaps explain better."[36] In the same *Raupen* book, she included two other species that have caterpillars of different colors, making it clear that she was curious about these variations, which were not the norm for most of the insects that she studied.[37]

Another species with variable larvae that Merian investigated, the red admiral butterfly (*Vanessa atalanta*), is featured in her second *Raupen* book, published in 1683. She describes three distinct types of caterpillar coloration in her text: Black with a yellow band and white spots, sulfur yellow with flesh- or rose-colored head and "claws" (*Klauen*), and brown or "hair-colored" (*Haarfarb*) with a red stripe (Fig. 4.13).[38] The red admiral appeared decades later with variously colored larvae in the entomological books of Eleazar Albin, Jan Christiaan Sepp, August Johann Rösel von Rosenhof, and Moses Harris, making an interesting case study for comparison of images and text. All of these authors described their observations, and the pigment variations in red admiral caterpillars can be seen in hand-colored copies of their plates. Albin addressed the differently colored caterpillars in this species almost forty years after Merian, and he described only two colors – a black morph with a yellow stripe and a yellow morph (Fig. 4.14).[39] In the three copies of Albin's Plate 3 that I have examined, the lower caterpillar is painted a light greenish yellow.[40] Yellow morphs of the red admiral caterpillar do exist and were included in the volumes by Sepp and Rösel von Rosenhof as well as Merian's. Sepp writes that he found this type of caterpillar primarily in black, but also in many other colors, emphasizing the fact that all of these caterpillars produced the same type of butterfly after pupation.[41] The meticulously painted Plate 1 in Sepp's first insect book, *Beschouwing der wonderen Gods* (1762), shows four color morphs (Fig. 4.15).

Rösel von Rosenhof credits Merian's 1683 *Raupen* book as the source of crucial information on the food plant (nettles) of the red admiral caterpillars,

36 "Woher nun solche ungleichen Raupen, wiewol einerley Art von Vögeln kommen, werden die Naturkündiger vielleicht besser auszusinnen wissen." Merian 1679, 36. Translation from Etheridge 2021, 219.

37 Merian 1679, 87 and 101. For an example of a tropical caterpillar that changes color, see Merian 1705, Plate 33 and accompanying text regarding the sphinx moth (*Pachylia ficus*).

38 Merian 1683, 81–82 and Plate 41. By "claws," Merian is referring to the prolegs of the caterpillars. The image in Fig. 4.13 is from a later Dutch edition, Merian 1712.

39 Albin 1720, text with Plate 3.

40 Two copies that I own and one at Oak Spring Garden Library Foundation, Upperville, VA.

41 Sepp 1762, 10 and Plate 1.

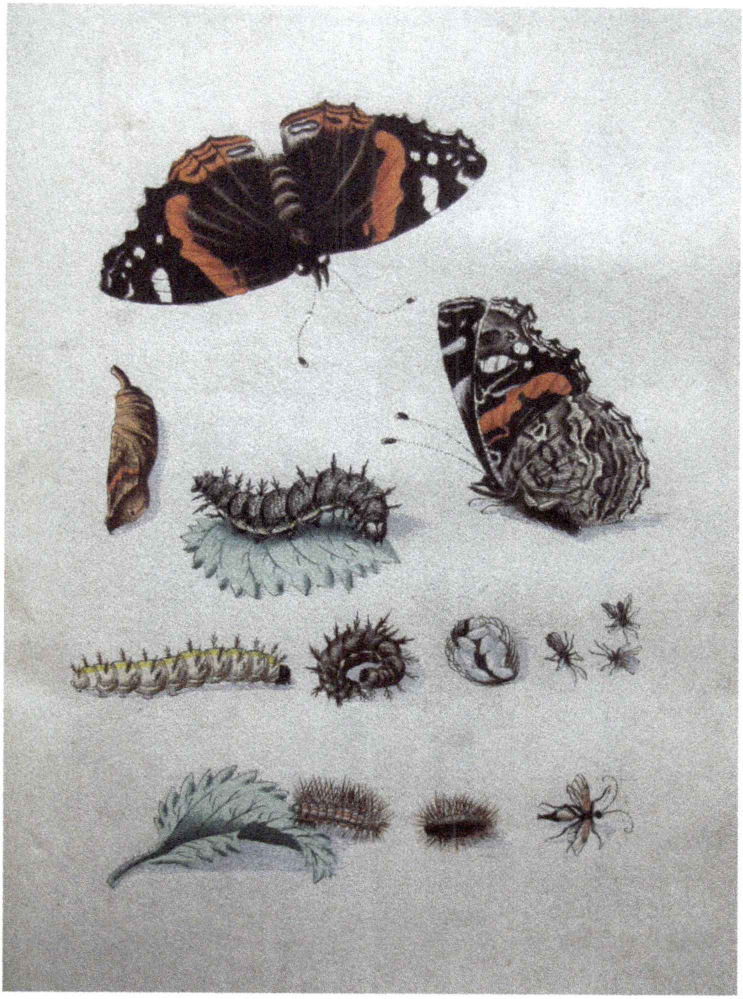

FIGURE 4.13 Maria Sibylla Merian, larvae, pupa, and adults of a red admiral butterfly (*Vanessa atalanta*) with parasitic flies and wasp in: Merian, Maria Sibylla, *Der rupsen begin, voedzel en wonderbaare verandering*, 1712, vol. 2, Plate 41, counterproof on paper, watercolor and body color, Amsterdam, Universiteit van Amsterdam, Artis Bibliotheek van het Allard Pierson, AB Legkast 019.02

which allowed him finally to raise them through metamorphosis after earlier failures.[42] He writes in great detail regarding the color morphs of the red

42 Rösel von Rosenhof 1746, 35 and Plate 6.

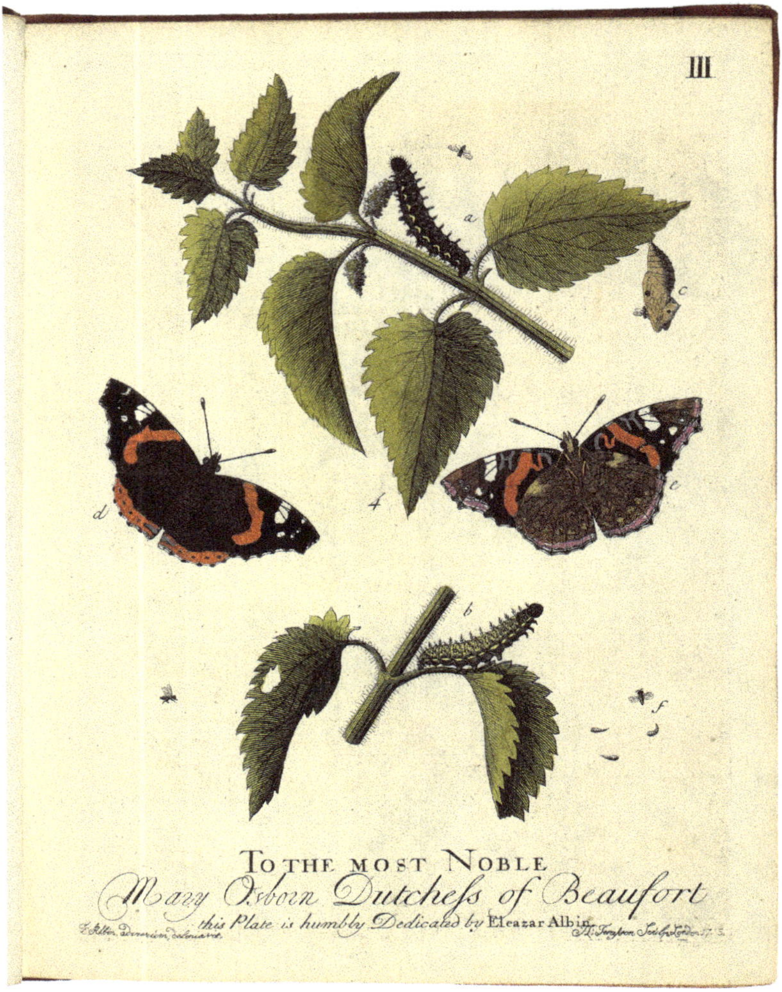

FIGURE 4.14 Eleazar Albin, larvae, pupa, and adults of a red admiral butterfly (*Vanessa atalanta*) with parasitic wasps, in: Eleazar Albin and William Derham, *A Natural History of English Insects*, 1720, Plate 3, etching on paper and watercolor, Upperville, Virginia, Oak Spring Garden Foundation

admiral caterpillar, emphasizing the importance of the fact that these are very different looking caterpillars that all become the same type of adult. He meticulously describes not only the overall color of each morph, but also the colors of stripes and dots on each one, ensuring that his readers could more readily identify these highly variable caterpillars as all being those of the red admiral butterfly. He found four major color types similar to those depicted by

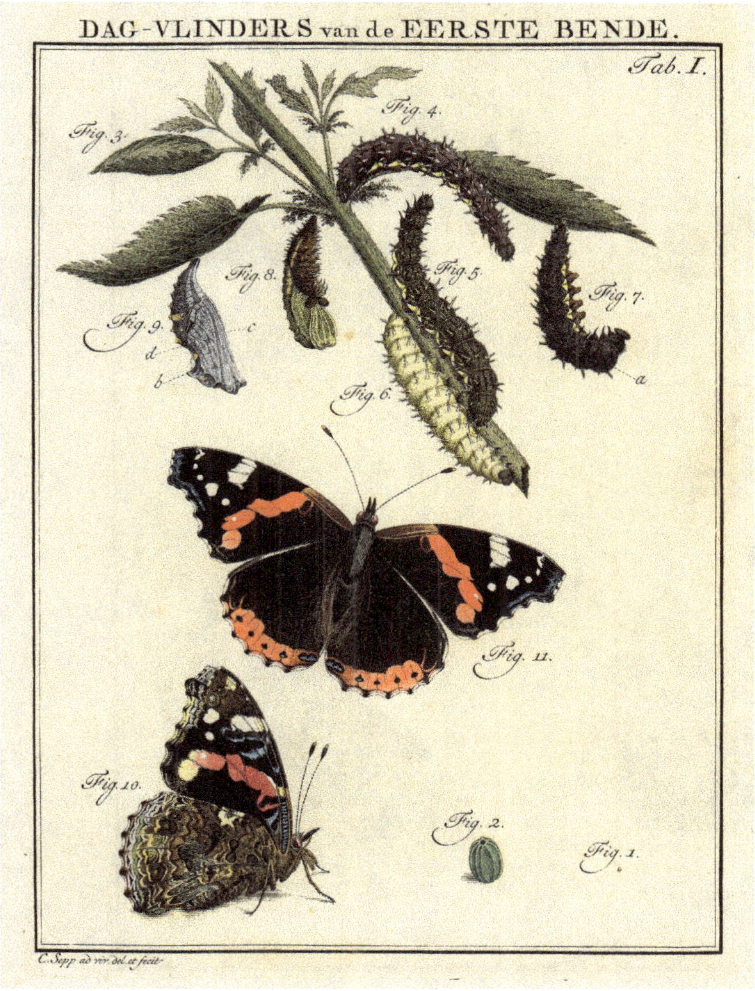

FIGURE 4.15 Jan Christiaan Sepp, eggs, larvae, pupae, and adults of a red admiral butterfly (*Vanessa atalanta*), in: Jan Christiaan Sepp, *Beschouwing der wonderen Gods, in de minstgeachte schepzelen*, 1762, Plate 1, etching on paper and watercolor, private collection © KAY ETHERIDGE

Sepp and depicted them in a plate devoid of the host plant, as was his habit (Fig. 4.16). Regarding this plate, Rösel von Rosenhof writes:

> I have presented four separate images of a single caterpillar on this page, simply for the purpose of showing the different colors of their skin for illuminating; were it not for that, a single image would have been quite

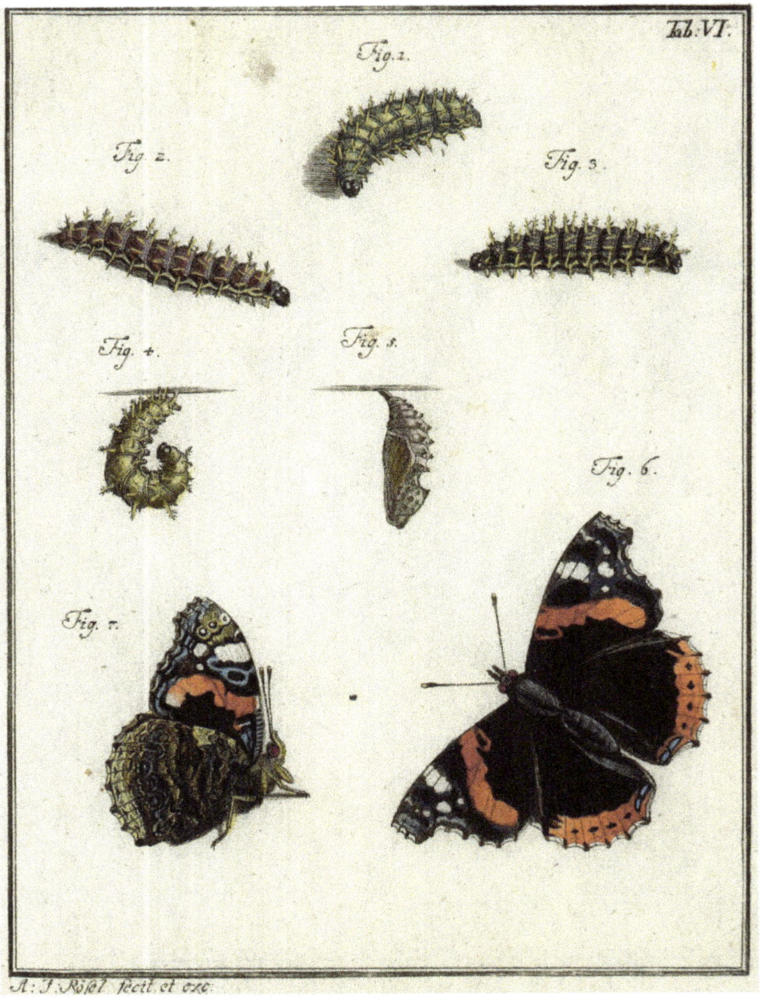

FIGURE 4.16 Johann August Rösel von Rosenhof, larvae, pupa, and adults of a red admiral butterfly (*Vanessa atalanta*), in: August Johann Rösel von Rosenhof, *Der monatlich-herausgegebenen Insecten-Belustigung*, 1746, vol. 1, Plate 6, etching on paper and watercolor, private collection

sufficient to describe the caterpillar. … I wish to note briefly the important thing about their different colors. The caterpillar in Figure 1 has an overall pale green color and a pair of yellow stripes running down both sides the length of its body, the lower stripe being somewhat wider and of a prettier color than the upper one. Figure 2 shows a crawling, stretched-out caterpillar with skin a reddish-brown color and adorned

> with just a single yellow stripe on each side. On this caterpillar one can see very clearly the above-mentioned dots on all the segments. But these dots show much more beautifully on the black caterpillars, one of which is pictured in Figure 3, the dots being yellow and thus standing out better against the black background. Thus also the yellow side stripes appear to be somewhat lighter than on the other caterpillars. Above these stripes is a row of especially yellow dots that, while somewhat larger, are not as raised as the others. In Figure 4 there appears one of our caterpillars in sulfur-yellow attire, and among this kind, too, one can find some that are a little lighter and others that are darker, just as among the other three types described here. All of the just-described types change into a single species of butterfly, of this I am certain.[43]

Naturalists who published on insects after Rösel von Rosenhof did not necessarily emulate his detailed descriptions. Moses Harris, who worked in England, was trained as an artist and printmaker. He became interested in raising and observing moths and butterflies, and published the beautifully illustrated 1766 *Aurelian, or Natural History of English Insects*, which offers relatively concise descriptions of the insects. Harris's text on the red admiral butterfly is similar to that of Albin in that it is brief and describes only two morphs, the yellow and black caterpillar, both of which appear in his Plate 6 (Fig. 4.17).[44]

43 "Daß ich auf unserer Tabelle vier besondere Figuren von einerley Raupen entworfen, ist blos deswegen geschehen, damit im Illuminiren die verschiedenen Farben ihrer Haut können ausgedrücket werden; auserdem hätte ich mich in meiner Beschreibung gantz wol mit einer einigen Figur behelfen können. ... so will ich vorhero mit wenigem noch von ihren verschiedenen Farben das nöthige erinnern. Die Raupe in der 1. Figur hat eine blaß-grüne Grund Farbe, und führet an jeder Seite ein paar gelbe Linien, so nach der gantzen Länge des Leibes hin laufen, davon die untere etwas breiter und schöner von Farbe ist als die obere. Die 2. Figur stellet eine im Kriechen ausge¬streckte Raupe vor, deren Haut von Roth-brauner Farbe und nur mit einer einigen gelben Linie an jeglicher Seite bezieret ist. An dieser Raupe siehet man die oben erwehnten Knöpflein auf allen Absätzen sehr deutlich; doch lassen sich dieselben an denen schwartzen Raupen, deren eine in der 3. Figur abgebildet ist, noch viel schöner beobachten, indeme sie hier von gelber Farbe sind und also auf dem schwartzen Grunde desto besser in das Gesicht fallen; so scheinen auch die gelben Seiten Linien deswegen etwas heller, als bey anderen, zu seyn. Über diesen Linien stehet eine Reihe besonderer gelber Puncte, die zwar etwas gröser, jedoch nicht so erhoben als die übrigen sind. In der 4. Figur zeiget sich eine unserer Raupen in Schwefel gelben Gewande, unter welcher Sorte man sowol, als unter denen 3 vorhergehenden, einige, die etwas heller, andere, die dunckeler sind, antrifft. Alle erst beschriebene Gattungen verwandeln sich nun zu einerley Papilions, und das weiß ich gewiß." Rösel von Rosenhof 1746, 36–37. Translated by Michael Ritterson.

44 Harris 1766, 11 and Plate 6.

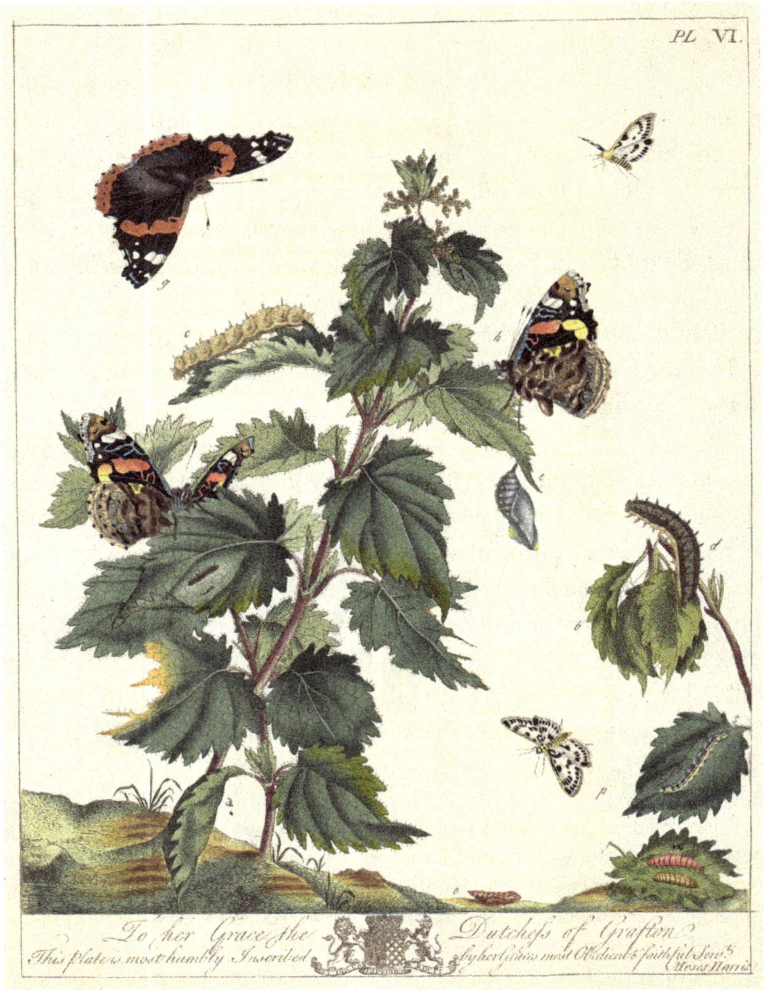

FIGURE 4.17 Moses Harris, larvae, pupa, and adults of a red admiral butterfly (*Vanessa atalanta*), in: Moses Harris, *The Aurelian*, 1766, Plate 6, etching on paper and watercolor, Upperville, Virginia, Oak Spring Garden Foundation

Harris was intensely interested in color theory as evidenced by his development and publication of a color wheel, and it may be that his focus on artistic representation of his subjects took precedence over describing the details of their natural history for his readers.[45]

45 See Harris *c.*1769.

The plates in this chapter featuring the adults, larvae, and pupae of the red admiral (Figs. 4.13 through 4.17) do not show evidence of copying, although the later naturalists most likely would have consulted earlier volumes on insects. Each author indicated that they raised their own specimens and made direct observations. By reading their introductory material and the text of several entries from each work, it seems clear that this is the case. They describe their methods to various degrees and sometimes their challenges and failures.[46] Moses Harris even details the types of nets he used to capture insects.[47] Comparisons of the red admiral plates herein to widely available photographs of the caterpillars of this species indicate that the colors of the larvae as portrayed are reasonably accurate. However, hand-colored copies can vary a great deal from volume to volume in some cases, particularly in the case of Merian's work.[48] She sold a fair number of uncolored copies of her books in her lifetime, and there were a number of post-mortem editions that appeared well into the eighteenth century, many of which have inaccurate coloring of both plants and insects; these poorly colored volumes had a negative effect on her reputation in the nineteenth century.[49] On the other hand, all copies that I have seen of the red admiral plates of the other naturalists discussed here are similar in coloration.[50] My observations have necessarily been of an admittedly small sample size of three or four copies of each, however the similarity in coloring holds for other plates within each edition and not just for the red admiral. It may be that the relative color consistency in works like that of Rösel von Rosenhof indicates that these volumes were colored 'in-house' by the artist and his family members, and that few if any uncolored copies were sold.[51]

46 See for example Rösel von Rosenhof 1746, 33–35, where he describes his initial difficulties in finding the caterpillars of the red admiral and further challenges in obtaining successful metamorphosis into pupae and adults.

47 Harris 1766, xii–xv.

48 Etheridge 2021, 99–101 and 128–130. Merian marketed three versions of her books: uncolored, handcolored with watercolor and bodycolor, and volumes with painted counterproofs on paper (watercolor and bodycolor). The images from Merian's 1679 *Raupen* book used in this chapter as well as in my 2021 book are counterproofs that were likely painted in Merian's workshop; thus, the colors of the insects are based on her studies and tend to be highly accurate.

49 Etheridge 2021, 117–125. It is impossible to know who colored most of the copies of Merian's published work.

50 Two copies, for instance, are kept at the Oak Spring Garden Foundation Library, Upperville, VA, in Albin's *A Natural History of English Insects* (1720), and Harris's *L'Aurelien* (1794).

51 See the chapter by Giulia Simonini in this volume.

The textual description of larval colors, the number of color morphs described, and their manner of depiction varies among these naturalists who raised the red admiral through metamorphosis. Over a span of years from 1683 (Merian) to 1766 (Harris), there is little expansion in the level of detail of their recorded observations other than that by Rösel von Rosenhof. Influence is difficult to trace unless an author cites every source that they have seen, but it seems likely that in most cases these naturalists were aware of the work that preceded theirs. As with the sexually dimorphic traits, the tendency for all of them was to describe what they saw and not to enter into speculation as to reasons or explanations for variations in larval color. Again, it was not until naturalists investigating the evolution of traits such as color began to conduct empirical studies, rooted in these earlier observations, that ideas regarding the causes of color variation in larvae were put forward.

8 Conclusion

Insect colors have evolved through natural selection to increase the probability of surviving to reproduce and to enhance reproductive success. Observations by early modern naturalists such as those discussed herein laid a foundation for our current understanding of the effect of sex, diet, habitat, and other factors on insect coloration. This knowledge in turn expanded our comprehension of natural selection and evolution in insects and other organisms. Sexual dimorphism in animals including humans has long been evident, but it was not until 1871 that sexual selection, the evolutionary mechanism behind the development of such traits, was put forward by Darwin.[52] The fact that insects, the most prevalent type of animal on the planet, exhibit this trait provided further evidence to support ideas on natural selection that are now central to biological studies.

The triggers for color changes in caterpillars were not investigated until more than two centuries after Merian pondered her miracle worms, the variously colored lappet moth larvae. In 1903, the naturalist Edward Poulton (1856–1943) published his experiments on the effects of raising this species of caterpillars on lichen-covered branches of different colors. The lappet moth larvae responded by developing colors and patterns that closely matched their branches (Fig. 4.18), and several of the images in his published photographs align closely with the colors of the caterpillars depicted in Merian's 1679 plates

52 Darwin 1871.

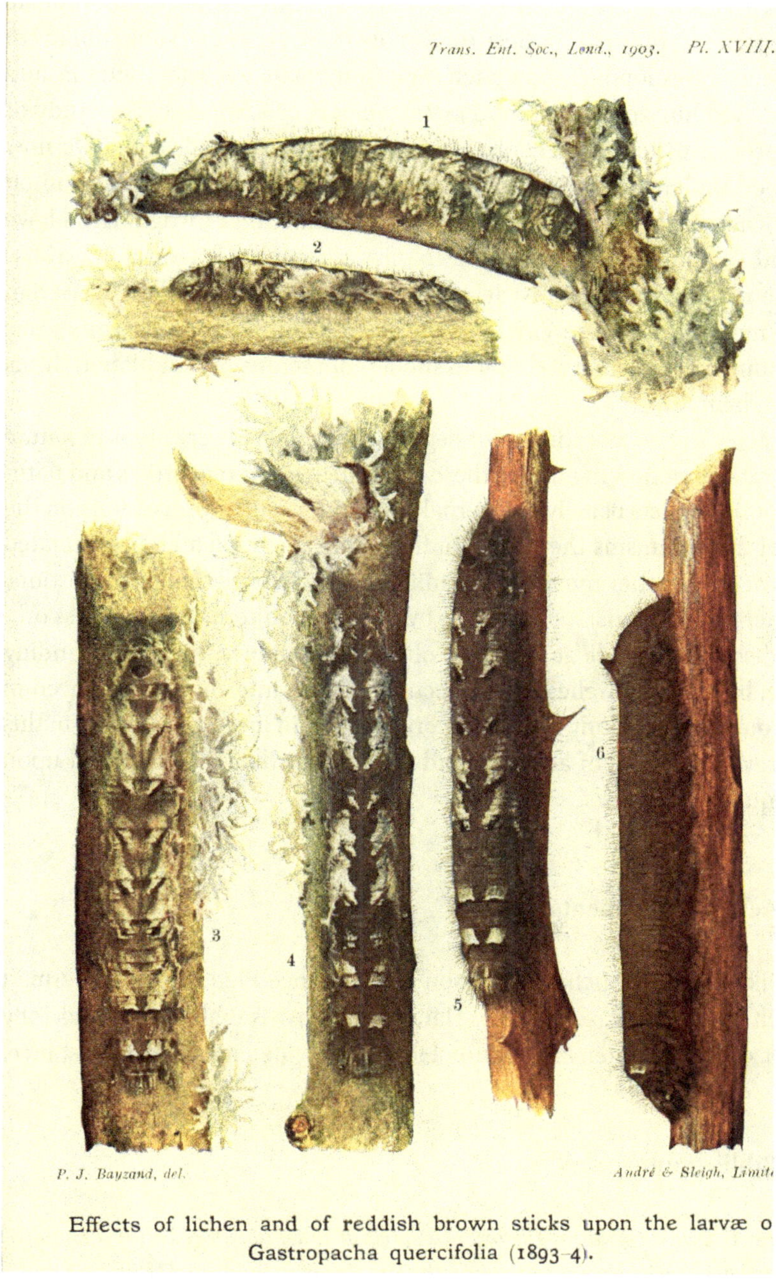

FIGURE 4.18 P.J. Bayzand (illustrator), Lappet moth caterpillars (*Gastropacha quercifolia*) grown on different colored lichen and branches, in: Edward B. Poulton, "Experiments in 1893, 1894, and 1896 upon the colour-relation between lepidopterous larvæ and their surroundings …," 1903, Plate 18, lithography. Public domain

(Figs. 4.11 and 4.12).[53] Poulton's experiments on both the pupae and larvae of lepidopterans were the earliest to demonstrate the remarkable ability of some insects to respond to and match their immediate environment; this and other work led him to publish in 1890 *The Colours of Animals*.[54] The study of every aspect of color in insects has been ongoing for decades, and we now have a good understanding of the role of natural selection and evolution on this physical trait that has such a wide variety of functions. Intriguingly, I have yet to find any publication on the mechanisms behind color variation in larvae of the red admiral, and it is clear that "the naturalists" invoked by Merian still have much to discover. With over one and a half million species of insects and over one hundred and sixty thousand lepidopterans on our planet, this is understandable.

The process of documenting color begins with careful observations of nature and accurate depiction in images; these steps are common to artists and naturalists, but naturalists usually add verbal descriptions and observations on the biology of the organisms they have studied. Once recorded and disseminated by publication or other means, these initial observations stimulate questions and further observations, such as those by Darwin, Poulton, and countless others. The discoveries made at each step ultimately enhance our understanding of nature, but progress relies on the accuracy of the information conveyed in the previous steps. The tiny subset of work on color in insects examined in this chapter reveal the kind of accuracy and consistency that could be built upon by subsequent naturalists.

Acknowledgements

I would like to thank Michael Ritterson and Florence Pieters for the German and Dutch translations, respectively. Hans Mulder (Artis Bibliotheek) and Tony Willis (Oak Spring Garden Foundation Library) provided invaluable assistance.

Bibliography

Albin, Eleazar. 1720. *A Natural History of English Insects: Illustrated with a Hundred Copper Plates, Curiously Engraven from the Life*. London: Published by the author.

53 See Poulton 1903, Plate 18.
54 Poulton 1887, 1890, 1892.

Badejo, Oluwatobi, Skaldina, Oksana, Gilev, Aleksei, and Jouni Sorvari. 2020. "Benefits of Insect Colours: A Review from Social Insect Studies." *Oecologia* 194, 27–40. https://doi.org/10.1007/s00442-020-04738-1.

Bates, Henry Walter. 1862. "Contributions to an Insect Fauna of the Amazon Valley: Lepidoptera: Heliconidae." *Transactions of the Linnean Society of London* 23: 495–566.

Crees, Logan D., Phil DeVries, and Carla M. Penz. 2021. "Do Hind Wing Eyespots of Caligo Butterflies Function in Both Mating Behavior and Antipredator Defense? (Lepidoptera, Nymphalidae)." *Annals of the Entomological Society of America* 114 (3): 329–37. https://doi.org/10.1093/aesa/saaa050.

Darwin, Charles, 1871. *The Descent of Man, and Selection in Relation to Sex*. London: John Murray.

Debat, Vincent, Serge Berthier, Patrick Blandin, Nicolas Chazot, Marianne Elias, Doris Gomez, and Violaine Llaurens. 2018. "Why Are Morpho Blue?" *Biodiversity and Evolution*, Philippe Grandcolas, and Marie-Christine Maurel (eds.). Amsterdam: Elsevier, 139–74. https://doi.org/10.1016/B978-1-78548-277-9.50009-7.

Eacock, Amy, Hannah M. Rowland, Nicola Edmonds, and Ilik J. Saccheri. 2017. "Colour Change of Twig-Mimicking Peppered Moth Larvae Is a Continuous Reaction Norm That Increases Camouflage against Avian Predators." *PeerJ* 5: e3999. https://doi.org/10.7717/peerj.3999.

Etheridge, Kay. 2020. *The Flowering of Ecology: Maria Sibylla Merian's Caterpillar Book*. Leiden: Brill.

Exnerová, Alice, Pavel Štys, Eva Fučíková, Silvie Veselá, Kateřina Svádová, Milena Prokopová, Vojtěch Jarošík, Roman Fuchs, and Eva Landová. 2007. "Avoidance of Aposematic Prey in European Tits (Paridae): Learned or Innate?" *Behavioral Ecology* 18 (1): 148–56. https://doi.org/10.1093/beheco/arl061.

Goedaert Johannes, and Johannes de Mey, Johannes. 1662. *Metamorphosis et historia naturalis insectorum. Autore Joanne Goedartio. Cum commentariis d. Joannis de Mey ... & duplici ejusdem appendice, una hemerobiis, altera de natura cometarum, & vanis ex iis divinationibus*. Middleburgh: J. Fierenes.

Greene, Erick. 1996. "Effect of Light Quality and Larval Diet on Morph Induction in the Polymorphic Caterpillar Nemoria Arizonaria (Lepidoptera: Geometridae)." *Biological Journal of the Linnean Society* 58 (3): 277–85.

Harris, Moses. 1766. *The Aurelian: or, Natural History of English Insects; Namely Moths and Butterflies*. London: Printed for the author, and sold by J. Edwards.

Harris, Moses. c.1769. *The Natural System of Colours*. Licester Fields: Printed for the author.

Heard, Kate. 2023. "One of the Most Curious Performances That Ever Was Published" Maria Sibylla Merian's Drawings in the British Royal Collection and the *Metamorphosis Insectorum Surinamensium*. In *Maria Sibylla Merian: Changing the*

Nature of Art and Science, Bert Van de Roemer, Florence Pieters, Hans Mulder, Kay Etheridge and Marieke van Delft (eds.). Tielt: Lannoo, 131–42.

Hossie, Thomas John, and Thomas N. Sherratt. 2013. "Defensive Posture and Eyespots Deter Avian Predators from Attacking Caterpillar Models." *Animal Behaviour* 86 (2): 383–89. http://dx.doi.org/10.1016/j.anbehav.2013.05.029.

Jonston, Johannes and Matthias Graus. 1660. *I. Jonstons Naeukeurige beschryving van de natuur der vier-voetige dieren, vissen en bloedlooze waterdieren, vogelen, kronkel-dieren, slangen en draken*. Amsterdam: Jan Jacobsz Schipper.

Merian, Maria Sibylla. 1679. *Der Raupen wunderbare Verwandelung und sonderbare Blumen-Nahrung: worinnen durch eine gantz-neue Erfindung der Raupen, Würmer, Sommer-vögelein, Motten, Fliegen, und anderer dergleichen Thierlein Ursprung, Speisen und Veränderungen samt ihrer Zeit, Ort und Eigenschaften, den Naturkündigern, Kunstmahlern und Gartenliebhabern zu Dienst, fleissig untersucht, kürtzlich beschrieben, nach dem Leben abgemahlt ins Kupfer gestochen und selbst verlegt*. Nuremberg: M.S. Merian.

Merian, Maria Sibylla. 1712/1714. *Der rupsen begin, voedzel en wonderbaare verandering*. Volumes 1 and 2. Amsterdam: Maria Sibylla Merian.

Merian, Maria Sibylla. 1705. *Metamorphosis Insectorum Surinamensium*. Amsterdam: G. Valck.

Merian, Maria Sibylla, van Delft, Marieke and Hans Mulder. 2016. *Maria Sibylla Merian. Metamorphosis Insectorum Surinamensium = Verandering der Surinaamsche insecten = Transformation of the Surinamese Insects: 1705*. Tiel; The Hague: Lannoo.

Petiver, James. 1717. *Papilionum Britanniae*. London: Royal Society.

Poulton, Edward B. 1887. "XIV. An Enquiry into the Cause and Extent of a Special Colour-Relation between Certain Exposed Lepidopterous Pupœ and the Surfaces Immediately Surround Them." *Philosophical Transactions of the Royal Society of London.(B.)*, no. 178: 311–441.

Poulton, Edward B. 1890. *The Colours of Animals: Their Meaning and Use, Especially Considered in the Case of Insects*. Vol. 67. New York: D. Appleton.

Poulton, Edward B. 1892. "XIX. Further Experiments upon the Colour-Relation between Certain Lepidopterous Larvœ, Pupœ, Cocoons, and Imagines and Their Surroundings." *Transactions of the Royal Entomological Society of London* 40 (4): 293–487.

Poulton, Edward B. 1903. "Experiments in 1893, 1894, and 1896 upon the Colour-Relation between Lepidopterous Larvæ and Their Surroundings, and Especially the Effect of Lichen-Covered Bark upon *Odontopera bidentata, Gastropacha quercifolia*, etc." *Transactions of the Royal Entomological Society of London*, 51(3), 311–374.

Rösel von Rosenhof, August Johann. 1746–61. *Der monatlich-herausgegebenen Insecten-Belustigung*. Four Volumes. Nuremberg: Johann Joseph Fleischmann.

Rutowski, Ronald L., Joseph M. Macedonia, Darrell J. Kemp, and Laura Taylor-Taft. 2007. "Diversity in Structural Ultraviolet Coloration among Female Sulphur

Butterflies (Coliadinae, Pieridae)." *Arthropod Structure & Development* 36 (3): 280–90. https://doi.org/10.1016/j.asd.2006.11.005.

Sepp, Jan Christiaan. 1762–1860. *Beschouwing der wonderen Gods: in de mintsgeachte schepzelen of Nederlandsche insecten ... naar 't leven naauwkeurig getekent, in 't koper gebracht en gekleurd.* Amsterdam: J.C. Sepp.

Schmidt-Loske, Katharina and Kay Etheridge. 2022. "The Transition State, Merian's Studies of Pupa, Chrysalis and Cocoon. In *Maria Sibylla Merian: Changing the Nature of Art and Science*, Bert Van de Roemer, Florence Pieters, Hans Mulder, Kay Etheridge, and Marieke van Delft (eds.). Tielt: Lannoo, 98–110.

Stavenga, Doekele G., Atsuko Matsushita, and Kentaro Arikawa. 2015. "Combined Pigmentary and Structural Effects Tune Wing Scale Coloration to Color Vision in the Swallowtail Butterfly Papilio Xuthus." *Zoological Letters* 1 (1): 14. https://doi.org/10.1186/s40851-015-0015-2.

Steinbrecht, Rudolf Alexander. 1985. "Fine Structure and Development of the Silver and Golden Cuticle in Butterfly Pupae." *Tissue and Cell* 17 (5): 745–62.

Sugumaran, Manickam. 2009. "Complexities of Cuticular Pigmentation in Insects." *Pigment Cell & Melanoma Research* 22 (5): 523–25. https://doi.org/10.1111/j.1755-148X.2009.00608.x.

Tutt, James William. 1896. *British Moths*. London: G. Routledge.

Vane-Wright, Richard I. 2020. "James Petiver's 1717 Papilionum Britanniae: An Analysis of the First Comprehensive Account of British Butterflies (Lepidoptera: Papilionoidea)." *Notes and Records* 74 (2): 275–302. https://doi.org/10.1098/rsnr.2019.0014.

Walton, Olivia C, and Martin Stevens. 2018. "Avian Vision Models and Field Experiments Determine the Survival Value of Peppered Moth Camouflage." *Communications Biology* 1 (1): 1–7. https://doi.org/10.1038/s42003-018-0126-3.

Yumnam, Tarunkishwor, Birupaksha Banerjee, and Ullasa Kodandaramaiah. 2021. "Pupal Colour Plasticity in the Butterfly *Catopsilia pomona* (Lepidoptera: Pieridae)." *Biological Journal of the Linnean Society* 134 (2): 331–41. https://doi.org/10.1093/biolinnean/blab087.

CHAPTER 5

Painting by Numbers and Insect Illustrations in the Eighteenth Century: Jacob Christian Schäffer and Stephan Loibel

Giulia Simonini

1 Introduction

This paper investigates how entomological plates were colored by hand in the eighteenth century, a subject which is almost *terra incognita*. It is especially dedicated to a case study, namely the implementation of the painting-by-numbers technique within the workshop of the naturalist Jacob Christian Schäffer in the late 1750s and early 1760s.[1] The chapter examines the popularity of the technique and asks to what extent the study of insects played a role in its development and dissemination. In attempting to answer these questions, I will present at first the *dramatis personae* of our story: Schäffer, the author of *Icones insectorum circa Ratisbonam indigenorum coloribus naturam referentibus expressae = Natürlich ausgemahlte Abbildungen Regensburgischer Insecten* (*Images of insects from Regensburg with their colors painted after nature*, 1766–1769), and Stephan Loibel, the painter who was commissioned to prepare the illustrations for the three-volume publication.[2] Unusually, Schäffer's *Icones insectorum* gave much more prominence to the images it contained than to the description and identification of the various species of insects. As we shall see, the motives behind this editorial choice were closely linked to Schäffer's attempts to optimize the process by which natural history plates, and especially those depicting insects, were colored by hand. I will then sketch how the hand coloring of printed plates was executed in comparable editorial projects, before comparing this widespread approach to the one Schäffer introduced in his own workshop to standardize paint mixtures: a sort of painting by numbers that greatly resembles the well-known coloring pastime. This will lead me to discuss the extent to which Loibel was responsible for the implementation of this method. As will become clear, other painters, too, resorted to similar

1 The most valuable publications on this subject are Nickelsen 2006a; 2006b; 2016; Goedings 2015; Eiermann 2018. For further literature, see below.

2 Hereafter abbreviated as *Icones insectorum*.

 | DOI:10.1163/9789004684553_006

color notation strategies to communicate with apprentices and collaborators in their workshops. Finally, I will present how the technique of painting by numbers was not only implemented for the plates in Schäffer's publications, but also employed and theorized by other authors, draftspeople, and illustrators, often in the context of entomological illustrations. The latter thus put into practice what Schäffer had advocated as early as in 1769: The use of color charts in the production of hand-colored natural history plates.

2 The Regensburg Insects of Jacob Christian Schäffer and Stephan Loibel

Jacob Christian Schäffer (1718–1790) was a Protestant parson, naturalist, and inventor based in the southern German city of Regensburg.[3] From the 1750s onward, his main interest was the study of insects and fungi,[4] as can be seen from a portrait by the painter Anton Graff (1736–1813), which shows the proud sitter next to three of the hand-colored plates that appeared in his books on the two subjects (Fig. 5.1).[5] While studying insects, Schäffer realized the significance of color for their classification and identification, and, consequently, the importance of their depiction. In *Erläuterte Vorschläge zur Ausbesserung und Förderung der Naturwissenschaft* (*Explained Suggestions for the Improvement and Advancement of Natural Science*, 1763), Schäffer argued that specimen descriptions must be neither too short nor too long and, most importantly, accompanied by illustrations,[6] with the latter being "indispensably necessary for clarity, because they make things perceivable and represent them."[7]

3 Today, Schäffer is mainly known for having invented the first washing machine (Roloff 2010, 168–169) and for his paper samples, which he produced from all kinds of materials (Szalay 2019).

4 His first works on insects were published already in 1752 under the titles *Apus pisciformis, insecti aquatici species noviter detecta. Adiecta est tabula aenea, insectum repraesentans* and *Nachricht von einer Raupe, so etliche Jahre her an manchen Orten in Sachsen vielen Schaden gethan: nebst einigen aus der Natur dieser Raupe hergeleiteten Vorschlägen, solche am leichtesten zu verringern und auszurotten*. Between 1752 and 1766, Schäffer published another 22 entomological papers with hand-colored illustrations.

5 After this portrait, the engraver Johann Jakob Haid (1704–1767) produced a mezzotint that was used as the frontispiece of *Icones insectorum*.

6 Hereafter abbreviated as *Erläuterte Vorschläge*. The same text with some amendments reappeared in the preface to Schäffer's *Abhandlung von Insektenlehre* (*Essay on Entomology*, 1764). Schäffer 1763, 20.

7 "sind zur Deutlichkeit unentbehrlich nothwendig, weil sie die Sachen sinnlich machen und darstellen." Schäffer 1763, 20. All translations, if not otherwise specified, are mine.

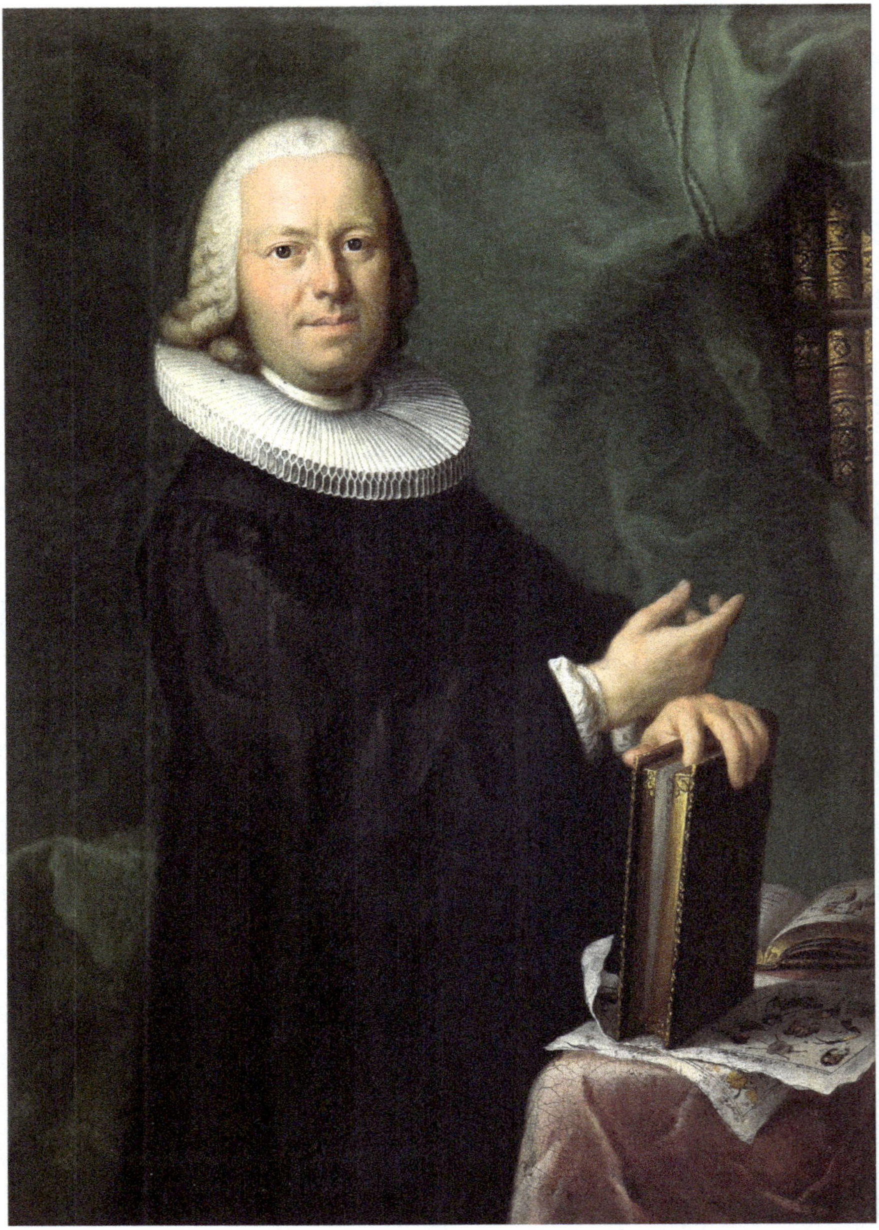

FIGURE 5.1 Anton Graff (signed), *Portrait of Jacob Christian Schäffer*, 1771, oil on canvas
© REGENSBURG, NATURKUNDEMUSEUM OSTBAYERN. PHOTO CREDITS: STEFAN EFFENHAUSER

Images, he declared, could easily illustrate what systems and lexica developed by naturalists could only explain through many words, thus performing a very helpful explicative function.[8] In order to maximize their beneficial effect, it was necessary for illustrations to be "enhanced through proper coloring or illumination, thus assigning to them their right and true value. Black prints must be completely banned."[9] Accordingly, Schäffer took care to include hand-colored illustrations in all his publications.

It is interesting to note that the content of *Icones insectorum* differed markedly from that of other natural history books of the same period in that its three volumes featured a total of 282 plates with many hand-colored illustrations of insects found in the vicinity of Regensburg, but provided very little in the way of written explanation. In the introduction to the volumes, Schäffer explained the reason "why even the initially promised short description [was] now omitted":[10] He considered the images in *Icones insectorum* to be self-explanatory, which is why he deemed it sufficient to identify each depicted insect with a name in Latin and German on the adjacent page – for instance, Figure Two of the eyed ladybug (*Anatis ocellata*) in Plate One is simply named "Coccinella prima. Erster Blattlaufkäfer" (Fig. 5.2). This peculiarity was noticed by a contemporary reviewer, who, while praising the "accuracy and beauty" of the images,[11] criticized the "arbitrary names, which, apart from the names of the genus, which are sometimes Linnaean, sometimes of own invention, consist only in numbers."[12] The same reviewer also found fault with the insects' arrangement on the individual plates, where they were represented in fairly neat lines, but without the host plant and in no clear taxonomical order.[13]

All 282 plates contained in the three volumes of *Icones insectorum* carry the signature "Loibel pinxit Ratisbona"[14] – the plates were thus made after the watercolor originals of one Loibel. A Stefan Loibel signed the plates of other works by Schäffer for the first time in 1759 as the engraver, and in 1760 as the

8 Schäffer 1763, 3.

9 "durch die Ausmalung, oder das Illuminieren, erst recht erhöhet werden, und damit ihren rechten und wahren Wert erhalten. Schwarze Abdrücke müssen gänzlich verbannet werden." Schäffer 1763, 14.

10 "warum sogar die anfänglich versprochene kurze Beschreibung jetzt weggeblieben ist" (Schäffer 1766a, 1:Vorbericht [n.p.]).

11 "Genauigkeit und Schönheit," L. 1770, 281.

12 "willkürliche Benennungen, die außer den Geschlechtsnamen, die bald Linnaeisch, bald von eigener Erfindung sind, nur in Nummern bestehen," L. 1770, 281.

13 L. 1770, 282.

14 Some are signed "Loibel pinx. Rat." or "Loibel p. R."

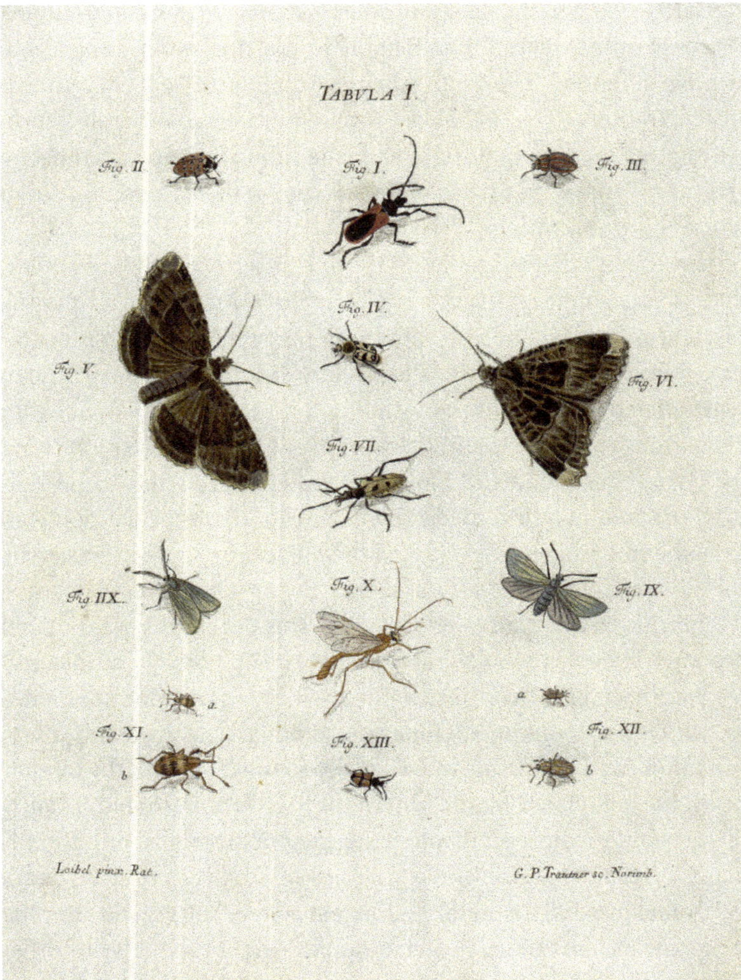

FIGURE 5.2 Stephan Loibel (illustrator), Gustav Philipp Trautner (engraver), Depiction of "Coccinella prima. Erster Blattlaufkäfer," the eyed ladybug (*Anatis ocellata*), in: Jacob Christian Schäffer, *Icones insectorum*, 1766, vol. 1, Plate 1, engraving on paper, watercolor and body color, Göttingen, Niedersächsische Staats- und Universitätsbibliothek, RMAG <4 ZOOL VI, 3205:1,1>. Public domain HTTP://RESOLVER.SUB.UNI-GOETTINGEN.DE/PURL?PPN3869 35742

draftsman.[15] Not much is known about this Stefan Loibel (1725–1775). The *Regensburgisches Diarium oder wöchentliche Frag- und Anzeige-Nachrichten* of June 1775 recorded the death of "Stephan Loibl" on June 15 and the burial two

15 Schäffer 1759; 1760.

days after of "Stephan Loibel" in the churchyard of St. Ruprecht.[16] Despite the slightly different spelling of the decedent's surname, there can be no doubt that the *Diarium* was referring to the same person: His profession is given as "painter and resident here, 50 years old" and as "miniature painter here," respectively.[17]

In *Erläuterte Vorschläge*, Schäffer disclosed additional details about Loibel, albeit without explicitly naming him. All told, Schäffer dedicated several pages to the vicissitudes he claimed to have experienced with his painters. Since there were no artists who specialized in natural history in Regensburg at the beginning of the 1750s, he resorted to training "destitute and poor" and "in part not very polite people" to prepare the illustrations for his publications, from the drawings to the hand coloring of the plates.[18] While the coloring was carried out by many anonymous colorists, Schäffer named some of the artists who produced the watercolor originals and performed the engraving or etching of the plates. At first, he trained a young goldsmith with a talent for painting named Johann Georg Beez.[19] In 1754, before leaving his position in Schäffer's workshop, "the young Petz … trained his sister in illumination,"[20] but the young woman in question – Susanna Sophia Beez – did not last long in Schäffer's employ: She unexpectedly died after having painted almost one hundred fungi.[21] After her sudden death, Schäffer decided to hire Stephan Loibel, a "proper so-called *Bildermahler*."[22] Like the Beez siblings before him, Loibel was at first unable to produce natural history illustrations to the desired standard; he was, however, a painter by profession: While his burial record identified him as a miniature painter (*Miniaturmahler*), Schäffer referred to him as a "Bildermahler."[23] In *Erläuterte Vorschläge*, Schäffer distinguished between "Mahler" (painters), "Illuministen" (illuminators), and "Bildermahler" (print colorists). Susan Dackerman has argued that the latter two are synonyms for *Briefmaler*, namely professional print colorists whose existence

16 Anonymous 1775, 195.

17 "Mahler und Beysitzer allhier, 50 Jahre alt" and "Miniaturmahler allhier," Anonymous 1775, 195.

18 "dürftig und arm;" "zum Teile eben nicht gar höfliche Leute," Schäffer 1763, 23; 25.

19 Schäffer (1763, 22) called him "Herr Peetz" (or Petz, see note 20 below), he himself used the signatures "Ioh. Geo. Bez," "I.G. Bez," or only "Beez." Beez is recorded in Thieme and Becker 1999, 487, and a family of goldsmiths in Augsburg named Betz is mentioned in Meißner and Beyer 1995, 271. See also Hünniger 2021, 192.

20 "Der junge Petz allhier hat seine Schwester zum illuminieren abgerichtet." Schäffer 1754. She signed her name "Beezin," "Soph. Beezin," or "Susanna Sophia Betzin."

21 Schäffer 1763, 23. On her untimely demise, see Hünniger 2021, 192.

22 "ordentlich sogenannten Bildermahler," Schäffer 1763, 23.

23 Schäffer 1763, 22–23. Hereafter, the modern German spelling (*Maler*) will be used.

is recorded as early as the sixteenth century.[24] This is supported by the fact that Schäffer lamented that *Bildermaler* "could paint little colorful pictures well enough. But when I put before them a cabbage, a worm, a fly, and the like to be painted, then nothing ever came of it."[25] In other words: While they were fairly proficient in the use of pigments and dyes, they proved incapable of producing accurate botanical or zoological outline drawings. Schäffer's text seems to hint that there was no sharp distinction between the professions of painter – likely working with water and body colors – and print colorist, with both sometimes being employed as natural history draftspeople. Their wages probably depended on their skill level, and we may assume that painters were more expensive than print colorists.[26] Thus, it would appear that Schäffer (as did other naturalists) hired, and frankly exploited, less specialized artists, and trained them to prepare the drawings for his publications in order to cut down on his expenses.[27] According to Schäffer, Loibel's training cost him a lot of effort and money, and even though the results were initially unsatisfactory, the latter often had to be paid in advance so he could feed his wife and children.[28] Loibel, though, "had the rare quality of being happy to be told what to do, and, without the slightest impatience, willingly changed a thing as often as I asked for it."[29] Schäffer proudly concluded this story by claiming to "have trained him [Loibel] in painting and engraving after nature."[30]

In summary, Loibel prepared the watercolor originals for the plates of *Icones insectorum* and likely supervised a whole number of unknown colorists, who hand colored the printed plates. In the following section, I will discuss why

24 Dackerman 2002, 17ff.

25 "Bildergen konnten sie artig und buntscheckig genug mahlen. Wenn ich ihnen aber ein Kraut, einen Wurm, eine Fliege, und dergleichen, vorlegte, um solche abzumahlen; so kam überall nichts heraus." Schäffer 1763, 22.

26 Nickelsen 2006b, 35–38.

27 One notable example is the Nuremberg physician Christoph Jacob Trew (1695–1769), who taught Georg Dionysius Ehret (1708–1770), a gardener who never underwent a formal education as a painter, how to properly execute botanical illustrations. See Nickelsen 2006b, 23–26, 32.

28 Schäffer 1763, 23–24.

29 "dieser mensch die seltene eigenschaft hatte, dass er sich gerne sagen lies, und ohne die geringste ungeduld eine sache so oft und willig abänderte, als ich es verlangte." Schäffer 1763, 23.

30 "wie ich ihm in Mahlen und Stechen nach der Natur abgerichtet habe." Schäffer 1763, 25. Dominik Hünniger (2021, 192) points out that the verb *abrichten* suggests the training of animals. However, while *abrichten* is indeed strongly pejorative in reference to humans in modern German usage, this was not necessarily the case in the eighteenth century (see Adelung 1793, 83–84). Loibel did not prepare the copperplates for *Icones insectorum*, but he did so for other works by Schäffer.

Icones insectorum, despite its idiosyncrasies, is a key publication for the implementation of paintingby numbers in entomological illustrations, and how the specific hand-coloring strategy employed in its creation differs from the usual method.

3 Hand-Colored Insect Illustrations

At the beginning of the establishment of entomology as an independent discipline, hand-colored natural history plates were widely regarded as superfluous. Depictions of insects were often deliberately left black and white like the ones in Jan Swammerdam's (1637–1680) *Bybel der natuure* (1737) – in fact, Swammerdam felt it would have been a pity to "deface" his accurate drawings with paints.[31] Black-and-white illustrations were appreciated for their clarity, which made them well-suited to illustrate the anatomical details of such tiny creatures.

Another important reason for the omission of hand-colored plates was economic in nature: When colorists were involved, the price of a book could easily double or even quadruple.[32] Yet despite the considerable expense, many eighteenth-century naturalists were dissatisfied with the quality of colored illustrations. Several factors contributed to this state of affairs: First, since there were no specialized schools for aspiring draftspeople in the field, the artistic and aesthetic rules for depicting true-to-nature specimens had to be defined by the naturalists themselves, who then instructed the artists on what details to highlight or leave out.[33] Second, while the coloring of epistemic images did form part of the training regime of many Enlightenment artists, the process of true-to-nature coloring could easily go awry during the many steps it took to produce each image: The artist began by preparing the original hand-colored drawing using a specimen (or another illustration) as their model; then, the copperplate was carved or etched (usually by another hand) and printed in many exemplars, of which one was hand colored by the artist and inscribed with color notations; finally, print colorists or illuminators colored the other printed plates by hand according to this model.[34] Third, print colorists were

31 "commaculare/bekladden" (Swammerdam 1737, 66). See also the introduction of this volume.

32 Examples of this are Aldrovandi's *De animalibus insectis* (1602) and Moffett's *The Theatre of Insects* (1634). See Etheridge 2021, 27–28.

33 Daston and Galison 2007, 59; Nickelsen 2006a, 77, 171; 2006b, 33.

34 Nickelsen 2006b, 62; 2006a, 8–14.

often members of socially and economically disadvantaged groups (women, children, etc.) who had not received proper training in this *métier*, and thus had to be supervised by the artist in charge of the project.[35] Since colorists were paid in piece rates, they often worked up to sixteen hours a day, sometimes by candlelight and not in the uniform and neutral northern light necessary to match the colors of the model[36] – no surprise, then, that book authors often lamented the sloppy execution of the coloring.[37] Schäffer himself denounced the illustrations in an earlier work on natural history as particularly hideous, and berated their colors as anomalous and absurd.[38] This was a commonly acknowledged problem, and likely the reason why Carl Linnaeus (170–1778), for instance, exhorted natural historians to "prevent their illustrators from rendering accidental traits, like color."[39]

Despite all these issues, much effort was expended on the publication of hand-colored insect illustrations. The miniature painter Johannes Goedaert (1617–1668) sold his three-volume *Metamorphosis naturalis* (1660–1669) either with black-and-white or hand-colored plates. The coloring was done by himself or his family, but no details regarding the process have been established so far.[40] The artist and entomologist Maria Sibylla Merian (1647–1717), too, offered her books either with colored or black-and-white illustrations. Merian painted separately the insects and plants on which they feed, and then combined them together in a drawing that served as the model for the copper plates.[41] How she and her daughters proceeded with the hand coloring of the printed plates remains unknown.[42] Some scholars have speculated that they may have used master copies, as did the miniature painter and entomologist August Johann Rösel von Rosenhof (1705–1759).[43] Master copies, also called master plates or pattern plates, were hand-colored prints with color notations written by the

---

35 Nickelsen 2006b, 62. On the role of women in the field of natural history illustrations, see also Tongiorgi Tomasi 2008.

36 Niekisch 2009a, 26. On the preferability of northern light, see for instance Beal 1984, 71–72.

37 Cockerell 1922, 75; Nickelsen 2006b, 63–65; Kusukawa 2012, 76.

38 The work in question is *Histoire naturelle des oiseaux* (1750), the re-print of Eleazar Albin's *A Natural History of the Birds* (1731) by the Dutch printing house De Hondt. Schäffer 1763, 7.

39 Daston and Galison 2007, 59.

40 Goedings 2019, 26.

41 Schmidt-Loske 2010, 7; Etheridge 2021, 97.

42 The coloring often differs from copy to copy, but in many cases we cannot tell if the plates were colored in Merian's workshop or by the purchasers themselves. Etheridge 2021, 100–101.

43 Grebe and Sauer 2017, 72.

leading painter of a project. This instrument was meant to expedite the print colorists' labor as well as ensure accuracy and uniformity among all exemplars of the same publication.[44] Specifically, the marginal color notations written on the master copy were intended to enable colorists to immediately determine the correct pigments and dyestuffs for washing the plate, thus sparing them the arduous task of testing different mixtures until they found the right one. This approach ensured, first, an accurate coloring and, second, a well-selling product.[45] Rösel von Rosenhof's son-in-law, Christian Friedrich Carl Kleemann (1735–1789), employed master copies, too. Manfred Niekisch has examined and published several of the still extant master plates of Kleemann's *Historia naturalis ranarum nostratium* (1758).[46] Intriguingly, some of the 23 plates, now held in the collection of the Naturhistorische Gesellschaft Nürnberg, were signed and dated by Kleemann himself – one example is labelled as "C.F.C. Kleemann ad pict[uram] prototyp[us]" ("C.F.C. Kleemann's original for painting") and "Bestes Muster 1770" ("best sample 1770").[47] The master plates are covered with copious handwritten notes on how their coloration was achieved: For example, "the cherry brown of the lung is a mixture of carmine with glair and very little black, while the reddish gleam is a mixture of vermilion and carmine. The green of the little ball in between is a mixture of lily green and white shaded with lily green."[48] Kleemann and his wife, Rösel's daughter Catharina Barbara (1741–1804), also produced master copies for *Insecten-Belustigung* (1740–1755), which are dated between 1766 and 1783.[49] Here, too, one can glean detailed information regarding the pigments they employed. On the master copy for Plate Ten depicting the metamorphosis of the painted lady (*Vanessa cardui*), for instance, Kleemann wrote: "the brown [of the body] mixed from burnt red ochre, with Berlin yellow and some dark yellow ochre, saffron, and very little black" (Fig. 5.3).[50] It is unclear what the exact purpose of these master

44 Nickelsen 2006b, 65.

45 Niekisch 2009a, 25; Ogilvie 2013, 85, 89.

46 Niekisch 2009a, 25–26; 2009b.

47 Niekisch 2009b, 17, 22.

48 "Die weichselbraune Lunge ist von Carmin mit Eiweiß und mit sehr wenig Schwarz vermischt, der röthliche Schimmer aber von Zinnober mit Carmin vermischt. Das grüne Kügelgen dazwischen ist Lilliengrün mit Weiß vermischt und mit Lilliengrün schattiert." Niekisch 2009b, 23. Also known as iris green, lily green was produced with the juice of various plants (Eastaugh 2008, 198).

49 Niekisch 2009a, 26. The master plates are kept at the Stadtbibliothek Nürnberg, Nor. H. 934 = Rar.

50 "Das braun von rothgebranntem Ocker, mit Berlinergelb und etwas dunkelgelb Ocker, Safran und sehr wenig Schwarz vermischt." Stadtbibliothek Nürnberg, 75 Nor. H. 934 = Rar, Plate 10. My transcription and translation.

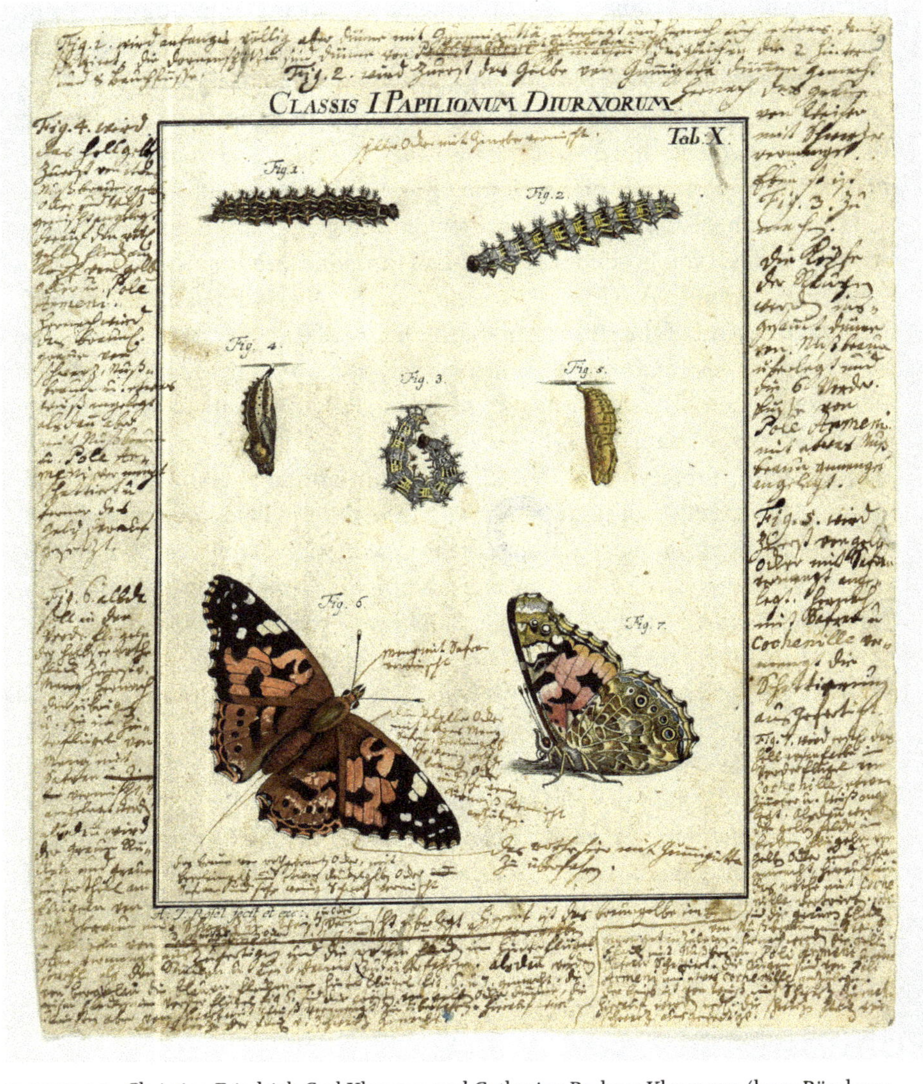

FIGURE 5.3 Christian Friedrich Carl Kleeman and Catharina Barbara Kleemann (born Rösel von Rosenhof), the stages of the metamorphosis of the painted lady (*Vanessa cardui*), master copy for Plate 10 of August Johann Rösel von Rosenhof's *Insecten-Belustigung*, 1746, vol. 1, engraving on paper, watercolor and body color, Nuremberg, Stadtbibliothek, 75 Nor. H. 934 = Rar 1, fol. 20
© STADTBIBLIOTHEK NÜRNBERG

plates was given that they were all hand colored after the publication of both *Insecten-Belustigung* and *Historia naturalis ranarum nostratium*. Niekisch has hypothesized that Rösel's family kept coloring the plates of unsold exemplars, or that the color notations were compiled to preserve this particular knowledge for future use.[51] Whatever Kleemann's original intentions may have been, his master plates were likely used as reference for the second edition of *Historia naturalis ranarum nostratium* in 1814.[52]

Another example of extant master copies are the ones personally hand colored by James Curtis (1791–1862) for his *British Entomology* (1824–1839). The collection is now housed at the Natural History Museum in London, and comprises 770 pattern plates, which, according to Pamela Gilbert, "are covered with streaks of colour, where the colourist has obviously tried to match the colours. These plates are rather dirty and greasy, not from age or poor storage but from constant handling by the colourists."[53] Like Kleemann, Curtis wrote down coloring instructions such as "N.B. use prussian blue & carmine lake" on the master copy of Plate Eight, which depicts the ruby-tailed wasp (*Chrysis fulgida*) (Fig. 5.4.). In the printed book, meanwhile, Curtis described the vibrant metallic and iridescent colors of this wasp as "bright blue variegated with green and purple ... bright crimson reflecting gold and green"[54] – instead of attempting to reproduce with pigments the greenish and purple gleam visible on the thorax next to the blue and crimson, he simply described this effect with words. Compared to Kleemann's coloring instructions, those by Curtis are rather cursory, if not perfunctory. From the colorists' perspective, though, sparse instructions actually made their task easier. Kleemann's notes, for example, were like sprawling clouds of handwritten text that engulfed the illustrations – how were colorists supposed to strictly follow such intricate instructions? Besides the inherent complexity of notes that attempted to describe every tiny visual detail of the master copy, another problematic issue was the absence of exact indications of the quantity of the respective paints in a given mixture. As the color proofs on Curtis's pattern plates reveal, colorists had to test several mixtures before they happened across the correct one. As widespread as this practice apparently was, it hardly prevented colorists from making mistakes, which led to other strategies being explored by authors and publishers.

That an alternative approach was chosen for the plates of *Icones insectorum* becomes clear from the afterlife of the book's publication material. In 1798, the Erlangen publisher Johann Jacob Palm (1750–1826) acquired from Schäffer's heirs the rights to all his natural history publications as well as the

51 Niekisch 2009a, 27–29.

52 Niekisch 2009a, 25.

53 Gilbert 2000, 6. On the coloring of these plates see French 2022.

54 Curtis 1824, 1:[plate 8].

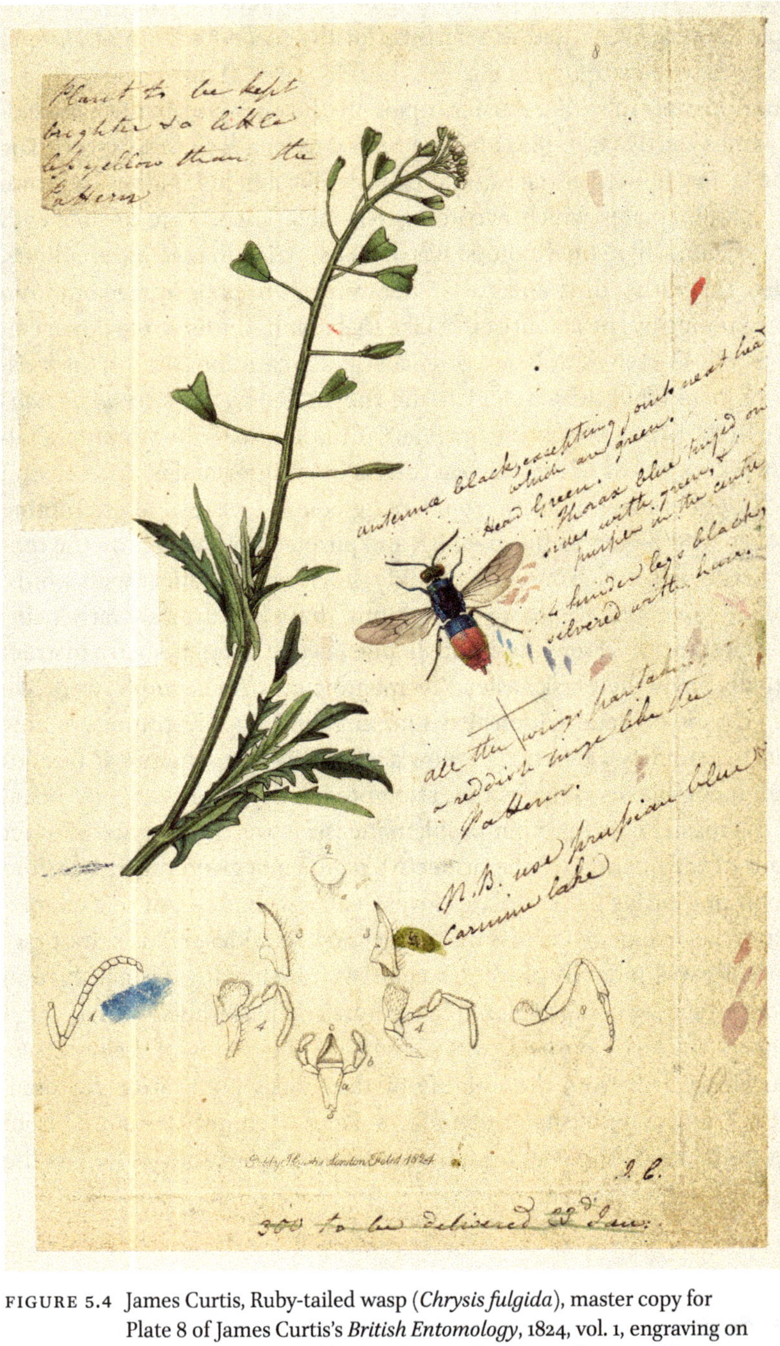

FIGURE 5.4 James Curtis, Ruby-tailed wasp (*Chrysis fulgida*), master copy for Plate 8 of James Curtis's *British Entomology*, 1824, vol. 1, engraving on paper, watercolor and body color, London, Natural History Museum, Entomology Rare Books, SB 0 C.11, page 8.

corresponding body of material,[55] which included the watercolor originals made by Stephan Loibel and the copperplates of *Icones insectorum*.[56] Palm then commissioned the physician and natural historian Georg Wolfgang Franz Panzer (1755–1829) to compile a new critical nomenclature of the insects depicted in the work. In furtherance of this editorial project, Palm provided Panzer with the "watercolor originals of the Regensburg insects" in the hope of facilitating the latter's task of identifying the specimens that had been so cursorily described by Schäffer.[57] When the critical edition was published in 1804, Panzer's preface praised the "clean, delicate, accurate, faithful, and evocative" watercolor originals as "almost consistently excellent"[58] – in fact, Panzer went as far as to state that "the majority" of the illustrations in question "could, more or less, replace the natural specimens."[59] While Panzer maintained that the watercolor originals had "not been produced by one hand alone," we now know that Loibel created all of them.[60] Unfortunately, the whereabouts of Loibel's watercolors for *Icones insectorum* – and of all the other material used for Schäffer's publications – are presently unknown, which made it impossible to assess their pictorial quality for the present essay. Comparing the plates of *Icones insectorum* with the watercolor originals, Panzer observed that the quality of the former did not match that of the latter. He explained this qualitative imbalance with the method employed by Schäffer's colorists (referred to by Panzer as *Illuministen*) in washing the plates:

> However, this apparent contradiction can easily be explained, if one knows that to obtain at least somewhat tolerable illuminated plates, one must not present the watercolors to the illuminators for copying, but rather the so-called master copies (*Musterblätter*), which, if possible, must be produced by the painter himself with great diligence. This did not happen with Schäffer's images of the Regensburg insects, though. The watercolor originals replaced the so-called colored master copies, and

55 Christine Glas (1988, 93–94) has suggested that the sale might have been arranged by Schäffer's nephew Johann Ulrich Gottlieb Schäffer (1753–1829), who studied at the University of Erlangen, of which Palm was the official publisher.

56 Glas 1988, 92.

57 "*Originalmahlereyen* der regensburgischen Insekten" Panzer 1804, xiii. Italics in the original.

58 "fast durchaus ... vortrefflich" and "rein, zart, treffend, getreu und characterisierend," Panzer 1804, xiii.

59 "die meisten ... konnten also so ziemlich die Stelle natürlicher Originale vertreten." Panzer 1804, xiii.

60 "nicht von einer Hand allein gefertigt worden sind," Panzer 1804, xiii.

> the colorists of Schäffer's plates could not cope with the many different ways of selecting and mixing colors, and therefore picked the colors at their own discretion, which led to the disfiguration of many an image.[61]

Panzer's observation confirms, on the one hand, that the usage of master copies was commonplace in the eighteenth century. On the other hand, it substantiates the hypothesis that Schäffer's colorists were not supplied with the usual master copies, but only with the watercolor originals. This raises the question of how and where the pigments, dyestuffs, and paint mixtures for the plates of *Icones insectorum* were standardized and set down.

4 Color Charts for *Icones insectorum*

As briefly mentioned above, Schäffer was a staunch advocate of hand-colored plates despite all the problems associated with their production. In his *Fernere Zweifel und Schwürigkeiten welche in der Insectenlehre annoch vorwalten* (*Further Doubts and Difficulties Which Still Prevail in Entomology*, 1766), Schäffer argued that "with some insects, the colored illustrations" were "more than necessary."[62] As he also pointed out, however, the color of insects could vary within the same species, and was liable to change with age and after their death. Moreover, naturalists lacked a vocabulary rich enough to describe the innumerable hues of insects, and draftspeople and print colorists were often unable to match their real-life colors with pigments because the latter, too, tended to change their properties or deteriorate over time.[63] As if matching the natural color of specimens was not demanding enough in and of itself, the work of both painters and colorists was rendered even more difficult by issues such as the lack of lightfastness of mixtures and the unpredictable chemical

61 "Indessen lässt sich dieser anscheinende Widerspruch gar wohl erklären, wenn man weiss, dass man, um nur etwas erträgliche illuminirte Blätter zu erhalten, den Illuministen nicht die Mahlereyen zum copiren vorlegen muss, sondern sogennate Musterblätter, die von dem Mahler – wo möglich, selbst mit grossem Fleisse gefertigt seyn müssen. Diess ist aber bey den Schäfferschen Abbildungen regensburgischer Insekten nicht geschehen. Die Originalmahlereyen vertraten die Stelle der sogenannten illuminirten Musterblätter, und die Illuministen der Schäfferschen Tafeln konnten sich in die so verschiedenen Arten und Weisen, Farben zu wählen und zu mischen, nicht finden, wählten daher nach eigenem Gutdünken ihre Farben, wodurch dann manches Bild entstellt worden," Panzer 1804, xiii–xiv.

62 Hereafter abbreviated as *Fernere Zweifel*. "Bey manchen Insecten [sind] die ausgemalten Abbildungen übernotwendig" Schäffer 1766c, 11.

63 Schäffer 1766c, 17.

reactions to which they were prone. In *Fernere Zweifel*, Schäffer rhetorically asked: "Who, alas, does not know that it is utterly impossible to prevent some colors, especially those that are exceedingly fresh and lively, from decaying over time and becoming dull?"[64] Another source of problems were the people who wielded the pigments. Schäffer maintained that "at first glance, illuminators may appear to be of the least importance; in fact, however, they are the ones who create the most issues, virtually everywhere"[65] – an almost inevitable result given that the colorists were often overworked and had to paint in the wrong light, not to mention cases where the master copy was lost.[66] All these challenges notwithstanding, Schäffer left no doubt that he considered the hand coloring of the plates an essential step in the production of true-to-nature images that would be of real benefit to fellow entomologists. In *Fernere Zweifel*, Schäffer noted that he had been pondering possible ways to standardize color terms and color mixtures for entomological plates for some time.[67] Given that the first volume of *Icones insectorum* was published in the same year (1766), this thought process must have been close to completion. In 1769, Schäffer finally presented the result of his deliberations under the title *Entwurf einer allgemeinen Farbenverein* [sic] *oder Versuch und Muster einer gemeinnützlichen Bestimmung und Benennung der Farben* (*Plan for a Universal Combination of Colors, or Attempt at and Model for Determining and Naming Colors in a Way that is Useful to the General Public*).[68] The little book listed a number of fundamental rules, which, needless to say, were illustrated with two hand-colored plates very likely executed by Loibel. The first plate, entitled *Farbenmuster der sieben einfachen und natürlichen Hauptfarben* (*Color Chart of the Seven Simple and Natural Main Colors*), displays 29 of the most frequently used colors,[69] which are divided into seven main groups: Red, yellow, blue, green, brown, white, and black (Fig. 5.5). The second, called *Muster und Register der rothen Farben* (*Model and Register of the Red Colors*), presents 33 samples of red colors.

64 "Ja, wer weis nicht, daß es bey manchen Farben, sonderlich, die gar zu frisch und lebhaft seyn, auf keine Weise zu verhüten ist, daß sie nicht mit der Zeit abstehen und matt werden sollten?" Schäffer 1766a, 17.

65 "Illuministen mögten, dem ersten Ansehen nach, am wenigsten zu bedeuten haben; und sie sind gleichwohl wirklich diejenigen, die fast überall das meiste zu schaffen machen." Schäffer 1763, 21.

66 Nickelsen 2006b, 62–63; 2006a, 5–6; 2016, 111–112.

67 Schäffer 1766c, 17.

68 Hereafter abbreviated as *Farbenverein*.

69 Hereafter abbreviated as *Farbenmuster*. In the text, Schäffer described only 28 colorants, of which one blue, the twenty-ninth sample in the chart, is left unnamed. Schäffer 1769, 20.

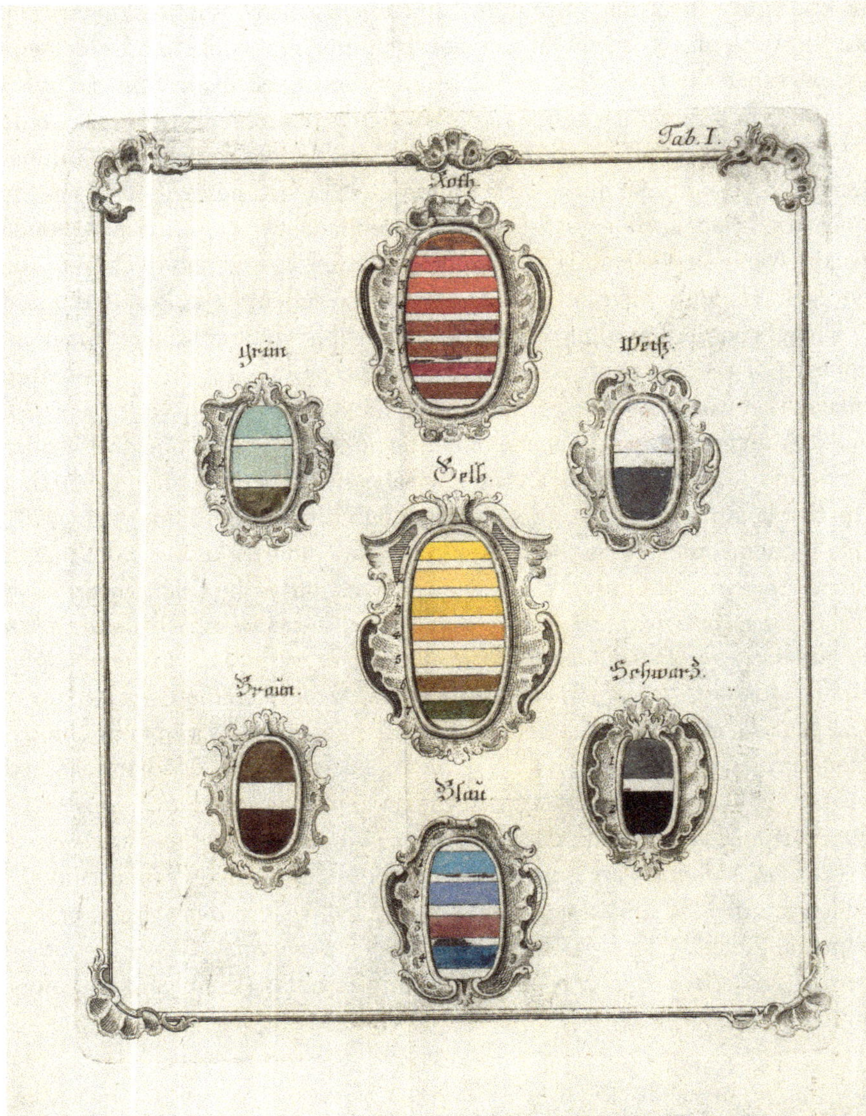

FIGURE 5.5 Stephan Loibel?, *Farbenmuster der sieben einfachen und natürlichen Hauptfarben*, in: Jacob Christian Schäffer, *Entwurf einer allgemeinen Farbenverein*, 1769, Plate 1, engraving on paper, body color, Zurich, Zentralbibliothek, Alte Drucke, NNN 821,5. Public Domain Mark 1.0
HTTPS://DOI.ORG/10.3931/E-RARA-51617

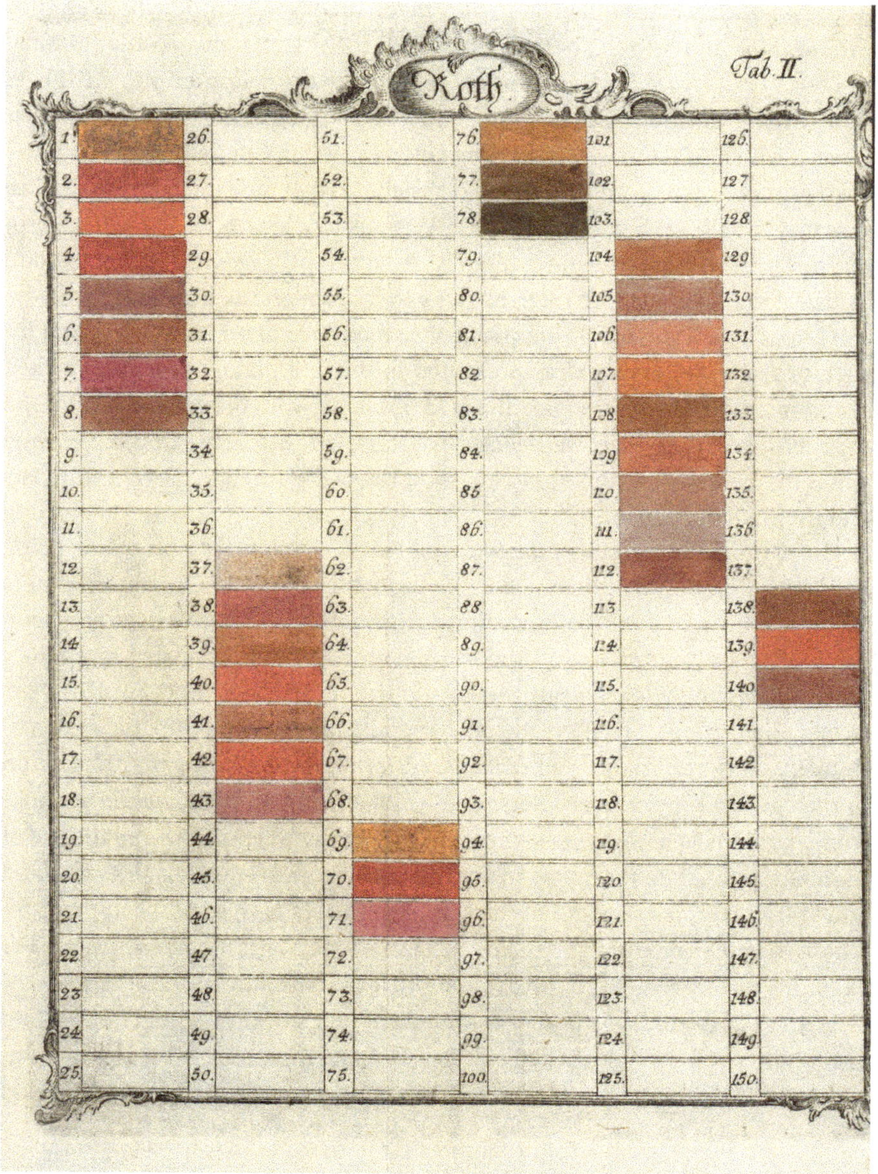

FIGURE 5.6 Stephan Loibel?, *Muster und Register der rothen Farben*, in: Jacob Christian Schäffer, *Entwurf einer allgemeinen Farbenverein*, 1769, Plate 2, engraving on paper, body color, Zurich, Zentralbibliothek, Alte Drucke, NNN 821,5. Public Domain Mark 1.0
HTTPS://DOI.ORG/10.3931/E-RARA-51617

Arranged within a grid-like color chart made up of 150 numbered boxes, they extend the gamut of red colorants shown in the first plate with an additional 25 hues of red (Fig. 5.6). The numbers assigned to the color samples are linked to a register entitled *Erklärung der Kupfertafeln* (*Explanation of the Copper Plates*), which lists the corresponding color names and ingredients (Fig. 5.7) – color number 71, for instance, is named *Rosenroth* (rose red) and is created from a mixture of cochineal and white lead.[70] With its 117 blank boxes, Schäffer's prototypical color chart was quite literally an "attempt" (*Versuch*) at, or "model" (*Muster*) of, what Schäffer referred to as *Farbenverein*, the final and complete version of which was to be comprised of a chart and corresponding register for each of the seven main colors presented in the first plate. As Schäffer lacked both the time and the necessary funds to publish a work of such magnitude, he expressed his hope that "well-versed painters, colorists, etc." would continue where he had left off and complete the project for the benefit of naturalists and their employees.[71]

Once the final *Farbenverein* was published, Schäffer envisaged it being employed as a double reference: On the one hand, the samples and registers it contained could guide colorists and draftspeople in the process of hand coloring drawings and plates; on the other hand, they also provided naturalists with standardized color terms they could use in their descriptions. However, as explained above, Schäffer sought to reduce the amount of textual information to a bare minimum – in *Icones insectorum*, he provided no description of the depicted specimens whatsoever, colors included. While he did propose some methods to name color samples in the chart, he clearly favored the use of numbers to indicate the colors of specimens:[72] "name only the main color and add to it the number or digit of the box from the color chart and color register where the color mixture you mean and want to be represented is located. For example, the elytra are red (no. 12) and the feet are yellow (no. 26)."[73]

With a complete *Farbenverein* at their disposal, naturalists could quickly and accurately visualize a described color, or find the right terms to describe their specimens.[74] Even more importantly, the comprehensive color chart provided

70 Schäffer 1769, 21. In *Farbenmuster*, cochineal (*Cochenille*) and carmine (*Carmin*) are classified as two different pigments, but they were likely extracted from the same insect.

71 "von kunstverständige Mahlern, Illuministen etc." Schäffer 1769, 6.

72 Simonini 2018, §21.

73 "Man nenne nur die Hauptfarbe, und setze zu solcher die Nummer oder Ziffer desjenigen Faches aus dem Farbenmuster und Farbenregister, wo diejenige gemischte Farbe sich befindet, welche man meynet und anzeigen will. Z. B. die Flügeldecken sind roth (No. 12) die Füße gelb (No. 26)." Schäffer 1769, 15.

74 Schäffer 1769, 17.

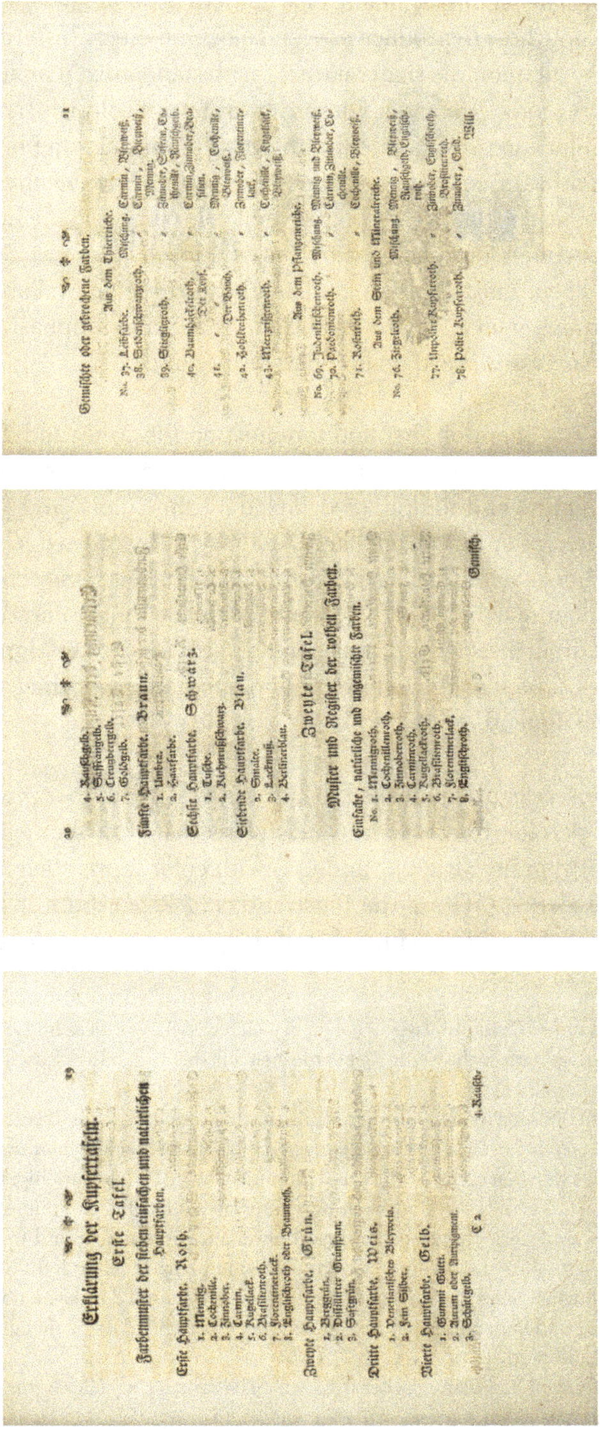

FIGURE 5.7 *Erklärung der Kupfertafeln*, in: Jacob Christian Schäffer, *Entwurf einer allgemeinen Farbenverein*, 1769, pp. 19–21, Zurich, Zentralbibliothek, Alte Drucke, NNN 821,5. Public Domain Mark 1.0
HTTPS://DOI.ORG/10.3931/E-RARA-51617

"painters, and especially colorists," many of whom, in Schäffers words, "[did] not immediately know how to produce a given color, and often waste[d] not a little time until they hit upon the right mixture," with standardized mixtures to wash natural history plates.[75] Schäffer's description of print colorists trying out several different color mixtures before finding the appropriate one brings to mind the stained and worn master copies used for Curtis's *British Entomology*, which, as briefly mentioned above, display several color proofs made by colorists next to the specimen hand colored by Curtis himself (Fig. 5.4). Schäffer evidently did not deem master copies sufficiently accurate and reliable, and considered the *Farbenverein* a superior solution. As Schäffer emphasized, his certitude on the matter was based on first-hand experience:

> Since I focus my attention first and foremost on the study and description of insects, I have only presented the natural and simple colors from which my painters and illuminators mixed all the colors, and painted and illuminated after nature the large quantity of insects, which occur in my insect books published so far; from which I believe I may conclude that this number of colors is sufficient and adequate for the painting and illumination of all insects, and consequently also for the creation of the color samples as well as the determination and naming of the colors of insects, which depends on it.[76]

In all likelihood, "insect books published so far" was a reference to *Icones insectorum*: Instead of naming the three last hues of red listed in the register of *Farbenverein* (color numbers 138, 139, and 140) with color terms such as "rose red," they were labeled as "[f]rom the Illustrations of Regensburg Insects."[77] Schäffer even revealed the precise numbers of the plates and figures from the

75 "Mahlern, und sonderlich Illuministen"; "die nicht gleich wissen, wie diese und jene Farbe herauszubringen, und oft nicht wenig Zeit verderben, bis sie die gehörige Mischung treffen" Both quotes Schäffer 1769, 18.

76 "Da ich [...] nur fördersamt auf die Kenntniß und Beschreibung der Insecten mein Augenmerk habe, so habe ich auch nur diejenigen natürlichen und einfachen Farben vorstellen lassen, aus welchen meine Mahler und Illuministen alle Farben derjenigen großen Menge der Insecten gemischet, und nach der Natur gemahlet und illuminiret haben, welche in meinen bishero ausgegebenen Insectenwerken vorkommen; und woraus ich glaube schließen zu dürfen, daß diese Anzahl der Farben zum Mahlen und Illuminiren aller Insecten, folglich auch zur Anlegung der Farbenmuster, und der davon abhangenden Bestimmung und Benennung der Insectenfarben, hinreichend und zulänglich sind." Schäffer 1769, Anmerkung [n.p.].

77 "Aus den Abbildungen Regensburgischer Insecten" Schäffer 1769, 22. The German title of *Icones insectorum* reads *Natürlich ausgemahlte Abbildungen Regensburgischer Insecten*.

second part of the first volume of *Icones insectorum* where the three mixtures had been deployed: Red number 138 for "Tab. LV. Fig. 1," red number 139 for "Tab. LXIX. Fig. 4," and red number 140 for "Tab. LXX. Fig. 1."[78] We can therefore assume that Loibel kept a record of the colorants and mixtures he employed for that project. He possibly sampled each new color in a color chart, which was in turn linked to a register of mixing instructions as described by Schäffer in his *Farbenverein*. In all likelihood, the color notations were not communicated in the manner of Kleemann and Curtis. As discussed above, Panzer, who had the opportunity to study the original material of Schäffer's *Icones insectorum*, was convinced that "the watercolor originals [had] replaced the so-called colored master copies" – had Loibel written color notations for the print colorists directly on the watercolors, Panzer would surely have noticed them. From Panzer's assertion that Schäffer's colorists had "picked the colors at their own discretion," we can instead deduce that the watercolors were unannotated. Although Schäffer did not explicitly state that painters were to mark master copies or watercolor originals with a color code to guide the colorists, both his explanation on how to use the *Farbenverein* and the names he gave to the red mixtures employed for washing the plates of *Icones insectorum* suggest that Loibel wrote down the hypernym of the respective color (such as blue, green, or white) along with the number of the corresponding sample. Where Loibel's color notations were located remains to be ascertained; if color charts or sheets with color-related shorthand annotations were among the original material of Schäffer's publications, Panzer did not recognize them as such. What is clear, however, is that the mixtures used to wash the plates were recorded somewhere, perhaps in a notebook with several color charts whose whereabouts are unknown today.

In light of all this, it seems plausible to infer that Schäffer's employees implemented a kind of painting-by-numbers technique that helped them to uniformly wash all the printed plates of *Icones insectorum*. But since Panzer judged the result as less accurate than Loibel's watercolors, this approach was possibly first put into practice for *Icones insectorum* and may have come into being when Loibel started working for Schäffer around 1759 or slightly earlier. It would appear, then, that Loibel, a miniature painter, contributed in some form to the development of the technique that Schäffer then presented as his brainchild in *Farbenverein*.

78 Schäffer 1769, 22. All plates are signed "Loibel p.[inxit] Rat.[isbona]," that is, "painted by Loibel in Regensburg."

5 Painting by Numbers and Color Notation Methods

This brings us to the question of the extent of Loibel's contribution. For one, the color notation method described in Schäffer's *Farbenverein* has many points of contact with a technique employed by three successful plant and animal illustrators, the brothers Joseph Anton (1756–1831), Franz Andreas (1758–1840), and Ferdinand Lucas Bauer (1760–1826): They, too, used a color chart with numbered samples to color-code their pencil drawings. The chart was likely linked to a register, where they wrote down the recipes needed to reproduce each sample. The three brothers appear to have started using this technique while working on a florilegium called *Liber regni vegetabili* in the early 1770s, when they were young boys.[79] Of the three, only Ferdinand used it throughout his life. He developed at least three different color codes and corresponding charts, of which only the first is known to be extant today (Fig. 5.8). Ferdinand's method of color-coding his field sketches was described by the botanist David John Mabberley as "painting by numbers."[80] In what follows, I shall demonstrate that many other practitioners made use of color notation methods.

Painting by numbers was a pastime for adults commercialized in the 1940s by Dan Robbins (1925–2019) with the promise to turn everyone into a Rembrandt.[81] It consisted of black-and-white drawings with numbered areas. These were sold along with premixed and ready-to-use pencils, watercolor cakes, or oil tubes labeled with a unique number that could be used to fill in the corresponding numbered area in the drawing. Robbins's painting kits were advertised with slogans like "I painted it myself!" or "You paint with numbered colors. A color for every number" (Fig. 5.9).

79 See Lack and Ibáñez 1997; Lack 2003a. Mulholland (2019; 2020) argues that the color chart now preserved in Madrid, Archivo del Real Jardín Botánico, was not used by Ferdinand Bauer and his brothers. Jane Jelley (2022) has hypothesized that Ferdinand Bauer used a numerical system that was based on his paint box and not on a color chart. However, Ferdinand's last color chart was used after his death by the Austrian painter Theodor Franz Zimmermann (1808–1880) to realize the colored version of a color-coded outline drawing of a parrot by the former's hand (Ridl-Dorn and Ridl 2019, 123). The collection of 2,748 bodycolor paintings is known today as *Codex Liechtenstein* and is housed in Vienna, Historische Bibliothek der Sammlungen des Regierenden Fürsten von Liechtenstein, inv. nr C I I. The first volume of *Codex Liechtenstein* is dated 1776 and the last 1805. See Lack 2015, 30. Many attempts have been made to explain how three teenage boys could have developed a technique that has been described as "a specialty of the Bauer brothers." Lack 2015, 37.

80 Mabberley 2017, ix.

81 Langer 2019; Seelye 2019.

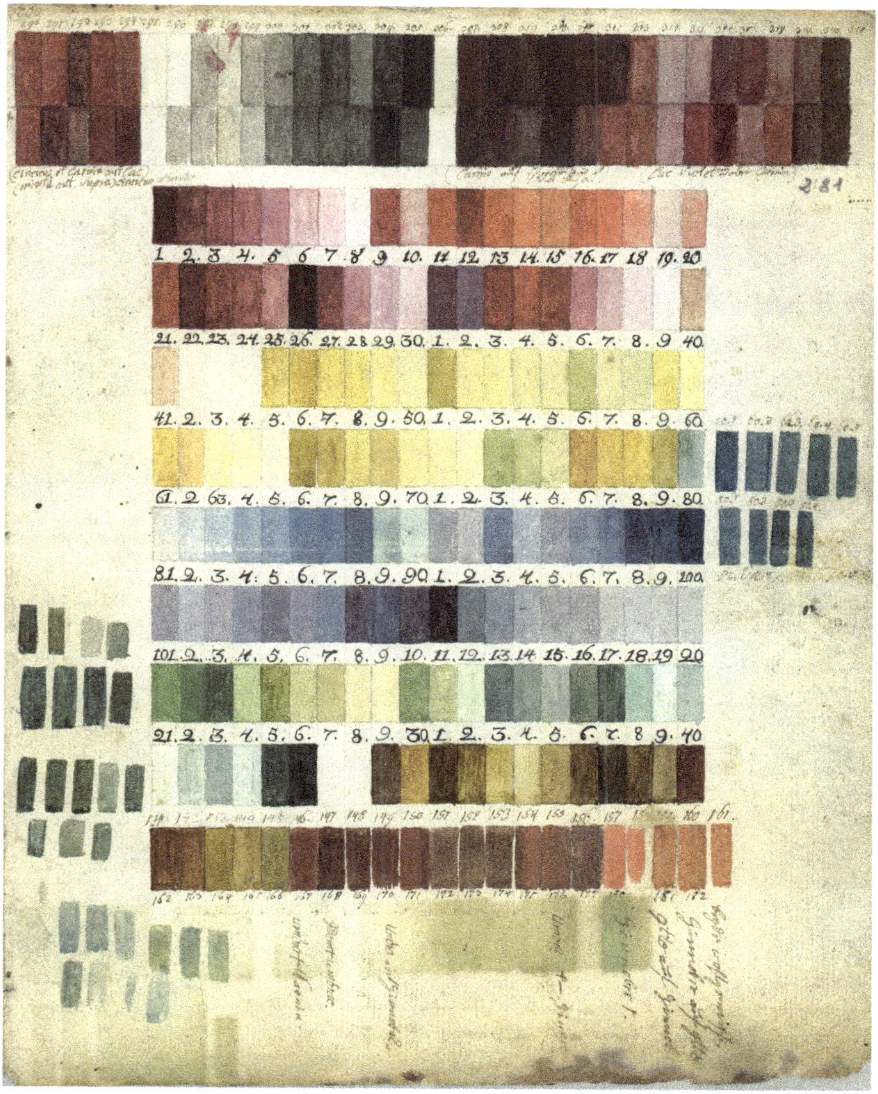

FIGURE 5.8 Franz Ferdinand and Joseph Bauer, color chart for *Liber regni vegetabili*, c.1770, in: Thaddäus Haenke, *Systema Colorum tabulare atque comparativum pro expeditiori plantarum cum vivis coloribus adumbratione in itinere cum hispanis navibus circa globorum terraqueum annis 1789–1793*, watercolor and body color on paper, Madrid, Archivo del Real Jardín Botánico, CSIC, Div. VI, 3, 2, fol. 281r

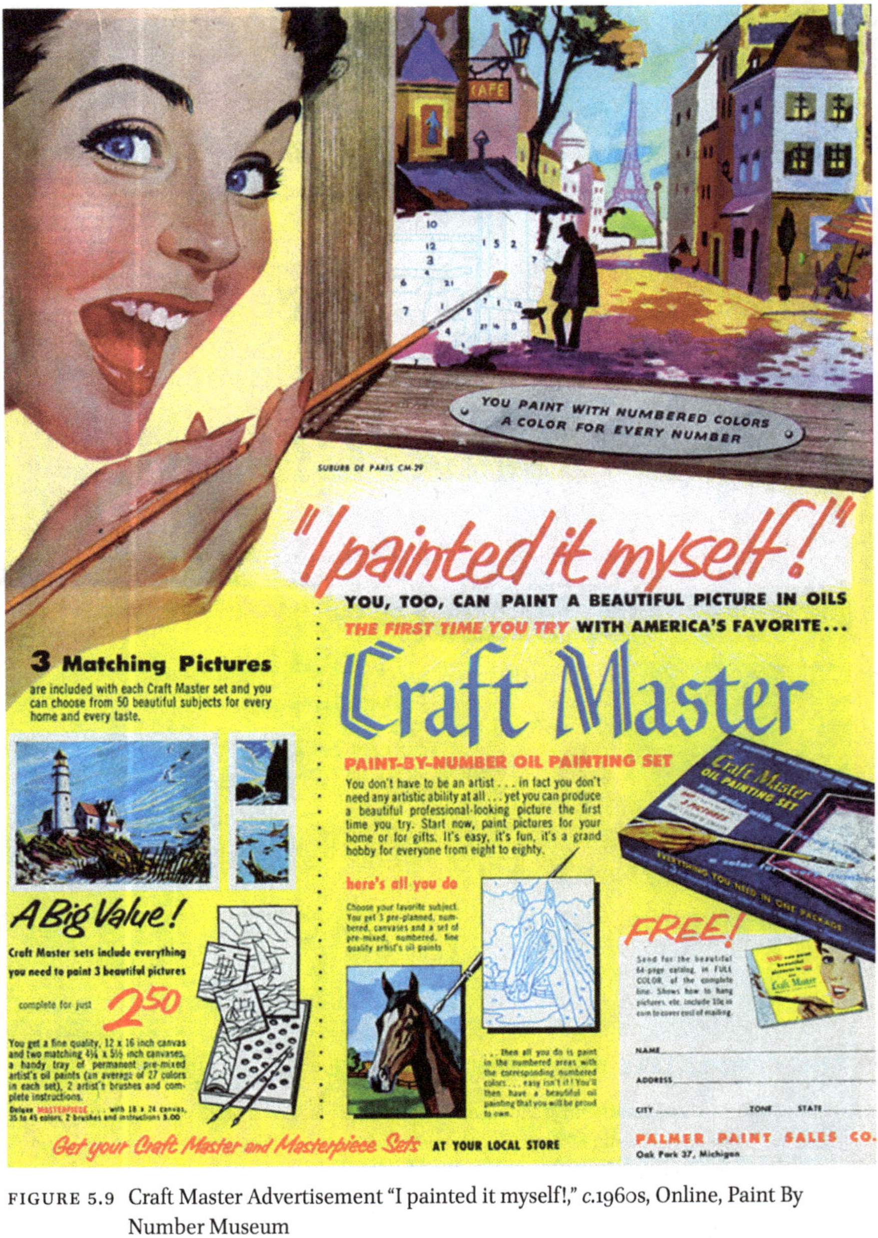

FIGURE 5.9 Craft Master Advertisement "I painted it myself!," *c.*1960s, Online, Paint By Number Museum

Robbins himself stated that he got the idea from Leonardo da Vinci (1452–1519), who supposedly supplied his own apprentices with numbered patterns indicating where certain colors should be used.[82] Although there is no actual evidence of this alleged practice of Leonardo's, color-coding methods were widespread among artists. For instance, it was common for medieval illuminators and early modern painters to resort to abbreviations or entire words to record color-related information in their works, a method that is evident in marginal notes in manuscripts, in sketches, and in the underdrawings of paintings.[83] Illuminators used letters like "a" for *azurus* (blue), "r" for *rubeus* (red), and "v" for *viridis* (green) to indicate the color of drop capitals,[84] and initials such as "p" for *purpur* (purple) and "gr" for *grün* (green) were often employed by glass painters in the drawings for windows and other glass artworks.[85]

The Flemish painter Jan van Eyck (*c.*1395–1441) added a legend of color notations to the draft for his portrait of Cardinal Albergati, which specified the ochre-grey of the hair, the purplish hue of a wart, and the blood-red color used for the nose. Van Eyck made use of the initials of these colors in the silverpoint portrait, where at least two letters "r" (for reddish) have been detected.[86] Similarly, an unknown southern German painter inserted in his draft for a triptych some letters and even a full word to indicate the color of the robes.[87] Albrecht Dürer (1461–1528) set down terms like *rot*, *grün*, and *gell* in his pen drawing *Ottoprecht*.[88] The workshop of Daniele da Volterra (1509–1566) likewise used abbreviations and initials of color terms for the fabrics in the sketch for the *Assumption of the Virgin*.[89] Anthony van Dyck (1599–1641) recorded

---

82 Langer 2019.

83 For an overview, see Seccaroni 2008.

84 Today, these color notations are rarely visible because they were either erased, or are now concealed under paint. Petzold 1990, 18, 20.

85 Ganz 1966, 14.

86 Dierick 2000, 79–80. Jan van Eyck, *Portrait Drawing of Cardinal Niccolò Albergati* (1435), silverpoint on prepared paper, 21.4 × 18.1 cm, Dresden, Kupferstichkabinett, inv. C 775. Ibid., *The Portrait of Cardinal Niccolo Albergati* (1438?), oil on wood, 32.5 × 25.5 cm, Vienna, Kunsthistorisches Museum.

87 One "w" for *weiß* (white), one "g" perhaps for *gelb* (yellow), one "s" for *schwarz* (black), one "b?" for *blau* (blue), and the word *grien* for green. Winkler 1939.

88 The terms can be translated as red (*rot*), green (*grün*), and yellow (*gell*). Albrecht Dürer, *Ottoprecht* (1516), pen drawing on paper, 22.5 × 15.6 cm, Berlin, Kupferstichkabinett, Inv. KdZ 26812. See Koreny 1989, 133–134.

89 "A" for *azzurro* (blue), "G" for *giallo* (yellow), "R" for *rosso* (red), "b" for *berrettino* (purple grey), and the phrase "te gialla abru," which stands for "terra gialla abbruciata" (burnt yellow ochre). Workshop of Daniele da Volterra, pencil drawing on paper, 29.3 × 28.6 cm, Florence, Gabinetto Disegni e Stampe degli Uffizi, Inv. 203 S. See Romani 2003, 104–105.

color names in a sketch he made after Titian's painting *Jacopo Pesaro and Pope Alexander VI before Saint Peter*, noting down *rosso*, *giallo*, *nero*, and *disegno d'oro color verde*.[90] Infrared reflectography has revealed that such initials or color words were sometimes written in the underdrawings of paintings. Examples of this practice are *The Court Jester Gonella*, where we can see several initials and color names such as "b" and "blanc" for white and "rouge" for red,[91] and Vittore Carpaccio's (c.1460/66–1525/26) *The Sermon of St. Stephen*, which features color names such as *beretin*, *roxeta ganzante*, and *azuro*.[92] Here and in similar cases, color-related annotations operated as loose chromatic reminders, and were not necessarily associated with specific colorants. Moreover, they generally indicated the colors of fabrics – only in the case of the Albergati portrait did they refer to aspects of human physiognomy.

While initials and entire words were in widespread use when it came to the coding of colors, numbers seem to have been much less common – so much so, in fact, that the number-based system employed by the Dutch engraver and painter Jacques de Gheyn II (*c.*1565–1629) was described by his biographer as "unusual":[93]

> He divided a panel into as many as a hundred squares, which he marked and numbered in a little book, and painted these squares with various colors – various greys, greens, yellows, blues, reds, flesh colors, and other mixtures giving to each as best as he could its own saturation, and he noted each one in the little book, as described.[94]

As proposed by Claudia Swan, De Gheyn possibly used the color charts contained in the "little book" to produce true-to-nature depictions of flowers and animals, which then served as models for his oil paintings.[95] Whether

90 Adriani 1965, 36; Brown 2015, 86.

91 Kreidl 1981; Pächt 1981. Jean Fouquet (attributed), *The Court Jester Gonella* (*c.*1440–1445), oil on wood, 36.1 × 23.8 × 0.8 cm, Vienna, Kunsthistorisches Museum.

92 The terms can be translated as purple grey, shimmering pink, and azure, respectively. Vittore Carpaccio, *The Sermon of St. Stephen* (?1514), oil on canvas, 148 × 194 cm, Musée du Louvre, Paris, Inv. 181. See LeChanu 1998, 47–48.

93 "oonghewoonlijk" Mander 1604, 294v.

94 "hy bedeelde een Penneel wel in hondert viercanten, en teyckendese met Cijfer ghetalen aen in een Boecxken, en beschilderde dese viercanten met verscheyden coleuren, verscheyden graeuwen, groenen, ghelen, blaeuwen, rooden, carnatien, en ander vermengselen, ghevende soo veel hy mocht elck zijn eygen diepsel, en teyckendet bysonderlijck 'in't Boecxken aen, als geseyt is." Mander 1604, 294v. Translation from Swan 2005, 73.

95 Swan 2005, 73–74. For instance: *Three Butterflies and a Stag Beetle* (1604), watercolor and gouache, 22.7 × 17.5 cm, Paris, Fondation Custodia, Collection Frits Lugt, Lugt Album fol. 1r, inv. nr 5655.

he marked his drawings with color notations is unknown. If so, however, De Gheyn may well have adopted numbers instead of color terms: What his work in the fields of botany and zoology called for were not vague chromatic reminders, but precise information regarding the colors of the respective specimens. Given the bewildering chromatic variety of his subjects, it is unlikely that verbal annotations would have been up to the task.

De Gheyn was not the only artist to employ color charts. Several examples used by manuscript illuminators and map colorists have come down to us,[96] which is no surprise given that they were an ideal tool to ensure uniformity among the products of a workshop. Even amateur painters sampled the paint mixtures they used for their works. A particularly fitting example in our context are the color notes – sometimes accompanied by samples – used by the naturalist and amateur painter William Jones (1745–1818) for his depictions of butterflies (Figs. 5.10 and 1.10).[97]

As I would argue, the fact that Loibel's burial record referred to him as a miniature painter makes it very likely that he, too, had learned to take advantage of color sampling during his apprenticeship. It is thus quite probable that it was Loibel who started using a color chart while producing the watercolor originals for *Icones insectorum*, which in turn inspired Schäffer to propagate this method as a solution that solved two problems at the same time: Not only did it assist painters and print colorists with the uniform coloring of printed illustrations, but it supplied naturalists with a much-needed standardized chromatic vocabulary. If my hypothesis is correct, Schäffer's contribution consisted, first, in the popularization of the painting-by-numbers technique, and second, in his suggestion to enhance the color charts used by painters with color terms, thereby turning them into a sort of color encyclopedia; the latter project, however, did not progress beyond the stage of a prototype in the form of his *Farbenverein*.

96 Examples of this are the color samples in the *Kölner Musterbuch*, a model book used by an illumination workshop, Cologne, Historisches Archiv, Inv. 7010-293; and in *Aantekening betreffende Schilderen* (1679), a painting manual written by the map colorists Jan Dirksz. Zoutman (1617/1618–1697) and Simon Eikelenberg (1663–1738), Regionaal Archief Alkmaar, Collectje Aanwinsten 390.

97 Jones's Paint Samples and Mixes, WJ/3/0/001, University of Oxford, Museum of Natural History, Vane-Wright 2021, 23. Another earlier example are the color blots in *Ricordi per belli colori*, a manuscript on landscape painting by the botanist Gherardo Cibo (1512–1600), Vatican City, Biblioteca Apostolica Vaticana, Urb.lat.1280.

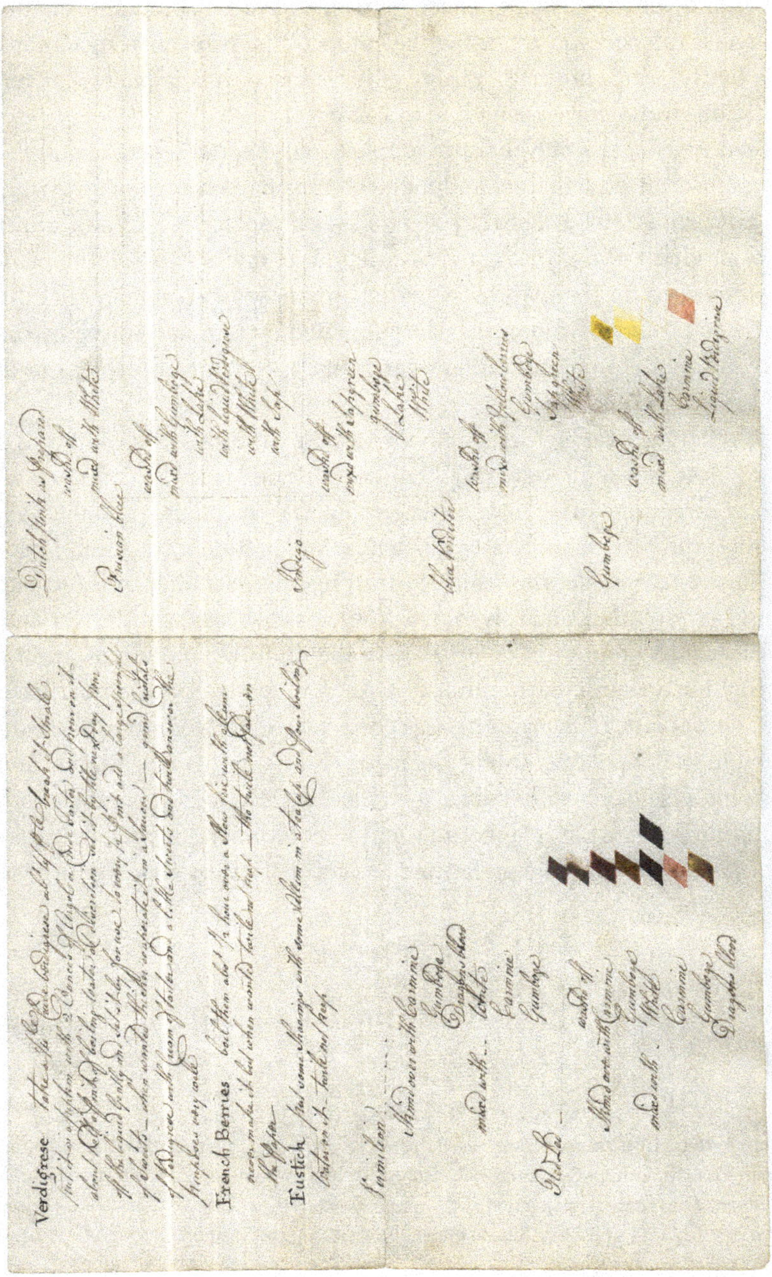

FIGURE 5.10 William Jones, paint samples and mixes used for his *Icones*, no date, ink and watercolor on paper, Oxford University Museum of Natural History, WJ/B/3/0/001, fol. 4

6 Reception of Loibel's Technique and Schäffer's *Farbenverein*

Schäffer's *Farbenverein* came off the press shortly before the Bauers started working on *Liber regni vegetabili*. As of yet, the resemblance between Schäffer's ideas and the Bauers' coloring method has been only cursorily noticed, but never fully investigated.[98] It is my contention that Loibel's painting-by-numbers technique as described in Schäffer's *Farbenverein* was adopted by the three young Bauer brothers in the early 1770s. Indeed, the chart that they presumably used (Fig. 5.8) is astonishingly similar to Schäffer's *Muster und Register der rothen Farben* (Fig. 5.6): Featuring 140 color samples to Schäffer's 150, it only has to be rotated about 90 degrees clockwise to make the two charts look almost exactly alike – the only noteworthy difference is that the Bauers' also contains colors other than red.

While the Bauers mainly focused on botanical subjects, they also painted zoological illustrations, including some of insects. In *Liber regni vegetabili*, for example, the plants are often embellished with insects such as the nine-spotted moth (*Amata phegea*) and the longhorn beetle (perhaps a *Prionus coriarius*) that adorn the illustrations of a white lily and a calla lily, respectively.[99] Regrettably, these insects are not depicted in the color-coded pencil drawings used as models for the bodycolor paintings in *Liber regni vegetabili*. In the 1780s, Franz Bauer painted the title page of an exemplar of the second edition of *Selectarum stirpium americanarum historia* for its author, the botanist Nicolaus Joseph von Jacquin (1727–1817) (Fig. 5.11),[100] decorating it with a plethora of stunningly lifelike butterflies and moths.[101] Given that Franz was simultaneously employed by Jacquin and Father Norbert Boccius (1731–1806), the commissioner of *Libri regni vegetabili*, it is quite possible that he first drew the lepidopterans in color-coded sketches. If extant, these have not emerged from any collection; however, Franz must have kept some sort of notebook

98 Nickelsen 2006b; 2006a; Lack 2015; Nickelsen 2016.

99 Many have been identified by Lack (2003a, 103). See also Mabberley 2017, 4.

100 Lack 1998. Franz Bauer, *Title Page of Nikolaus Joseph von Jacquin's Selectarum Stirpium Americanarum Historia* (c.1780), bodycolor, Vienna, Österreichische Nationalbibliothek, 177687-E.

101 In Franz's wreath, one can identify a dead common fly (*Musca domestica*), a common brimstone (*Gonepteryx rhamni*), two swallowtails (*Papilio machaon*), two peacock butterflies (*Aglais io*), a large emperor moth (*Saturnia pyri*), an African death's-head hawk-moth (*Acherontia atropos*), an elephant hawk-moth (*Deilephila elpenor*), five red admirals (*Vanessa atalanta*), a small tortoise shell (*Nymphalis urticae*), a large white (*Pieris brassicae*), and a spurge hawk-moth (*Hyles euphorbiae*).

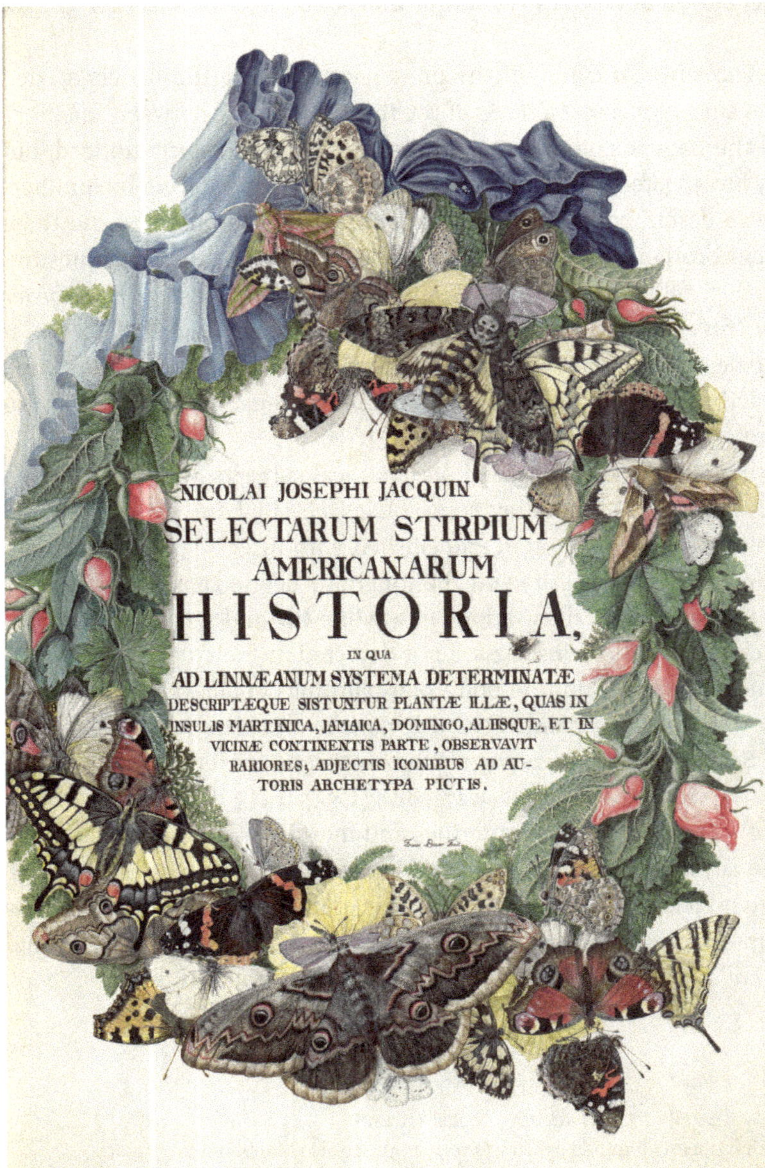

NICOLAI JOSEPHI JACQUIN
SELECTARUM STIRPIUM
AMERICANARUM
HISTORIA,
IN QUA
AD LINNÆANUM SYSTEMA DETERMINATÆ
DESCRIPTÆQUE SISTUNTUR PLANTÆ ILLÆ, QUAS IN
INSULIS MARTINICA, JAMAICA, DOMINGO, ALIISQUE, ET IN
VICINÆ CONTINENTIS PARTE, OBSERVAVIT
RARIORES, ADJECTIS ICONIBUS AD AU-
TORIS ARCHETYPA PICTIS.

FIGURE 5.11 Franz Bauer, wreath with several butterfly and moth species, bodycolor on paper, in: Nikolaus Joseph von Jacquin, *Selectarum Stirpium Americanarum Historia*, c.1780, title page, Vienna, Österreichische Nationalbibliothek, 177687-E. Public Domain Mark 1.0
HTTPS://DIGITAL.ONB.AC.AT/REPVIEWER/VIEWER.FACES?DOC=DTL_8846113&ORDER=1&VIEW=SINGLE

with the sketches of these insects, because the red admiral (*Vanessa atalanta*) reappeared in a miniature on vellum that he painted in 1798.[102]

His brother Ferdinand, too, depicted zoological subjects: During a journey that took him to Australia and the neighboring Norfolk Island, for instance, Ferdinand sketched and color-coded several animals, among them the Christmas spider (*Austrachanta minax*), which, in the early nineteenth century, was still not fully discriminated from insects.[103] Of this spider, Ferdinand produced a separate color-coded study and a colored drawing, and he may well have done the same for the insects depicted next to it on the sheet that is now kept at the Linnaean Society of London.[104] Ferdinand must have executed many more such color-coded drawings over the course of his career, which are now presumably lost. Unfortunately, none of the three brothers shared their method publicly, nor did they publish a color encyclopedia as Schäffer had hoped, which consigned this early reception of Schäffer's *Farbenverein* (and thus of Loibel's strategy to sample color mixtures) to more than two centuries of obscurity.

Not only was the technique itself adopted, as in the case of the Bauer brothers – the description and possible applications of painting by numbers set forth in Schäffer's *Farbenverein* reverberated widely in German-language publications of the time. As discussed above, Schäffer suggested that the colors of insects be named using a hypernym and the number of the sample which came closest to the coloration of the specimen in question. Moreover, his idea to sample mixtures for the coloring of natural history plates, although not clearly communicated as painting by numbers, surely was understood as such, as the example of the Bauer brothers demonstrates. The technique was also picked up by the astronomer and mathematician Johann Heinrich Lambert (1728–1777), who authored *Beschreibung einer mit dem Calauschen Wachse ausgemalten Farbenpyramide* (1772) with the help of the Berlin court painter Benjamin Calau (1724–1785).[105] Lambert was familiar with Schäffer's *Farbenverein*, as he held a copy in his personal library and described it in

102 Franz Bauer, *Flower Piece* (1798), bodycolor on vellum, Cambridge, Fitzwilliam Museum, PD. 58–1975.

103 Ferdinand Bauer, *Austrachant minax* (Christmas spider), 1804, pencil drawing, Vienna, Archiv des Naturhistorischen Museums; Ferdinand Bauer, *Austrachant minax* (Christmas spider), 1804–1805, bodycolor and watercolor, London, Natural History Musem, South Kensington, Zoology Special Collection, RBR Shelf 169–174.

104 Ferdinand Bauer (?), *Australian Insects and Arachnids*, bodycolor and watercolor, London, Linnaean Society, MS no 622, No. 6, Drawer 37. See Mabberley 2017.

105 Hereafter abbreviated as *Farbenpyramide*.

his own book.[106] Advancing a theory on the mixing of colors, Lambert's *Farbenpyramide* comprised two illustrations, a colored pyramid made of 112 color samples and a numerical counterpart that assigned to the latter a total of 69 numbers (Fig. 5.12).[107] In the book, Lambert set forth how each of the samples could be replicated using just three colorants. He also conjectured possible uses of the two illustrations, one of which is highly pertinent to the topic of this paper:

> Those who wish to engage in the drawing after nature of flowers, fruits, animals, insects, or entire landscapes begin with the drawing itself … Here, one must not worry about hitting the colors. One looks them up in the pyramid and notes the number down. One can then return home, certain that one will not forget the colors, but will be able to find them again exactly.[108]

What Lambert suggested, then, was for his readers to draw in the field with a copy of his *Farbenpyramide* at hand. The colors of the depicted specimen were to be compared with those in the pyramid, and their respective numbers noted down in the drawing. The readers could then head home with the color-coded drawing, and prepare the colorants needed following Lambert's instructions. Interestingly, from 1773, Calau sold a color box which contained 28 prefabricated watercolor cake colors mixed after the instructions in Lambert's book and numbered according to its numerical counterpart. Owners of Lambert's book and Calau's color box could thus wash their color-coded drawings using the cakes provided in the box[109] – a true forerunner of Robbins's painting-by-numbers kit. Lambert was the first to provide an explicit description of how the painting-by-numbers technique could be applied to insects, which strongly suggests that Schäffer's *Farbenverein* inspired one of the applications of his color-mixing pyramid. Lambert, however, developed a color chart that was very different in terms of its premises and configuration from the one envisaged by Schäffer.

106 Anonymous 1778, 15–16; Lambert 1772, 27–28.

107 See Dijksterhuis 2015; Simonini 2023.

108 "Wer sich üben will, Blumen, Früchte, Thiere, Insecten, oder auch ganze Landschaften nach der Natur zu zeichnen, der fängt bey der Zeichnung selbst an … Um die Farben zu treffen darf man hiebey nicht besorgt seyn. Man sucht sie in der Pyramide auf und schreibt die No. bey. Dann kann man nach Hause gehen, und man ist sicher, daß man die Farben nicht vergessen, sondern genau wieder finden wird." Lambert 1772, 110.

109 Simonini 2023, 128–133.

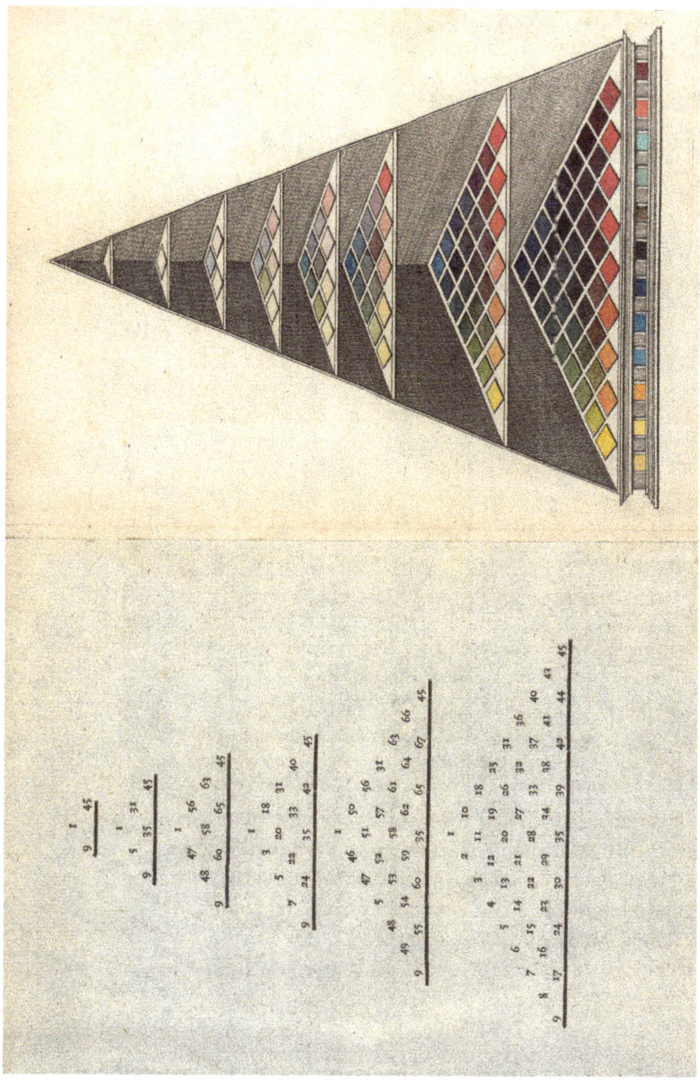

FIGURE 5.12 Johann Heinrich Lambert and Benjamin Calau, color pyramid and numerical counterpart, in: Johann Heinrich Lambert, *Beschreibung einer mit dem Calauschen Wachse ausgemalten Farbenpyramide*, 1772, engraving on paper, watercolor, Zurich, ETH-Bibliothek, Rar 5100. Public Domain Mark 1.0 HTTPS://DOI.ORG/10.3931/E-RARA-3770

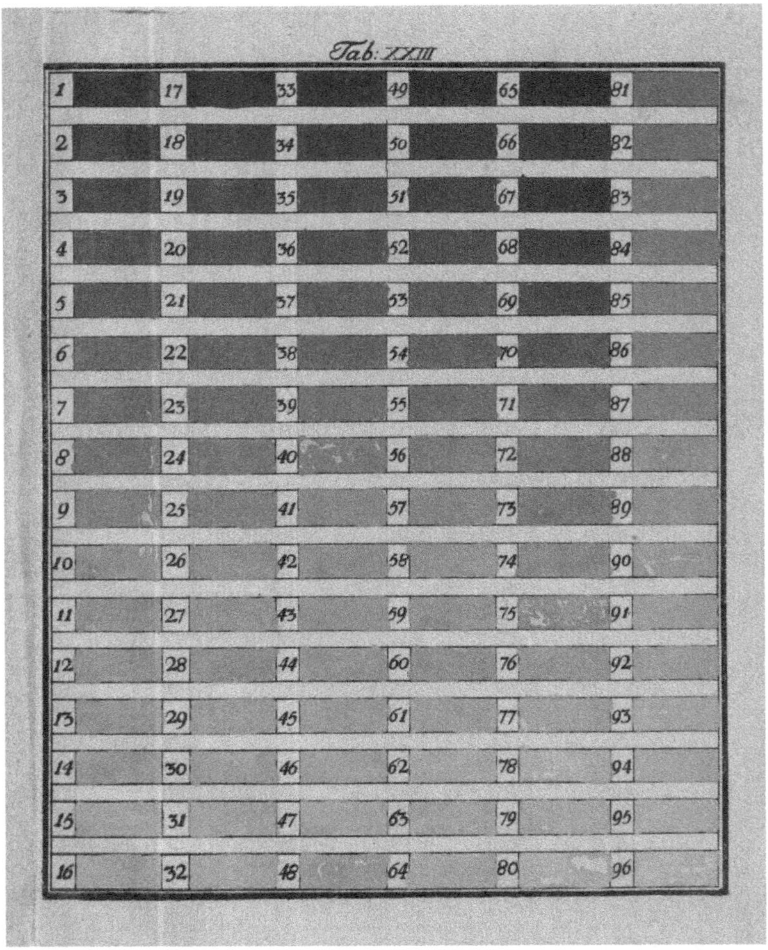

FIGURE 5.13A Christian Friedrich Prange?, samples of green colors (*Grüne Farben*), in: Christian Friedrich Prange, *Farbenlexicon*, 1782, Plate 23, engraving on paper, body color, Dresden, Sächsische Landesbibliothek – Staats- und Universitätsbibliothek, Technol.B.168.m. Public Domain Mark 1.0 HTTPS://DIGITAL.SLUB-DRESDEN.DE/WERKANSICHT/DLF/53059/1

Schäffer's hope that resourceful painters or print colorists would one day produce the universal color chart described in his *Farbenverein* was realized in 1782, when the artist and art teacher Christian Friedrich Prange (1752–1836) published the first color dictionary complemented by a painted sample for each color term, thereby following the model proposed by Schäffer in *Muster und Register der rothen Farben*. Prange's *Farbenlexicon* contains a total of 4,608

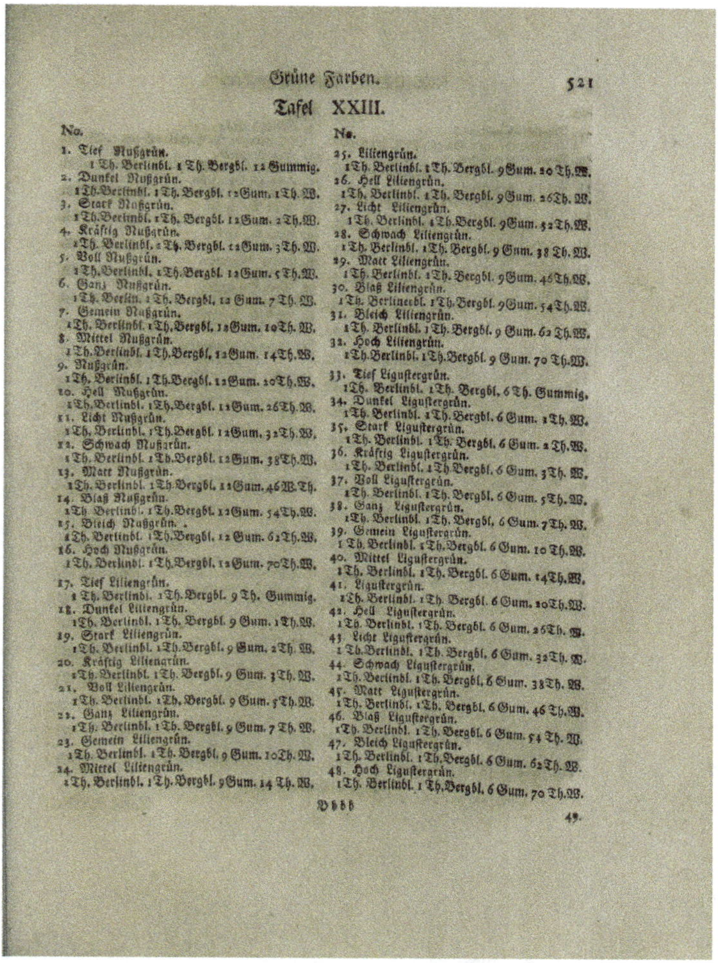

Grüne Farben. 521

Tafel XXIII.

No.
1. Tief Nußgrün.
1 Th. Berlinbl. 1 Th. Bergbl. 12 Gummig.
2. Dunkel Nußgrün.
1Th. Berlinbl. 1 Th. Bergbl. 12Gum. 1Th. W.
3. Stark Nußgrün.
1Th. Berlinbl. 1Th. Bergbl. 12Gum. 2Th. W.
4. Kräftig Nußgrün.
1Th. Berlinbl. 1Th. Bergbl. 12Gum. 3Th. W.
5. Voll Nußgrün.
1Th. Berlinbl. 1Th. Bergbl. 12Gum. 5Th. W.
6. Ganz Nußgrün.
1Th. Berlin. 1Th. Bergbl, 12 Gum. 7Th. W.
7. Gemein Nußgrün.
1Th. Berlinbl. 1Th. Bergbl, 12Gum. 10Th. W.
8. Mittel Nußgrün.
1Th. Berlinbl. 1Th. Bergbl, 12Gum. 14Th. W.
9. Nußgrün.
1Th. Berlinbl. 1Th. Bergbl. 12Gum. 20Th. W.
10. Hell Nußgrün.
1Th. Berlinbl. 1Th. Bergbl. 12Gum. 26Th. W.
11. Licht Nußgrün.
1Th. Berlinbl. 1Th. Bergbl. 12Gum. 32Th. W.
12. Schwach Nußgrün.
1Th. Berlinbl. 1Th. Bergbl. 12Gum. 38Th. W.
13. Matt Nußgrün.
1Th. Berlinbl. 1Th. Bergbl. 12Gum. 46W. Th.
14. Blaß Nußgrün.
1Th. Berlinbl. 1Th. Bergbl. 12Gum. 54Th. W.
15. Bleich Nußgrün.
1Th. Berlinbl. 1Th. Bergbl. 12 Gum. 62Th. W.
16. Hoch Nußgrün.
1Th. Berlinbl. 1Th. Bergbl. 12Gum. 70Th. W.
17. Tief Liliengrün.
1 Th. Berlinbl. 1Th. Bergbl. 9 Th. Gummig.
18. Dunkel Liliengrün.
1Th. Berlinbl. 1Th. Bergbl. 9 Gum. 1Th. W.
19. Stark Liliengrün.
1Th. Berlinbl. 1Th. Bergbl. 9 Gum. 2Th. W.
20. Kräftig Liliengrün.
1Th. Berlinbl. 1Th. Bergbl. 9 Gum. 3Th. W.
21. Voll Liliengrün.
1Th. Berlinbl. 1Th. Bergbl. 9 Gum. 5Th. W.
22. Ganz Liliengrün.
1Th. Berlinbl. 1Th. Bergbl. 9 Gum. 7 Th. W.
23. Gemein Liliengrün.
1Th. Berlinbl. 1Th. Bergbl. 9 Gum. 10Th. W.
24. Mittel Liliengrün.
1Th. Berlinbl. 1Th. Bergbl. 9Gum. 14 Th. W.

No.
25. Liliengrün.
1Th. Berlinbl. 1Th. Bergbl. 9Gum. 20 Th. W.
26. Hell Liliengrün.
1Th. Berlinbl. 1Th. Bergbl. 9Gum. 26Th. W.
27. Licht Liliengrün.
1Th. Berlinbl. 1Th. Bergbl. 9Gum. 32Th. W.
28. Schwach Liliengrün.
1Th. Berlinbl. 1Th. Bergbl. 9 Gum. 38 Th. W.
29. Matt Liliengrün.
1Th. Berlinbl. 1Th. Bergbl. 9Gum. 46Th. W.
30. Blaß Liliengrün.
1Th. Berlinerbl. 1Th. Bergbl. 9Gum. 54Th. W.
31. Bleich Liliengrün.
1Th. Berlinbl. 1Th. Bergbl. 9 Gum. 62 Th. W.
32. Hoch Liliengrün.
1Th. Berlinbl. 1Th. Bergbl. 9 Gum. 70 Th. W.
33. Tief Ligustergrün.
1Th. Berlinbl. 1Th. Bergbl, 6Th. Gummig.
34. Dunkel Ligustergrün.
1Th. Berlinbl. 1Th. Bergbl. 6 Gum. 1Th. W.
35. Stark Ligustergrün.
1Th. Berlinbl. 1Th. Bergbl. 6 Gum. 2Th. W.
36. Kräftig Ligustergrün.
1Th. Berlinbl. 1Th. Bergbl. 6 Gum. 3Th. W.
37. Voll Ligustergrün.
1Th. Berlinbl. 1Th. Bergbl. 6 Gum. 5Th. W.
38. Ganz Ligustergrün.
1Th. Berlinbl. 1Th. Bergbl, 6 Gum. 7Th. W.
39. Gemein Ligustergrün.
1 Th. Berlinbl. 1Th. Bergbl. 6 Gum. 10 Th. W.
40. Mittel Ligustergrün.
1Th. Berlinbl. 1Th. Bergbl. 6 Gum. 14Th. W.
41. Ligustergrün.
1Th. Berlinbl. 1Th. Bergbl. 6 Gum. 20Th. W.
42. Hell Ligustergrün.
1Th. Berlinbl. 1Th. Bergbl. 6 Gum. 26Th. W.
43. Licht Ligustergrün.
1 Th. Berlinbl. 1Th. Bergbl. 6 Gum. 32Th. W.
44. Schwach Ligustergrün.
1Th. Berlinbl. 1Th. Bergbl. 6 Gum. 38Th. W.
45. Matt Ligustergrün.
1Th. Berlinbl. 1Th. Bergbl. 6 Gum. 46 Th. W.
46. Blaß Ligustergrün.
1Th. Berlinbl. 1Th. Bergbl. 6 Gum. 54 Th. W.
47. Bleich Ligustergrün.
1Th. Berlinbl. 1Th. Bergbl. 6 Gum. 62Th. W.
48. Hoch Ligustergrün.
1Th. Berlinbl. 1Th. Bergbl. 6 Gum. 70 Th. W.

Vddd

49.

FIGURE 5.13B Christian Friedrich Prange?, register of green colors (*Grüne Farben*), in: Christian Friedrich Prange, *Farbenlexicon*, 1782, p. 521, Dresden, Sächsische Landesbibliothek – Staats- und Universitätsbibliothek, Technol.B.168.m. Public Domain Mark 1.0
HTTPS://DIGITAL.SLUB-DRESDEN.DE/WERKANSICHT/DLF/53059/1

numbered color samples subdivided into 48 color charts, all of which are linked to a register disclosing each sample's name and the corresponding mixing instructions (Figs. 5.13a and b). In 1782, Prange also published a cheaper spin-off of the color lexicon, *Die Schule der Mahlerei*, which contained a reduced number of color samples distributed over six painter's palettes (Fig. 5.14). In both books, Prange provided some suggestions on how to use the color charts.

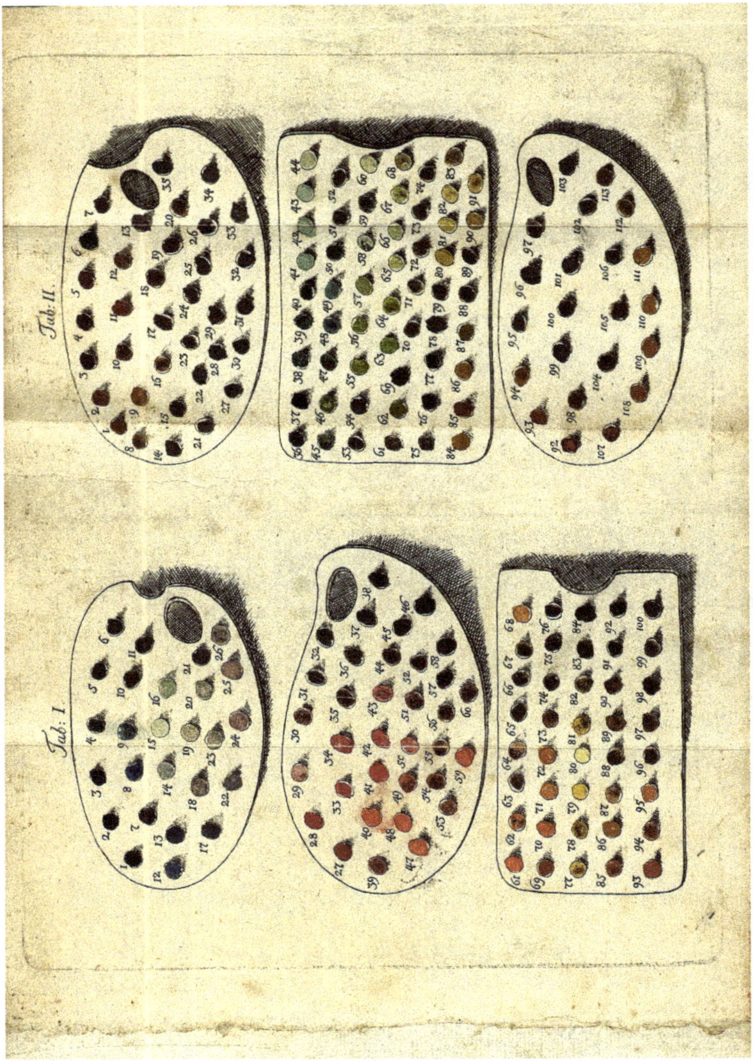

FIGURE 5.14 Christian Friedrich Prange?, color chart with 213 samples, in: Christian Friedrich Prange, *Die Schule der Mahlerey*, 1782, Plates 1 and 2, engraving on paper, body color, Göttingen, Niedersächsische Staats- und Universitätsbibliothek, DD90 A 33229. Public Domain Mark 1.0 HTTP://RESOLVER.SUB.UNI-GOETTINGEN.DE/PURL?PPN645205737

In doing so, Prange plagiarized with minor amendments Lambert's previously quoted description of painting by numbers:

> Thus, those who wish to engage in the drawing after nature of flowers, fruits, animals, insects, or entire landscapes begin with the drawing itself. In order to match the colors, one looks them up in the illuminated plates and notes the numbers down, as in the examples I have given in the second section [of the *Farbenlexicon*]. One can then return home, certain that one will not forget the colors, but will be able to find them again exactly.[110]

Here, too, artists were encouraged to head into the field equipped only with a sketchbook, a pencil, and a copy of the *Farbenlexicon*, which is quite compact and can be easily slipped into a bag; here, too, they were to mark the drawing with the numbers of the matching color samples, and to wash the drawing at home with the help of the mixing instructions supplied in the register. Although the thousands of samples in Prange's *Farbenlexicon* all carry fanciful names such as "birch green" (*Birkengrün*), "tiger yellow" (*Tiegergelb*), "bullfinch red" (*Tompfaffenroth*), and "cornflower blue" (*Kornblumenblau*), Prange apparently felt that using the corresponding numbers was more practical – following his instructions, one could simply write on the sketch the Roman numeral of one of the 48 color charts and the Arabic numeral of one of the 96 samples in the chart, which resulted in a code that was never longer than 8 digits.

The "second section" of the *Farbenlexicon* mentioned in the above-quoted passage contains a list in which Prange gave practical examples of how drawings of various motifs – mostly flowers, insects, human bodies, and landscapes – could be colored with the tints and shades cataloged in his dictionary.[111] A similar but longer list was included in *Die Schule der Mahlerey*, which added fruits, vegetables, metals, buildings, quadrupeds, birds, and fishes.[112] In explaining to his readers how the washing of drawings was to be executed, Prange employed what could be termed a descriptive version of painting by numbers: Instead of

110 "Wer sich also üben will, Blumen, Früchte, Thiere, Insekten oder auch ganze Landschaften nach der Natur zu zeichnen, der fängt bey der Zeichnung selbst an. Um aber die Farben zu treffen, sucht er selbige in den illuminirten Tafeln auf, und schreibt die No. bey, so wie ich einige Beyspiele davon im zweyten Abschnitt angegeben habe. Dann kann man nach Hause gehen und man ist sicher, daß man die Farben nicht vergessen, sondern genau wieder finden wird." Prange 1782b, vii.

111 Prange 1782b, 393–412; 434–435; 445–448.

112 Prange 1782a, 88–116.

providing color-coded illustrations, as Dan Robbins would do in the twentieth century, Prange quoted the numbers of the respective sample and the chart in which it was contained. It is interesting to note that the examples listed in the *Farbenlexicon* included eleven insects, mostly of the orders of Coleoptera and Lepidoptera.[113] First on the list was the stag beetle (*Lucanus cervus L.*), which had served as the motif of famous paintings by the likes of Albrecht Dürer (1505) and Joris Hoefnagel (*c.*1575), and which would also be skillfully rendered by one of the Bauer brothers, possibly Franz.[114] For the coloring of this coleopteran, Prange indicated only two "blackish red" samples (*Rothschwärzlich*), namely numbers 49 and 50 from plate XXXIX. The fourth example was the rose chafer (*Cetonia aurata*), a creature whose "shiny golden green" Prange described as ultimately "inimitable," but similar to sample number 17 from plate XXIII: "deep lily green" (*tief Liliengrün*, see Fig. 5.13a/b).[115] Prange's slender instructions alone were not enough to ensure the true-to-nature coloring of insect drawings: Neither did he specify the colors of anatomical details, nor did he explain how to produce the appropriate shadings. Clearly, he took for granted that his readers possessed the requisite knowledge and skills, and simply advised them to take their cues from similar images in *Insecten-Belustigung*.[116] However, while Prange's instructions were addressed to proficient amateurs who wanted to hand-color printed illustrations, his audience also included the underpaid women and children without formal artistic training whom naturalists like Schäffer so frequently recruited to color their illustrations.

In the latter context of reception, Prange's instructions and Loibel's color notations were not sufficient to ensure the accurate coloring of drawings or printed illustrations of insects – they had to be complemented with a well-executed model, such as Rösel von Rosenhof's hand-colored prints or Loibel's watercolor originals. Alternatively, real-life specimens could have been used, but, as Schäffer and other entomologists noted, the colors of insects

---

113 European stag beetle (*Lucanus cervus L*), rhinoceros beetle (*Scarabaeus nasicornis L.*), red dung beetle (*Aphodius fimetarius*), rose chafer (*Cetonia aurata*), common cockchafer (*Melolontha melolontha*), spring dumbledor (*Trypocopris vernalis*), summer chafer (*Amphimallon solstitiale*), common burying beetle (*Nicrophorus vespillo*), swallowtail (*Papilio machaon*), common brimstone (*Gonepteryx rhamni*), and red admiral (*Vanessa atalanta*). Prange 1782b, 411–412.

114 Neri 2011, 7–10; Lack 2003b, 483.

115 "Das goldglänzende Grün dieses Käfers ist zwar unnachahmlich" (Prange 1782b, 411). *Pace* Prange, Hoefnagel had managed to imitate the chafer's metallic sheen quite impressively with a particular painting technique that involved metal components. See Kimberly Schenck's and Stacey Sell's chapter in this volume.

116 Prange 1782b, 412.

tended to change or fade over time.[117] Provided a suitable model was available, painting by numbers enabled greater accuracy in scenarios where natural history plates were hand colored by people who "could not cope with the many different ways of choosing and mixing colors," to once again quote Panzer. In effect, it was a simple but efficient solution to the problem of inaccurately colored illustrations that anyone could implement, even the three young Bauer brothers working on *Liber regni vegetabili* in the early 1770s.

7 Conclusion

The most common eighteenth-century method to standardize the hand-coloring of natural history plates within a workshop, the use of so-called master copies, was not flawless. As the instructions written on these models were almost inevitably ambiguous, they often yielded unsatisfactory results. Although this practice was still implemented in the nineteenth century, in the eighteenth century the problem of inaccurate written instructions was partially overcome in the framework of a large-scale entomological publication project: Schäffer's *Icones insectorum*.

I argued in this essay that the artist commissioned for this project, the Regensburg miniature painter Stephan Loibel, sampled pigments, dyestuffs, and paint mixtures used in the watercolor originals for Schäffer's book. Sampling colors was commonplace in the workshops of miniature painters and other practitioners, as it ensured a uniform coloring of many identical copies of the same subject. Painters also used to write down coloring instructions on their works, which, for the most part, were loose chromatic reminders. Similarly, Loibel implemented a mainly numerical system of color notation to provide the colorists employed in the same project with less ambiguous coloring instructions, a system that was based on the colors he had previously sampled. Loibel and the colorists active in Schäffer's workshop thus implemented a sort of painting-by-number technique, which, given its novelty, did not always live up to Schäffer's expectations. Despite the many difficulties associated with this project, Schäffer recognized that Loibel's strategy could perfect the existing practice of using master copies to efficiently produce uniform and accurate hand-colored plates, the usage of which he preferred over long descriptive texts in natural history books.

117 See Beth Fowkes Tobin's chapter in this volume.

Moreover, Schäffer felt that using color charts with a register of mixing instructions would not only be of great benefit to painters and colorists – if each sample was named, the color chart would also allow naturalists to systematize their chromatic vocabulary. Schäffer therefore published a little book (*Farbenverein*) which presented Loibel's approach as his brainchild and illustrated it with a prototypical chart of red nuances, which he hoped well-versed painters could extend to all other colors. In turn, painting by numbers was thematized and popularized in a variety of ways in the publications of later German-speaking authors. As we have seen, the technique itself was used by the three young Bauer brothers to paint a florilegium in the 1770s, with the youngest of the brothers, Ferdinand Bauer, continuing to employ it for the rest of his life. The extent to which this technique was implemented among naturalists, natural history painters, and colorists after the publication of *Farbenverein* has yet to be fully explored. In the meantime, however, I hope that this paper has demonstrated the significance of this color-coding and painting practice for the accurate coloring of insect illustrations.

Bibliography

Manuscript Sources

Schäffer, Jacob Christian. 1754. Letter to the Nuremberg Apothecary Johann Ambrosius Beurer (1716–1754), dated September 18, 1754. Erlangen, Universitätsbibliothek, Briefsammlung Trew, H62/TREWBR SCHAEFFER_JACOB_CHRISTIAN[4.

Printed Sources

Adelung, Johann Christoph. 1793. *Grammatisch-kritisches Wörterbuch der Hochdeutschen Mundart mit beständiger Vergleichung der übrigen Mundarten, besonders aber der Oberdeutschen*. Leipzig: Breitkopf.

Adriani, Gert, ed. 1965 (= 1940). *Anton van Dyck: Italienisches Skizzenbuch*. Wien: Schroll.

Anonymous. 1775. *Regensburgisches Diarium oder wöchentliche Frag- und Anzeige-Nachrichten, in welchen die hiesigen Begebenheiten, als: Trauungen, Gebuhrten und Absterben beyderseits Religionverwandten Personen u. Ankunft und Abgang der Passagiers zu Wasser und zu Lande, obrigkeitliche Verordnungen, Intimationen, kleine gelehrte Abhandlungen und Bücher, desgleichen der Brod-, Bier-, Fleisch, und Getränke-Preis bekannt gemacht werden; Dann weiters von Sachen, die gestohlen, verlohren und gefunden worden, zum Verkauf und Vermiethung stehen, oder zu kaufen und zu miethen anverlanget werden; von Promotionen, Besonderungen und suchenden Bedienungen etc. Und überhaupt von allem dem, was sonst zu des Publici*

Wissenschaft, hier und in der Nachbarschaft, dienlich ist, hinlängliche Nachricht gegeben wird. Vol. 25. Regensburg: Zeidler.

Anonymous. 1778. *Verzeichniß der Bücher und Instrumente, welche der verstorbene Königl. Ober Baurath und Professor Herr Heinrich Lambert hinterlassen hat, und die den Meistbiethenden sollen verkauft werden.* Berlin: Winterschen Schriften.

Beal, Mary. 1984. *A Study of Richard Symonds: His Italian Notebooks and Their Relevance to Seventeenth-century Painting Techniques.* New York; London: Garland Publishing.

Brown, Beverly Louise. 2015. "Seeing Red: Was Titian Too Young to Know Better?" *Artibus et Historiae* 36 (72): 85–105.

Cockerell, T.D.A. 1922. "Dru Drury, an Eighteenth Century Entomologist." *The Scientific Monthly* 14 (1): 67–82.

Curtis, John. 1824. *British Entomology: Being Illustrations and Descriptions of the Genera of Insects Found in Great Britain and Ireland: Containing Coloured Figures from Nature of the Most Rare and Beautiful Species, and in Many Instances of the Plants upon Which They Are Found.* Vol. 1. London: Printed for the author.

Dackerman, Susan. 2002. "Painted Prints in Germany and the Netherlands." In *Painted Prints: The Revelation of Color in Northern Renaissance & Baroque Engravings, Etchings & Woodcuts*; edited by Susan Dackerman, 9–48. University Park (Pa.): Pennsylvania State University Press.

Daston, Lorraine, and Peter Galison. 2007. *Objectivity.* Cambridge, New York: Zone Books.

Dierick, A.L. 2000. "Jan van Eyck's Handwriting." In *Investigating Jan van Eyck*, Susan Foister, Sue Jones, and Delphine Cool (eds.). Turnhout: Brepols, 79–82.

Dijksterhuis, Fokko Jan. 2015. "Perception of Colours by Different Eyes." In *Colour Histories: Science, Art, and Technology in the 17th and 18th Centuries*, Magdalena Bushart and Friedrich Steinle (eds.). Berlin: De Gruyter, 23–41.

Eastaugh, Nicholas. 2008. *Pigment Compendium, a Dictionary and Optical Microscopy of Historical Pigments.* Second ed. Amsterdam: Elsevier, Butterworth-Heinemann.

Eiermann, Wolf (ed.). 2018. *Prachtvoll illuminirt: Das Handkolorit in der Druckgraphik (1493–1870).* Munich: Hirmer.

Etheridge, Kay. 2021. *The Flowering of Ecology: Maria Sibylla Merian's Caterpillar Book.* [Emergence of Natural History 3]. Leiden: Brill.

French, Anthony. 2022. "The colouring of John Curtis's British entomology (1834–1839): Joseph Standish and 'the paragon of perfection.'" *Archives of Natural History* 49 (1): 62–77. https://doi.org/10.3366/anh.2022.0758.

Ganz, Paul Leonhard. 1966. *Die Basler Glasmaler der Spätrenaissance und der Barockzeit.* Basel: Schwabe.

Gilbert, Pamela. 2000. *Butterfly Collectors and Painters: Four Centuries of Color Plates from the Library Collections of the Natural History Museum, London.* Jurong, Singapore: Beaumont.

Glas, Christine. 1988. *Johann Jacob Palm (1750–1826): Ein Erlanger Verleger und Buchhändler; Mit einer Verlagsbibliographie von 1780 bis 1826*. [Erlanger Studien 78]. Erlangen: Palm und Enke.

Goedings, Truusje. 2015. *'Afsetters en meester-afsetters': de kunst van het kleuren 1480–1720*. Nijmegen: Vantilt.

Goedings, Truusje. 2019. "De werking van zwart en wit. Kleur in Goedaerts Metamorphosis Naturalis." *Tijd-Schrift. Erfgoedpraktijk in Vlaanderen* 2019 (3): 21–37.

Grebe, Anja, and Sauer, Christine. 2017. *Maria Sibylla Merian: Blumen, Raupen, Schmetterlinge*. Aktualisierte Auflage. Ausstellungskatalog der Stadtbibliothek Nürnberg 110. Nuremberg: Stadtbibliothek Nürnberg.

Hünniger, Dominik. 2021. "Visible Labour? Productive Forces and Imaginaries of Participation in European Insect Studies, ca. 1680–1810." *Berichte zur Wissenschaftsgeschichte* 44 (2): 180–210. https://doi.org/10.1002/bewi.202100002.

Jelley, Jane. 2022. "A Puzzle in a Paintbox: A Painter's Solution to Ferdinand Bauer's Colour Code for the Flora and Fauna Graeca 1786–1794." *Art & Perception* 10 (4): 299–333. https://doi.org/10.1163/22134913/bja.10042.

Koreny, Fritz. 1989. "'Ottoprecht Fürscht' Eine unbekannte Zeichnung von Albrecht Dürer. Kaiser Maximilian I. und sein Grabmal in der Hofkirche zu Innsbruck." *Jahrbuch der Berliner Museen* 31: 127–48.

Kreidl, Detlev. 1981. "Le portrait de Gonella: Le dessin sous-jacent dans le tableau de Gonella." *Gazette des Beaux-Arts* 123 (VI/97): 5–8.

Kusukawa, Sachiko. 2012. *Picturing the Book of Nature: Image, Text, and Argument in Sixteenth-Century Human Anatomy and Medical Botany*. Chicago: University of Chicago Press.

L. 1770. "D. Iacobi Christiani Schäfferi Icones insectorum circa Ratisbonam indigenorum coloribus naturam referentibus expressae Vol. I. P. I. P. II. Natürlich ausgemahlte Abbildungen Regensburgischer Insecten. Regensburg gedruckt bey Heinrich Zenkel 1766. Zusammen 100 Tafeln in Gr. 4." *Allgemeine deutsche Bibliothek* 11 (2): 281–282.

Lack, Hans Walter. 1998. "Jacquin's 'Selectarum stirpium Americanarum historia' – The Extravagant Second Edition and Its Title Pages." *Curtis's Botanical Magazine* 15 (3): 194–214. https://doi.org/10.1111/1467-8748.00171.

Lack, Hans Walter. 2003a. *Ein Garten für die Ewigkeit: Der Codex Liechtenstein*. Second edition. Bern: Benteli.

Lack, Hans Walter. 2003b. "Ferdinand, Joseph und Franz Bauer: Testamente, Verlassenschaften und deren Schicksale." *Annalen des Naturhistorischen Museums in Wien (B Supp.)* 104: 479–551.

Lack, Hans Walter. 2015. *The Bauers: Joseph, Franz & Ferdinand; Masters of Botanical Illustration; an Illustrated Biography*. Munich: Prestel.

Lack, Hans Walter, and Maria Victoria Ibáñez. 1997. "Recording Colour in Late Eighteenth Century Botanical Drawings: Sydney Parkinson, Ferdinand Bauer and

Thaddäus Haenke." *Curtis's Botanical Magazine* 14 (2): 87–100. https://doi.org/10.1111/1467-8748.00074.

Lambert, Johann Heinrich. 1772. *Beschreibung einer mit dem Calauschen Wachse ausgemalten Farbenpyramide wo die Mischung jeder Farben aus Weiß und drey Grundfarben angeordnet, dargelegt und derselben Berechnung und vielfacher Gebrauch gewiesen wird … mit einer ausgemahlten Kupfertafel.* Berlin: Haude und Spener.

Langer, Emily. 2019. "Dan Robbins, Paint-by-Number Inventor Who Made 'Every Man a Rembrandt,' Dies at 93." *Washington Post*, April 6, 2019. Accessed September 22, 2023. https://www.washingtonpost.com/local/obituaries/dan-robbins-paint-by-number-inventor-who-made-every-man-a-rembrandt-dies-at-93/2019/04/05/6f80f412-574e-11e9-8ef3-fbd41a2ce4d5_story.html.

LeChanu, Patrick. 1998. "The Names of the Colors in the Painting 'Preaching of St. Stephen' by Vittorio Carpaccio." *Kermes. Arte e tecnica del restauro* 11 (32): 46–52.

Mabberley, David John. 2017. *Painting by Numbers: The Life and Art of Ferdinand Bauer*. Sydney: NewSouth Publishing.

Mander, Karel Van. 1604. *Het schilder-boeck waerin voor eerst de leerlustighe ieught den grondt der edel vry schilderconst in verscheyden deelen wort voorghedraghen*. Haarlem: voor Paschier van Wesbusch.

Meißner, Günter, and Andreas Beyer (eds.). 1995. "Betz, Dt. Goldschmiede-Familie." In *Saur Allgemeines Künstlerlexikon: Die Bildenden Künstler aller Zeiten und Völker – Bd. 10 Berrettini – Bikkers*. Berlin: De Gruyter.

Mulholland, Richard. 2019. "The Mechanism and Materials of Painting Colour 'Ad Vivum' in the Eighteenth Century." In *Ad Vivum?: Visual Materials and the Vocabulary of Life-Likeness in Europe before 1800*, Thomas Balfe, Claus Zittel, and Joanna Woodall (eds.). London: Brill, 328–355.

Mulholland, Richard. 2020. "Analysis of Watercolour Pigments Found in Thaddeus Haenke's Eighteenth Century Colour Chart and Attributed to Ferdinand Bauer." In *Ciencia y Arte VII. Ciencias y tecnologías aplicadas a la conservación del patrimonio*, Carolina Aguado Serrano, Asunción Guerrero García (eds.). Madrid: Ministerio de Educacion, Cultura Y Deporte, 96–121.

Neri, Janice. 2011. *The Insect and the Image: Visualizing Nature in Early Modern Europe, 1500–1700*. Minneapolis: University of Minnesota Press.

Nickelsen, Kärin. 2006a. "The Challenge of Colour: Eighteenth-Century Botanists and the Hand-Colouring of Illustrations." *Annals of Science* 63 (1): 3–23. https://doi.org/10.1080/00033790500151177.

Nickelsen, Kärin. 2006b. *Draughtsmen, Botanists and Nature: The Construction of Eighteenth-Century Botanical Illustrations*. Archimedes. [New Studies in the History and Philosophy of Science and Technology 15]. Dordrecht: Springer Science & Business Media.

Nickelsen, Kärin. 2016. "Illuminierungspraktiken. Zur Handkolorierung naturhistorischer Tafeln des 18. Jahrhunderts." In *Farre. Farbstandards in den frühen Wissenschaften*, André Karliczek and Andreas Schwarz (eds.). Jena: Salana, 101–129.

Niekisch, Manfred. 2009a. *August Johann Rösel von Rosenhof Künstler, Naturforscher und Pionier der Herpetologie; eine Einführung zum Reprint der "Historia Naturalis Ranarum Nostratium – Die natürliche Historie der Frösche hiesigen Landes."* Saarbrücken: Fines mundi.

Niekisch, Manfred. 2009c. "Neue Ergebnisse aus den Forschungen zur Geschichte der Familie Rösel-Kleemann sowie zu den Musterblättern und zur Illumination der 'Historia Naturalis Ranarum Nostratium.'" *Sekretär. Beiträge zur Literatur und Geschichte der Herpetologie und Terrarienkunde. Symposium zum 250. Todestag von August Johann Rösel von Rosenhof am 27. März 2009* 9 (2): 7–32.

Ogilvie, Brian W. 2013. "The Pleasure of Describing: Art and Science in August Johann Rösel von Rosenhof's Monthly Insect Entertainment." In *Animals on Display: The Creaturely in Museums, Zoos, and Natural History*, Liv Emma Thorsen, Karen A. Rader, and Adam Dodd (eds.). [Animalibus: Of Animals and Cultures]. University Park, Pennsylvania: The Pennsylvania State University Press, 77–100.

Pächt, Otto. 1981. "Le portrait de Gonella: le problème de son auteur." *Gazette des Beaux-Arts* 123 (VI/97): 1–4.

Panzer, Georg Wolfgang Franz. 1804. *D. Jacobi Christiani Schaefferi iconum insectorum circa Ratisbonam indigenorum enumeratio systematica opera et studio* [= *Systematische Nomenclatur über Weiland Herrn Dr. Jacob Christian Schäffers natürliche ausgemahlte Abbildungen Regensburgischer Insekten*]. Erlangen: Palm.

Petzold, Andreas. 1990. "Colour Notes in English Romanesque Manuscripts." *British Library Journal* 16 (1): 16–25.

Prange, Christian Friedrich. 1782a. *Die Schule der Mahlerey: mit zwey illuminirten Kupfertafeln*. Halle: Hendel.

Prange, Christian Friedrich. 1782b. *Farbenlexicon: worinn die möglichsten Farben der Natur nicht nur nach ihren Eigenschaften, Benennungen, Verhältnissen und Zusammensetzungen sondern auch durch die wirkliche Ausmahlung enthalten sind; zum Gebrauch für Naturforscher, Mahler, Fabrikanten, Künstler und übrigen Handwerker, welche mit Farben umgehen; mit 48 illuminirten Tafeln und einer großen Landschaft*. Halle: Hendel.

Ridl-Dorn, Christa and Mario Dominik Ridl. 2019. "Ferdinand Bauer or Johann and Joseph Knapp? A Rectification." *Gardens' Bulletin Singapore* 71 (suppl. 2): 123–142. https://doi.org/10.26492.gbs71(suppl.2).2019.10.

Roloff, Eckart. 2010. "Jacob Christian Schäffer: Der Regensburger Humboldt wird zum Pionier für Waschmaschinen, Pilze und Papier." In *Göttliche Geistesblitze: Pfarrer und Priester als Erfinder und Entdecker*. Erlebnis Wissenschaft. Weinheim: Wiley-VCH: 159–182.

Romani, Vittoria (ed.). 2003. *Daniele da Volterra: amico di Michelangelo*. Firenze: Mandragora.

Schäffer, Jacob Christian. 1759. *Isagoge in botanicam expeditiorem: iconibvs aeri incisis et pictis illvstrata*. Regensburg: Zunkel.

Schäffer, Jacob Christian. 1760. *Der Gichtschwamm mit grünschleimigem Hute*. Regensburg: Montag.

Schäffer, Jacob Christian. 1763. *Erläuterte Vorschläge zur Ausbesserung und Förderung der Naturwissenschaft*. Regensburg: Zunkel.

Schäffer, Jacob Christian. 1766a. *Icones insectorum circa Ratisbonam indigenorum coloribus naturam referentibus expressae = Natürlich ausgemahlte Abbildungen Regensburgischer Insecten*. Vol. 1 part 1. Regensburg: Zunkel.

Schäffer, Jacob Christian. 1766b. *Icones insectorum circa Ratisbonam indigenorum coloribus naturam referentibus expressae = natürlich ausgemahlte Abbildungen Regensburgischer Insecten*. Vol. 1 part 2. Regensburg: Zunkel.

Schäffer, Jacob Christian. 1766c. *Zweifel und Schwürigkeiten welche in der Insectenlehre annoch vorwalten. Nebst einer ausgemahlten Kupfertafel*. Regensburg: Montag.

Schäffer, Jacob Christian. 1769. *Entwurf einer allgemeinen Farbenverein oder Versuch und Muster einer gemeinnützlichen Bestimmung und Benennung der Farben: nebst zwey ausgemahlten Kupfertafeln*. Regensburg: Weiß.

Schmidt-Loske, Katharina. 2010. "Historical Sketch Maria Sibylla Merian – Metamorphosis of Insects." *Deutsche Entomologische Zeitschrift* 57 (1): 5–10. https://doi.org/10.1002/mmnd.201000001.

Seccaroni, Claudio. 2008. "L'uso di notazioni cromatiche di supporto alla realizzazione di un'opera." In *Dipinti tibetani: dalle spedizioni di Giuseppe Tucci; materiali e tecniche alla luce delle indagini non invasive*, Marisa Laurenzi Tabasso, Massimiliano A. Polichetti, and Claudio Seccaroni (eds.). [ENEA per i beni culturali]. Roma: De Luca, 149–169.

Seelye, Katharine Q. 2019. "Dan Robbins, Who Made Painting as Easy as 1-2-3 (and 4-5-6), Dies at 93 (Published 2019)." *The New York Times*, April 5, 2019, sec. Obituaries. Accessed October 15, 2023. https://www.nytimes.com/2019/04/05/obituaries/dan-robbins-dead.html.

Simonini, Giulia. 2018. "Organising Colours: Patrick Syme's Colour Chart and Nomenclature for Scientific Purposes." *XVII–XVIII. Revue de La Société d'études Anglo-Américaines Des XVIIe et XVIIIe Siècles* 75. https://doi.org/10.4000/1718.1327.

Simonini, Giulia. 2023. "Calau's Punic Wax, Lambert's Farbenpyramide (1772), and Prefabricated Watercolour Cakes." In *Colour Order Systems in 18th and Early 19th Century Europe*, Tanja Kleinwächter, Sarah Lowengard, and Friedrich Steinle (Eds.), 119–149. [International Archives of the History of Ideas / Archives internationales d'histoire des idées 244]. Cham: Springer Nature.

Swammerdam, Jan. 1737. *Bybel der natuure of historie der insecten / Biblia naturae sive, Historia insectorum*. Herman Boerhaave (ed.), translated in Latin by Hieronymus David Gaubius. Leiden: Severinus and Vander Aa.

Swan, Claudia. 2005. *Art, Science, and Witchcraft in Early Modern Holland: Jacques de Gheyn II (1565–1629)*. Cambridge: Cambridge University Press.

Szalay, Gabriella. 2019. "Paper Trials, Multiple Masculinities, and the Oeconomy of Honor." In *Working with Paper: Gendered Practices in the History of Knowledge*, Carla Jean Bittel, Elaine Leong, and Christine von Oertzen (eds.). Pittsburgh: University of Pittsburgh Press.

Thieme, Ulrich, and Felix Becker. 1999 (= 1931/32). "Peetz (Peez) J. G." In *Allgemeines Lexikon der Bildenden Künstler von der Antike bis zur Gegenwart, Bd. 26: Olivier – Pieris*. Leipzig: Seemann.

Tongiorgi Tomasi, Lucia. 2008. "'La femminil pazienza': Women Painters and Natural History in the Seventeenth and Early Eighteenth Centuries." In *The Art of Natural History: Illustrated Treatises and Botanical Paintings, 1400–1850*, Therese O'Malley and Amy W.R. Meyers (eds.). [Studies in the History of Art, Symposium Papers / Center for Advanced Study in the Visual Arts, 69.46]. Washington: National Gallery of Art; distributed by Yale University Press, 158–185.

Vane-Wright, Richard I. 2021. "Introduction. William Jones and the Birth of the Icones." In *Iconotypes. A Compendium of Butterflies & Moths; Jones's Icones Complete*, Oxford University Museum (ed.). Oakland: University of California Press, 10–35.

Winkler, Friedrich. 1939. "Ein spätgotischer Altarentwurf im Kupferstichkabinett." *Jahrbuch der Preußischen Kunstsammlungen* 60: 212–216.

CHAPTER 6

The Colors of Lepidopterans: Ignaz Schiffermüller's Caterpillar Watercolors and Their Iconographic Impact

Stefanie Jovanovic-Kruspel and Harald Bruckner

1 Introduction

This paper focuses on the work of the Austrian entomologists Michael Denis (1729–1800) and Ignaz Schiffermüller (1727–1806). Together, they are known as the authors of the anonymously published *Ankündung eines systematischen Werkes von den Schmetterlingen der Wienergegend* (Announcement of a Systematic Work on the Butterflies of the Viennese region, 1775), commonly referred to as the *Wiener Verzeichnis* (Viennese Catalog). Having been placed on the official list of works approved as available for zoological nomenclature in 1958, the *Wiener Verzeichnis* continues to be relevant for entomological taxonomical nomenclature to this very day.[1]

Besides his interest in lepidopterology, Ignaz Schiffermüller was also a trained painter and teacher of architectural drawing. In his book *Versuch eines Farbensystems* (Attempt at a System of Colors, 1771), he sought to develop a method to standardize the usage of color names in the fields of art and natural history. In what follows, Schiffermüller's expertise on colors will be of special interest. Colors were considered artistic tools, but also important traits in the context of the systematic determination of species, which made it necessary to define them as unambiguously as possible. As we shall see, many cross-references show that the *Wiener Verzeichnis* and *Versuch eines Farbensystems* stand in a close relationship with each other.

The *Wiener Verzeichnis* was based on Denis and Schiffermüller's extensive collection of lepidopterans from the surroundings of Vienna, which was later acquired by the imperial collections in Vienna. In 1848, a fire destroyed the entire collection with its very large number of valuable type specimens, many of whose names are still taxonomically valid today (more on this issue later).[2] The loss of the collection is compounded by the fact that the planned illustrated

1 Hemming 1958, 1–44; Melville and Smith 1987.
2 Kudrna and Belicek 2005, 28.

 | DOI:10.1163/9789004684553_007

edition of the *Wiener Verzeichnis* never materialized. In the absence of both collection and illustrations, the identification of the numerous species newly described by Denis and Schiffermüller rests exclusively on the rather meagre textual information provided by the *Wiener Verzeichnis*. An album of more than 400 watercolors of pupae, larvae, and plants painted by Schiffermüller in preparation for an illustrated reprint of the *Wiener Verzeichnis* is thus of particular interest: These depictions are now the only remaining evidence of the lost collection.

While images are not officially accepted as types in zoological taxonomy, Schiffermüller's illustrations can be interpreted as "iconotypes," as "proxy name bearers" that carry historical significance for the discipline of lepidopterology.[3] Moreover, as we will show, Schiffermüller's watercolors had a significant impact on later entomological illustrations. Made available to other scholars through the mediation of the Augsburg entomologist and illustrator Jacob Hübner (1761–1826), they continued to serve as iconographic templates more than a century later.

The details and colors of Schiffermüller's watercolors are immensely accurate, to the point that they allow modern lepidopterologists to determine the respective species. Given the lifelikeness of these illustrations, we have compiled a list of the current scientific names of the species they depict in order to render them accessible to entomological research (see Appendix).[4]

2 Denis and Schiffermüller: Friendship and Joint Research

Michael Denis and Ignaz Schiffermüller met in 1759, when they took up their posts as teachers at the *Theresianum* academy.[5] Both were members of the *Societas Jesu*,[6] then the responsible body for teaching and

3 Vane-Wright 2021, 11–35, esp. 34.

4 In the appendix and in this essay we use square brackets when the name of the taxon was published anonymously and/or the year was unclear. See recommendations 22A and 51A of International Commission on Zoological Nomenclature 1961, 22 and 51.

5 For Denis's own extensive description of the academy's history, purpose, and curriculum, see Denis 1797b, 220–229.

6 Schiffermüller's life has been the subject of various scholarly articles, but our main source remains an undated autobiography handwritten by an unknown and rather clueless scribe held in the archive of the Oberösterreichisches Landesmuseum in Linz (Anonymous *c.*1806). Written in part after Schiffermüller's death in 1806, the manuscript was partially transcribed and published by Christian Promitzer and Frantz Speta (1990, 62–66). Additional information on Schiffermüller's life can be found in Promitzer and Speta 1990, Speta 2003, and Hoffmann 1952. Retzer 1801 compiled Denis's literary estate, including many autobiographical texts. In

research in the Austrian Empire and a fertile environment for the study of natural history.[7]

Already before his appointment to the *Theresianum*, Schiffermüller had begun to collect and study insects (mainly Lepidoptera).[8] Denis's interest in insects also had roots in his childhood. In 1765, he started to study natural history, and began to breed caterpillars and collect insects in Schiffermüller's company.[9] Between 1770 and 1774, the two young scholars explored the Viennese surroundings together and went on excursions to the Viennese Alps,[10] using their free time to determine Lepidoptera species according to the thirteenth edition of Carl Linnaeus's *Systema naturae* (1767);[11] only if both agreed on a new species description was it included in their own system.[12]

3 The *Wiener Verzeichnis*

What is known today as the *Wiener Verzeichnis* first appeared in 1775 under the title *Ankündung eines systematischen Werkes von den Schmetterlingen der Wienergegend.* The book was republished in the following year as *Systematisches Verzeichniß der Schmetterlinge der Wienergegend* (Systematic Catalog of the Butterflies of the Viennese Region), almost without any alterations. In both cases, no author's name was indicated – we only learn on the respective title pages that the books were "edited by several teachers at the *Theresianum*."[13] In 1958, the International Commission on Zoological Nomenclature (ICZN) placed the 1775 edition on the official list of works approved as available for zoological nomenclature.[14] Established in 1895, it is the task of the ICZN to

---

addition to his contribution to the *Wiener Verzeichnis*, Denis published two books entitled *Lesefrüchte I* and *II* (Denis 1797a and b). Many of the short texts on a list of alphabetically ordered topics contained in the two volumes also provide autobiographical information.

7 Jesuit scholars like the founder of the Museum mathematicum, Joseph Franz (1704–1776), the astronomer Maximilian Hell (1720–1792), the entomologist Nikolaus Poda (1723–1798), and the mineralogist Franz Xaver Freiherr von Wulfen (1728–1805) were among the leading Austrian naturalists of the time and belonged to Schiffermüller's and Denis's intellectual network.

8 Anonymous *c.*1806, 10.

9 Denis 1797a, 201; Retzer 1801, 58.

10 Anonymous *c.*1806, 18.

11 The edition in question is Linnaeus 1767, 744–900. See Anonymous *c.*1806, 14.

12 Anonymous *c.*1806, 14.

13 "herausgegeben von einigen Lehrern am k.k. Theresianum." Denis and Schiffermüller 1775, title page; 1776, title page.

14 Hemming 1958, 1–44. See also Melville and Smith 1987, 314.

provide and regulate a uniform system of zoological nomenclature that ensures that every animal has a unique and universally accepted scientific name.[15] The *International Code of Zoological Nomenclature* (here, too, the initialism ICZN is used) takes Linnaeus's tenth edition of the *Systema Naturae* (1758) as the official starting point of zoological nomenclature.[16] The book commonly referred to as *Wiener Verzeichnis* and attributed to Michael Denis and Ignaz Schiffermüller, is one of only a handful of lepidopterological works from the eighteenth century that have been accepted by the ICZN.[17] While the publication history (1775 vs. 1776) as well as the authorship (Denis and/or Schiffermüller and/or others) of the *Wiener Verzeichnis* are still debated,[18] we agree with the ICZN in recognizing Denis and Schiffermüller as the authors of the 1775 edition.

The *Wiener Verzeichnis* is not merely a natural history treatise, but an intriguing combination of science, poetry, and graphic art, a *Gesamtkunstwerk* in the truest sense of the term – arguably more so than any other systematic entomological work of the time.[19] With regard to the illustrations, we know from a handwritten biography of Schiffermüller that there had been a plan to have the work extensively illustrated with images of all the life stages of lepidopterans.[20] The biography discusses this plan in three separate passages, with the first stating that Schiffermüller had already executed roughly 400 images of caterpillars, which were intended for a systematic juxtaposition of adult lepidopterans and their larval stages.[21] The second reads as follows:

> One had, as a matter of fact, ... given hope or even promised to exactly describe all species ... that had not been depicted in books so far, even to provide, as it were, for each of them a life history and illustration of the caterpillars and butterflies.[22]

---

15 For detailed information on the ICZN, see their website: https://www.iczn.org.

16 International Commission on Zoological Nomenclature 1961, 5.

17 Melville and Smith 1987, 313–6. The title *Wiener Verzeichnis* was first used in Kirby 1781, v. Other accepted eighteenth-century works were authored by Johan Christian Fabricius (1743–1808), Giovanni Antonio Scopoli (1723–1788), and Johann Caspar Füssli (1706–1782), among others.

18 In 2005, Otakar Kudrna and Joseph Belicek argued that Schiffermüller had been the sole author, while Klaus Sattler and Walter Gerard Tremewan (2009) maintained that there had been others. A similar position was taken by Balázs Tóth, Gergely Katona, and Zsolt Bálint (2020), who thought it highly probable that several Jesuits from the *Theresianum* had participated in the preparation of the book.

19 Cf. Fabricius 1775, Scopoli 1763, Füssli 1781–1786.

20 Anonymous *c.*1806, 19 f.

21 Anonymous *c.*1806, 13 f.

22 "Man hatte wirklich ... Hoffnung gegeben oder auch versprochen, wenigstens alle bis dahin in Büchern Abbildungen nicht bekannter Arten in nachfolgenden Ländern genau

From the biography's third reference to the project, an approximate timeframe for the preparations can be deduced: "Regarding copperplate engravings and illumination, all arrangements had been made when a harsh, all-destroying fate struck."[23] It can be assumed that this "fate" was the abolishment of the *Societas Jesu* in 1773, which means that most of the preparations must have been completed before this *terminus ante quem.*

In the end, only four images were incorporated into the *Wiener Verzeichnis*: A frontispiece, a vignette illustration before the appendix, and two tables (Tab. I. a & b).[24] The frontispiece and the two tables were executed by the copper engraver Ferdinand Landerer (1730 or 1746–1795).[25] All four images had both a decorative and an informative function. The vignette was meant to serve as a visual instruction indicating the terminology of the body parts of caterpillar, pupa, and lepidopteran.[26] The two tables – in some prints they are colored – as well as the frontispiece feature several different species and are each described in a dedicated chapter.[27]

A special feature of the *Wiener Verzeichnis* is the extensive integration of poetic texts. In addition to his taxonomical work, Michael Denis, who also published poetry under the anagrammatic pseudonym *Sined der Barde* ('Sined the Bard'), contributed eight poems and many literary references.[28] The sources of these literary adornments are highly diverse and range from authors of all ages to Denis's own lyrical oeuvre. One of the things the book conveys through this added poetic layer is the idea of physico-theology. Denis and Schiffermüller were not the only scholars of their time who approached their endeavors in the field of natural history from the perspective of physico-theological thinking.[29] Brian W. Ogilvie has plausibly argued that physico-theology contributed

zu beschreiben, ja von einer jeden derselben gewissermassen eine Lebensgeschicht und Abbildung der Raupen und der Schmetterlinge zu liefern," Anonymous *c.*1806, 19–20.

23 "in Betreff auf Kupferstiche und auf Illuminierung [hatte man] alle Anstalt getroffen-allein war, kam nun wieder ein wiedriges, alles zerstörendes Schicksall." Anonymous *c.*1806, 21.

24 Appendix (= Nachtrag) Denis and Schiffermüller 1776, 305. It is signed with "I. Schiffermüller sculp." Denis and Schiffermüller 1776, 8; the signature on Tab. I a reads "Land. f." (= Landerer fecit).

25 Landerer was also employed as art teacher at the *Savoyische Adelige Akademie*, or, as Denis and Schiffermüller referred to it, the *k. k. Cadettenschule*.

26 It seems likely that Denis was responsible for compiling the terminology; in *Lesefrüchte* 2, he translated a list of Linnaean terms from Latin into German under the title "Schmetterlinge" (Denis 1797b, 180–4).

27 Denis and Schiffermüller 1776, 236–244 and 244–303.

28 The poems can be found in Denis and Schiffermüller 1776, 7, 10, 16, 17, 18, 19, 26, 36. For a full list of Denis's poetic works, see Retzer 1801.

29 On physico-theology, see Blair and Von Greyerz 2020, 1–20.

significantly to the formation of entomology as a discipline towards the end of the eighteenth century.[30] The notion that the Creator's wisdom and omnipotence could be revealed through the observation of His minor works (such as insects) chimed with Denis and Schiffermüller's Jesuit mindset.[31] One of Denis's poems in the *Wiener Verzeichnis* is especially emblematic of this line of thinking:

> Before me opens up the book
> Of instructive nature.
> I, the pupil, lay eyes on it and already
> Already I read, stammeringly, God
> …
> God in the splendor of the butterfly,
> God in the bee's diligence,
> In the chirping of the cicada, God
> In the cricket's song, God.[32]

Here, Denis takes up the ancient metaphor of the 'book of nature,' which to him finds its parallel in the Bible: One has but to 'read' in it for God's plan to be unveiled.[33] To Denis and Schiffermüller, the study of the 'book of nature' and especially of entomology thus had a deep moral purpose,[34] an idea that goes back to the German theologian and historian Friedrich Christian Lesser (1692–1754), whose *Insecto-Theologia* (1738) was extremely popular.[35] The work's 1742 French edition, which contained additional remarks by the Dutch naturalist and engraver Pierre Lyonnet (1706–1789), is referred to several times in the *Wiener Verzeichnis*.[36] It was against the backdrop of this moral-theological purpose that Denis and Schiffermüller advised their readers to pay special attention to the appearance and behavior of caterpillars.[37]

---

30 Ogilvie 2020, 171–182.

31 See, for example, Denis and Schiffermüller 1776, 1, 16, 20, 23, 32.

32 "Vor mir eröffnet sich das Buch / Der lehrenden Natur. / Ich Schüler blick' es an und schon / Schon les' ich stammelnd Gott / … / Gott in der Pracht des Schmetterlings, / Gott in der Biene Fleiß, / Im Zwitschern der Cicade Gott, / Im Grillenliede Gott." Denis and Schiffermüller 1776, 36 in the footnotes.

33 On the metaphor of the 'book of nature,' see Harrison 2006, 1–26, esp. 5–6. The importance of this metaphor to Denis's thinking is also evident in Denis 1797a, 206–207.

34 Denis and Schiffermüller 1776, 32.

35 Oligivie 2020, 176.

36 Denis and Schiffermüller 1776, 7, 32, 35, 224. See also the references to Lyonnet in Lesser 1742 and a direct reference to Lesser 1742 in: Denis and Schiffermüller, 1776, 175.

37 Denis and Schiffermüller 1776, 32.

Denis and Schiffermüller's focus on caterpillars is evident from their approach to systematic scientific nomenclature, which is a key concern of the *Wiener Verzeichnis*: More than 164 out of a total of 322 pages are dedicated to naming the lepidopteran species described in the book.[38] While Denis and Schiffermüller generally followed Linnaeus's system, they sometimes perceived a need to refine it further.[39]

Before Denis and Schiffermüller, the primary way of determining lepidopteran species, as introduced by Linnaeus, was to describe traits of the imago form, focusing on details like the wings and antennae.[40] In the course of their endeavor to classify the diverse species of moths and butterflies around Vienna, the two Jesuit scholars realized that this approach required improvement.[41] To that end, they began to study the entire life cycle of lepidopterans including the larval stages,[42] and incorporated their findings into Linnaeus's system, which constituted an innovative step in zoological taxonomy.[43] Building on the work of the entomologists Etienne Louis Geoffroy (1725–1810) and Giovanni Antonio Scopoli (1723–1788), they studied lepidopterans from a holistic point of view, describing comprehensively the many traits of caterpillars that could be used to draw clearer distinctions between Lepidoptera species, such as the number of legs (which ranges between 10 and 16), the form of the body (deviations from a regular cylindrical shape in the form of humps, points, knobs, etc.), and whether the specimen was naked or hairy.

Denis and Schiffermüller's research left them convinced that "More visible and clearer characteristics will probably never be discovered in lepidopterans. Thus, a system of classification in which the caterpillars are included can certainly be a useful resource."[44] In order to resolve difficult taxonomic challenges, they succinctly recommended to keep "one eye on the butterfly, the other on the caterpillar"[45] – in fact, a whole chapter of the *Wiener Verzeichnis* was dedicated to this new approach.[46] Denis and Schiffermüller's integration of the

38 Denis and Schiffermüller 1776, 40–187, in the supplement: 305–322.

39 Denis and Schiffermüller 1776, footnote on page 194.

40 Linnaeus 1758, 458 ff. See also Beth Tobin's chapter in this volume.

41 Zilli 2021a, 265–274, esp. 273.

42 Prior to Linnaeus, Johannes Goedaert (1617–1668), Jan Swammerdam (1637–1680), and Maria Sibylla Merian (1647–1717) had also examined the larval stages.

43 Kudrna 2015, 8; Hemming 1937, 6; Higgins 1959, 49.

44 "Sichtbarere und deutlichere Kennzeichen wird man an den Schmetterlingen wohl niemal entdecken. So kann also eine Eintheilung, zu welcher auch die Raupen gezogen werden, noch immer ein dienliches Hilfsmittel seyn." Denis and Schiffermüller 1776, 32.

45 "Ein Aug auf den Schmetterling, das andere auf die Raupe." Denis and Schiffermüller 1776, 31 f.

46 Denis and Schiffermüller 1776, 27–35.

larval stage into taxonomic thinking contributed to the incipient development of an ecological view of nature as pioneered by Merian.[47] The division of caterpillars into families was intended to enable collectors to determine what kind of lepidopteran they would have metamorphosed into, information that could also contribute to a better outcome when breeding caterpillars for the purpose of setting up a collection.[48] For the latter reason, Denis and Schiffermüller also supported the idea of a German nomenclature of lepidopterans based on the food plants of their caterpillars, e.g. "Ligusterschwärmer, Weidenspinner, Rittersporneule, Birkenspanner, Eichenwickler" ("privet moth, willow moth, larkspur moth, birch moth, oak moth").[49] Interestingly, the inclusion of the food plant in the name is a phenomenon that is almost exclusively limited to Denis and Schiffermüller's German nomenclature, the development of which conformed to the general trend at the time to promote the use of vernacular languages in the field of natural history.

The importance Denis and Schiffermüller attached to the caterpillar stage is also evident from the fact that the descriptions of the caterpillars generally precede those of the adult stage in their systematic nomenclature.[50] According to Schiffermüller's biography, he and his collaborator described 1150 species of Lepidopterans – 700 more than Linnaeus.[51]

4 Denis and Schiffermüller's Collection of Lepidopterans

Denis and Schiffermüller's focus on the caterpillar stage had a major impact on the way they collected. Those of their readers who wished to assemble a collection of nocturnal and crepuscular moths were advised to breed them from caterpillars.[52] This was due to three reasons: First, unlike the mostly well-camouflaged adult moths, their caterpillars often display very lively colors,

47 Denis and Schiffermüller 1776, 28. See Etheridge 2021.

48 Denis and Schiffermüller 1776, 30.

49 Denis and Schiffermüller 1776, 24 and 26. Compared to Linnaeus 1758, where about one fifth of the Latin species names refer to food plants, this proportion is considerably lower in the *Wiener Verzeichnis*. The vast majority of Denis and Schiffermüller's Latin names refer to the color of the adult animals.

50 Denis and Schiffermüller 1776, 40–187. In Linnaeus 1758, the description of the adult stage is given first.

51 Anonymous *c.*1806, 16 f. This is confirmed by Hoffmann (1952, 59) and Sattler and Tremewan (1984, 7:3): 282. Today, genital preparation of adult specimens has largely replaced the examination of caterpillars in the identification of lepidoptera species.

52 Denis and Schiffermüller 1776, 29.

which makes them much easier to find;[53] second, breeding caterpillars supplies the collector with specimens "in their original beauty, with intact ornamentation of their pattern and colors";[54] and third, a collection of nocturnal moths was far more interesting from a taxonomic perspective than one featuring the already well-studied diurnal butterflies – in Denis and Schiffermüller's words, the number of moths "seems inexhaustible, with new species appearing almost daily. Since this genus predominates, we should pay more attention to it than to the others."[55]

With regard to how their collection was displayed, we can assume that Denis and Schiffermüller followed their own advice, which in turn mirrored the general collection practices of the time:[56] The specimens were needled on a board, with diurnal butterflies being mounted in such a way that both sides were visible, whereas nocturnal and crepuscular moths displayed the dorsal side of both sexes.[57] There is a plausible reason for this divergent practice: While the ventral side of diurnal butterflies often shows important determination characteristics, the display of both sexes of nocturnal and crepuscular moths showcases the species' sexual dimorphism.[58] All specimens were ordered systematically, and protected against dust and pests under glass lids.[59] The collection's layout was described by the Austrian zoologist Leopold Joseph Fitzinger (1802–1884) as follows: "The same was arranged in a strictly systematic manner, furnished with labels by Schiffermüller's own hand and separated according to natural groups in small blue-gray boxes, which were closed with a glass lid framed by gilded wooden bars."[60] Interestingly, there is no mention of dried caterpillars preserved in the collection.

53 Denis and Schiffermüller 1776, 29.

54 "In ihrer ursprünglichen Schönheit, mit unversehrtem Zierrathe ihrer Zeichnung und Farben …" Denis and Schiffermüller 1776, 29.

55 "Sie scheint unerschöpflich, indem man fast täglich neuer Arten gewahr wird. Diese Gattung überwiegt also, folglich ist auf sie mehr, als auf die anderen zu sehen." Denis and Schiffermüller 1776, 29. Denis's and Schiffermüller's understanding of the term *Gattung* (genus) differs from modern usage – more on this issue below.

56 Denis and Schiffermüller 1776, 19. See Hünniger 2018, 696–702 and Zilli 2021b, 138.

57 Denis and Schiffermüller 1776, 19.

58 Sexual dimorphism is also present in diurnal butterflies, but Denis and Schiffermüller apparently saw no need to display it.

59 Denis and Schiffermüller 1776, 19.

60 "Dieselbe war streng systematisch geordnet, mit Etiketten, von Schiffermüller's eigener Hand geschrieben, versehen und nach natürlichen Gruppen abgesondert, in kleinen blaugrau grundirten Kästchen aufgestellt, die mit einer von vergoldeten Holzstäben umrahmten Glasdecke geschlossen waren." Fitzinger 1868, 28.

After the dissolution of the *Societas Jesu* in 1773, Denis and Schiffermüller's entomological cooperation came to an end. In the same year, Denis was appointed head of the *Garellische Bibliothek* at the *Theresianum*, a post which required him to be physically present in the library and thus prevented him from going on excursions. In 1777, Schiffermüller was appointed director of the *Nordisches Collegium*, which entailed a move to Linz. Due to Schiffermüller's relocation, Denis decided to leave the entire collection to his friend and collaborator. Looking back at this decision, he later wrote: "Hence a splendid collection was formed, my share of which I relinquished to my friend when I took on the sedentary duty of a librarian and we were separated by his promotion."[61]

Schiffermüller took the collection to Linz, where he continued to make it accessible to interested naturalists: We know that the collection was studied by Franz de Paula von Schrank (1747–1835) in 1778, and by Johan Christian Fabricius (1743–1808) in 1784.[62] In particular, the above-mentioned Jacob Hübner was allowed to work intensely with Denis and Schiffermüller's collection of specimens: Schiffermüller hoped that Hübner's illustrations would render the *Wiener Verzeichnis* "useful as a system."[63] Conversely, the many references in volumes one and eight of Hübner's *Sammlung europäischer Schmetterlinge* (1796 & 1836) reflect the importance of the *Wiener Verzeichnis* to the Augsburg painter and entomologist.[64]

When the Jesuit monastery in Linz was suppressed in 1788, Schiffermüller obtained the deanship in a village named Waizenkirchen, but for unknown reasons his lepidopterans were left behind in Linz.[65] Upon Schiffermüller's death in 1806, Carl Schreibers (1775–1852), head of the Imperial Natural History Cabinet in Vienna, managed to convince Emperor Franz I (1768–1838) to purchase the collection. Never incorporated into the main collection, it was maintained in separate cabinets, which can be interpreted as a reflection of its special importance.[66] During the bombardment of the Hofburg Palace on October 31, 1848 the entire collection was destroyed, leaving behind a historical

61 "Daher entstand die herrliche Sammlung, deren Antheil ich meinem Freunde abtrat, als ich zur sitzenden Bibliothekarspflicht kam, und wir durch seine Beförderung getrennet wurden." Retzer 1801, 58.

62 Kudrna 2015, 4; Hemming 1937, 6 and 67 f.; Hope 1845, xi.

63 "… als System brauchbar …" Anonymous *c.*1806, 61.

64 Hübner 1796; 1836. For a list of Hübner's references to the *Wiener Verzeichnis*, see Sattler and Tremewan 2009, 7 f.

65 Fitzinger 1868, 27.

66 Fitzinger 1856, 46. The usual practice, which is followed to this day, was to incorporate acquired collections into the general collection, making it difficult for historians to trace the work of a specific collector.

gap which continues to be keenly felt by entomologists whose research is focused on taxonomy.[67]

5 The "Somber Colors" of Lepidopterans: Order and Names

The second field in which Schiffermüller developed extensive expertise was color. One of his duties as a teacher of architectural drawing at the *Theresianum* was to supply students with painting materials, which led him to a deeper understanding of the properties of colors and "gave him the courage" to create *Versuch eines Farbensystems*.[68] The work appeared in 1771 – four years before the *Wiener Verzeichnis* – with the same publisher, Augustin Bernardi (1753–1798); Schiffermüller was credited as its sole author. A second edition from 1772 shows only minor changes in its dedication.

In simplified terms, *Versuch eines Farbensystems* was aimed at two goals: First, the creation of a color system, and second, the standardization of color names. Here, too, Schiffermüller expressed his belief that behind the rich variety of colors lay "a completely divine harmony and therein undeniably the clearest traces of a most perfect Master."[69] He was convinced that close observation and diligent empirical research would reveal a classificatory system for colors like the one Linnaeus had created for plants, animals, and minerals, a system whose order was "not fabricated, not imposed by force, but founded entirely in nature."[70] The system proposed by Schiffermüller consisted of twelve colors, which he called "pure, bright, and delightful."[71] He depicted them in a color circle based on the system described by the French Jesuit priest and mathematician Louis-Bertrand Castel (1688–1775) in 1740.[72] Based on the position of each color in the circle, Schiffermüller formulated rules for their harmonious relationship. According to Schiffermüller, colors with only one hue between

67 See Russell et al. 2020, 449–472; Russell and Vane-Wright 2022, 279–294; Wikström et al. 2020, 337–364.

68 "den Muth eingeflösset haben mag", Anonymous *c.*1806, 6. See also Anonymous *c.*1806, 8.

69 "eine ganz göttliche Harmonie und hierinn unlaugbar [sic!] deutlichste Spuren eines vollkommensten Meisters," Schiffermüller 1772, 7.

70 "die nicht erdichtet, nicht gewaltig hergezogen, sondern in der Natur ganz gegründet ist," Schiffermüller 1772, 6.

71 "reinen, hohen und lieblichen Farben," Schiffermüller 1772, 10 (definition) and 19. The colors so classified were either pure or mixed from only two colors.

72 See Kemp 1990, 290. For more information on the topic of color systems, see Karliczek 2013 and 2016. According to Thomas Lersch (1984, 6), Schiffermüller thus created the earliest representation of a color circle in a German publication on color theory.

them were not a good match, neither in a dress nor in a painting.[73] When two colors had several others between them, their contrast ranged from "very delicate and dainty" (two colors) to "more vivid" (three colors) to "strong and powerful" (four colors); a difference of five colors "would be perceived as pleasant only by senses inured to being moved by none other than fairly powerful stimuli."[74] Schiffermüller's "system or doctrinal edifice" of colors was intended to serve painters and other practitioners of art, as well as natural historians.[75]

However, it had not gone unnoticed by Schiffermüller that the "pure, bright, and delightful" hues depicted in his circle were actually much rarer in nature than what he referred to as "somber" (*trübe*) colors, an observation that in turn caused him to ponder the counsel of the "wisest Creator."[76] Schiffermüller's conclusion was clear: God's purpose had been the creation of a harmonious whole.[77] The same thought is also found in the *Wiener Verzeichnis* in reference to the colors of lepidopterans:

> Admittedly, their colors are not always bright and shiny, like those of the rainbow, but deep, somber, dull; yet how grateful must the former be to the latter, how necessary are they to prevent our eye from being dulled, with what variety has nature placed them next to each other to prevent any violent transition, so that, in the painter's parlance, none of them infringes upon the others, all of them are in sweet harmony, and bring out the best in each other.[78]

Denis and Schiffermüller also addressed the fact that the subtle shadings of naturally occurring colors often do not obey the received laws of color harmony:

---

73 Schiffermüller 1772, 15.

74 "sehr zart und niedlich" (2), "lebhafter" (3), "stark und gewaltig" (4), "dann kann ihre Zusammensetzung insgemein nur jenen Sinnen gefallen, die allein von recht starken Gegenständen gerühret zu werden, gewöhnt sind," Schiffermüller 1772, 16–17.

75 "ein System oder Lehrgebäude," Anonymous *c.*1806, 6, 7ff.

76 "weiseste Schöpfer" Schiffermüller 1771, 19. There is an interesting analogy between this finding and the world of lepidopterans: nocturnal butterflies also outnumber their diurnal and crepuscular counterparts; see Denis and Schiffermüller, 1776, 29.

77 Schiffermüller 1771, 19.

78 "Freylich sind es nicht immer hohe und glänzende, wie jene des Regenbogens, sondern tiefe, düstere, matte Farben; allein wie viel haben diesen die ersteren zu danken, wie nötig sind sie ihnen unser Aug nicht zu stumpfen, mit welcher Auswahl hat die Natur sie nebeneinander gestellet, jeden gewaltsamen Abfall zu verhindern, so daß in der Malersprache keine der anderen wehe tut, alle lieblich aufeinander lassen, und sich wechselseitig empfehlen." Denis and Schiffermüller 1776, 37.

> According to accepted opinion, green and blue are almost the least compatible. That may be, if we are talking about the common green and blue; but one should consider the far from rare caterpillar of the moth that goes by the name of the great night peacock, and is probably the largest in Europe. One will behold with joy how pleasantly a bright sky blue with its perfect shading appears above a delicate parakeet, or rather olive, green.[79]

The caterpillar in question belongs to the most iconic species described by Denis and Schiffermüller: the Viennese emperor, *Saturnia pyri* ([Denis & Schiffermüller], 1775).[80] The *Wiener Verzeichnis* calls this species "*Bombyx Pyri.*" While the genus name was later changed to *Saturnia* (Schrank 1802), the specific epithet '*pyri*' is taxonomically still valid today. Vienna and its surroundings are thus considered the type locality of this species, the place where the first described specimen – the so-called type specimen – originated. In the *Wiener Verzeichnis*, the Viennese emperor is also referred to with the German common name *Birnspinner*, which is derived from one of its caterpillar's main food plants: the pear tree.[81] The distribution of this species is widespread, extending from North Africa through all of southern Europe and eastwards to Iran; in Central Europe, it is limited to the warm areas. In Austria, the Viennese emperor also occurs mainly in the southeastern parts of the country. Its wingspan reaches up to 16 cm, which makes it the largest moth in this geographical range. The hibernation stage is the pupa. In Austria, its Red List status is 'vulnerable' (VU).[82]

The sheer abundance of lepidopteran colors, combined with their apparent disregard for the rules of harmony that lay at the heart of Schiffermüller's color circle, created a serious problem for natural historians:

> Describing these colors presents no small difficulty to entomologists. Many people have no knowledge at all of some unusual colors. Others know

---

79 "Grün und Blau verträgt sich nach dem angenommenen Urtheile fast am wenigsten. Sey es, wenn vom gemeinen Grün – und Blauen die Rede ist; aber man betrachte die bey uns eben nicht seltene Raupe des Nachtschmetterlinges, der unter dem Namen des grossen Nachtpfauen geht, und in Europa wohl der größte ist. Man wird mit Vergnügen sehen, wie angenehm ein hohes Himmelblau mit seiner vollkommenen Schattirung über einem zarten Sittich- oder vielmehr Olivengrün stehe." Denis and Schiffermüller 1776, 37.

80 Other common names are large emperor or great peacock moth; the modern German common name is *Wiener Nachtpfauenauge or Großes Nachtpfauenauge.*

81 Denis and Schiffermüller 1776, 204. On the animal's diet, see Steiner et al. 2014, 69.

82 Huemer 2007, 2.

> the names, but they associate different concepts with them. Descriptions therefore generally remain obscure, leaving us in uncertainty.[83]

Denis and Schiffermüller were not alone in realizing that the incredible variety of colors displayed in all the life stages of lepidopterans necessitated a more specific terminology.[84] In art, too, the problem of naming was unresolved, and many color names were "undefined."[85] All this led to Schiffermüller's attempt to standardize color names in his *Versuch eines Farbensystems*. The terminology he employed reflects his wish to adopt Linnaeus's systematic approach for this endeavor, and it is certainly no coincidence that he designated colors as *Gattungen* (genera). Schiffermüller maintained that the naming of things should be as unambiguous as possible:

> The first degree of wisdom, says the knighted Linnaeus (*Systema Naturae* page 13, Viennese edition), is the knowledge of things. This knowledge consists in a true idea of objects, by means of which the similar are distinguished from the dissimilar by their features imposed on them by the Creator. Whoever wishes to communicate this knowledge to others must give all different things their proper names, which must never be confused. For with the name comes the knowledge of things.[86]

The many thematic similarities and direct cross-references between the *Wiener Verzeichnis* and *Versuch eines Farbensystems* show that both works were not only inspired by the same Linnean idea of standardizing and describing

---

83 "Diese Farben nun zu beschreiben fühlen alle Entomologen eine nicht geringe Beschwerniß. Viele Menschen haben von manchen sonderbaren Farben gar keine Kenntniß. Anderen sind zwar die Namen bekannt; allein sie verbinden verschiedene Begriffe damit. Beschreibungen bleiben also insgemein dunkel, lassen in der Ungewißheit." Denis and Schiffermüller 1776, 37–38.

84 Denis and Schiffermüller 1776, 39. A footnote refers to the terminological efforts of the entomologist Jacob Christian Schäffer and the botanist Georg Christian Oeder. Denis and Schiffermüller 1776, 37–38.

85 "unbestimmt," Schiffermüller 1771, 22.

86 "Der erste Grad der Weisheit, sagt der Ritter v. Linne (** Syst. Nat. pag. 13, Edit. Vien.), ist das Erkenntniß der Dinge. Dieses Erkenntniß besteht in einem wahren Begriff der Gegenstände, vermöge dessen Aehnliche von Unaehnlichen durch eigene vom Schöpfer ihnen aufgedrückte Kennzeichen unterschieden werden. Wer dieses Kenntniß anderen mittheilen will, muss allen verschiedenen Dingen eigene Namen geben, die niemals vermenget werden müßen. Denn mit dem Namen fällt auch das Erkenntniß der Dinge." Denis and Schiffermüller 1776, 23. The edition of *Systema Naturae* being referred to here is Linnaeus 1767.

nature, but were developed simultaneously.[87] Denis and Schiffermüller dedicated an entire chapter of the *Wiener Verzeichnis* to the topic of colors,[88] where they claimed that flowers only exhibited two or three colors at the same time, whereas lepidopterans displayed ten or more, most of them somber and muted.[89] Employing common color names like yellow, red, green, or brown was insufficient, because these terms did not cover the existing variety, and could be used in reference to very different tints and shades.

There are many passages in the *Wiener Verzeichnis* where its authors announce their intention to tackle the problem of color naming. For instance, the preface – dated March 16, 1771 – states that their goal was "to remedy any ambiguity" in the description of colors.[90] Elsewhere, Denis and Schiffermüller profess their desire to "present, as much as possible, the colors to the eye in a natural sequence."[91] This sequence was to be easy to remember, "requiring no other mechanism than that of the eye" to clarify the color terms.[92] Admitting that this undertaking was beyond the scope of the *Wiener Verzeichnis*, they signaled their willingness "to arrange for a special edition."[93] It is our contention that this "special edition" was Schiffermüller's *Versuch eines Farbensystems*.[94]

The above-mentioned "natural sequence" was converted into a tabular overview of blue color terms and a corresponding color chart.[95] In this overview and chart, Schiffermüller differentiated between twelve types of blue subdivided into three nuances each, resulting in a total of thirty-six shades. To all of them, he assigned German and, in many cases, also Latin and French names. Instead of inventing an entirely new nomenclature, Schiffermüller tried to collect names that were already common or had been employed in

87 See the cross-references in Denis and Schiffermüller 1776, 9 and 38–39. Together with the date of the preface of the *Wiener Verzeichnis* (1771), this allows us to speculate that the two works might have been intended for publication in the same year.

88 "Von den Farben der Schmetterlinge." Denis and Schiffermüller 1776, 35.

89 Denis and Schiffermüller 1776, 37.

90 "Ueber dieß haben wir noch ein Mittel auszufinden gesucht ... jeder Undeutlichkeit in den Beschreibungen abhelfen, und überhaupts den Begriff von den Farben festsetzen soll." Denis and Schiffermüller 1776, 9.

91 "Wir werden die Farben selbst, so viel möglich ist, in einer natürlichen Folge dem Auge vorlegen." Denis and Schiffermüller 1776, 38–39.

92 "kein anderer Mechanismus erfordert, als jener des Auges." Denis and Schiffermüller 1776, 38–39.

93 "eine besondere Ausgabe davon zu veranlassen." Denis and Schiffermüller 1776, 39.

94 As announced in the *Wiener Verzeichnis* (1776, 39), its layout is indeed very similar to that of *Versuch eines Farbensystems*.

95 The position of color circle, tabular overview, and color chart differs from edition to edition. In the 1772 edition held by the Naturhistorisches Museum, the tabular overview and chart can be found at the end of the book in the form of an annex.

the publications of his (mainly entomological) predecessors, such as Linnaeus, René-Antoine Ferchault de Réaumur (1683–1757), Pieter Cramer (1721–1776), and Johann Leonhard Frisch (1666–1743).[96] His goal was to identify each shade with a unique and standardized name for future use.[97] When describing the blue colors of lepidopterans in the *Wiener Verzeichnis*, Denis and Schiffermüller obviously drew on the names defined in *Versuch eines Farbensystems*: For instance, the terms 'full blue' (*Vollblau*), 'common blue' (*Gemeinblau*), 'high blue' (*Hochblau*), 'fine blue' (*Feinblau*), 'shiny blue' (*Blankblau*), and 'lazure blue' (*Lazurblau*) can be found in the descriptions of several species.[98] However, as a look at the many other descriptive texts in the book reveals, the standardized names were not even sufficient for the color blue. Further nuances like 'steel blue' (*Stahlblau*), 'blueish light grey' (*Blaulichtgrau*), and 'blue grey' (*Blaugrau*) had to be introduced, not to mention all the other colors that the authors needed to describe.

One way of dealing with this problem was to resort to comparisons. In many cases, Denis and Schiffermüller made recourse to the names of plants to define the nuance of a specific color. This practice not only served the purpose of defining the hue in question, but it also reminded their readers that plants and lepidopterans are closely linked through the phenomenon of mimicry:[99]

> Meanwhile, it is obvious enough from our *Verzeichnis* that the caterpillars, and not infrequently also the butterflies, resemble through their colors, and to some extent also through their shape, the blossoms of the plant or the leaves of the same, sometimes the tree bark, more often the twigs, and occasionally even the entire little plant.[100]

96 Denis and Schiffermüller 1776, 25.

97 Although Schiffermüller's belief in the intuitive accessibility of his color sequence proved overly ambitious, color charts did exhibit a certain potential for standardization. See Prange 1782, xvi; for details regarding the reception of Schiffermüller's color circle and color chart, see Simonini 2024.

98 Schiffermüller 1772, tabular overview on pp. 25–26. For example, these color names were used in the species descriptions in Denis and Schiffermüller 1776, 181 and 183.

99 On the impact of the surroundings on the coloration of caterpillars, see Kay Etheridge's essay in this volume.

100 "Unterdessen kann man aus unserem Verzeichnisse genugsam bemerken, daß die Raupen und nicht selten zugleich die Schmetterlinge durch ihre Farben oder einigermaßen auch durch ihre Gestalt, bald den Blüthen der Pflanze, bald den Blättern derselben, ein andersmal der Baumrinde, noch öfter den Aestchen, zuweilen auch den ganzen Pflänzchen ähnlich sehen" Denis and Schiffermüller 1776, 189.

As already mentioned, Denis and Schiffermüller often included the names of food plants in their German names for lepidopterans. However, it was the observation of mimicry that prompted them to describe the color shades of the caterpillar and sometimes also of the imago by referring to the respective plant. The species with the common name green-underside blue, *Glaucopsyche alexis* (Poda 1761), which was described in the *Wiener Verzeichnis* under the name "Wirbelkrautfalter *Papilio Damaetas*,"[101] is a striking example of this practice. Denis and Schiffermüller referred to its caterpillar with the German common name "wirbelkrautfarbene Raupe," which translates as "caterpillar colored like the plant *Wirbelkraut*."[102] The old German common name *Wirbelkraut* can be identified with the plant *Astragalus* spec., to whose blossoms the color of the caterpillar of *Glaucopsyche alexis* indeed bears a striking resemblance. Although the adult stage does not share this resemblance, its German name, too, was derived from the food plant of its caterpillar: *Wirbelkrautfalter*. This approach to nomenclature can be read as indicative of the importance Denis and Schiffermüller attached to a holistic perspective on lepidopterans that paid special attention to their caterpillars and the respective food plants.

6 Ignaz Schiffermüller's Watercolors

The relationship between caterpillars and food plants is also the subject of an album of watercolors painted by Schiffermüller, which is held by the Natural History Museum (NHM) in London.[103] Little is known about its provenance except that it was at one time in the possession of Baron Dr. Kurt von Rosen (1883–1946), a zoologist from Munich. Von Rosen's collection of manuscripts and zoological illustrations was acquired by the NHM in 1949.[104]

The album consists of 59 small quarto sheets, which all carry the caption 'Tab.' Five of them were left blank (Plates 2, 10, 20, 34, and 56). Each sheet features between one and up to twenty insects, which sometimes depict different larval stages of the same species. In total, the album comprises 454 watercolor drawings: 416 show larvae, many of them on their food plant, and 38 depict pupae. There are no images of eggs. Pencil notes on the sheets, which Higgins assumes to be by Hübner, give a specific scientific name and/or refer

101 The name *Papilio damaetas* ([Denis & Schiffermüller], 1775) is a synonym today.
102 Denis and Schiffermüller 1776, 183.
103 Schiffermüller before 1773.
104 Harvey, Gilbert, and Martin 1996, 174.

to an illustration number, presumably from Hübner's *Geschichte europäischer Schmetterlinge* (1793–1842).[105]

Plate 28 is an example of what a typical sheet from Schiffermüller's album looks like (Fig. 6.1). In this particular instance, Schiffermüller depicted seven caterpillars ("A," "B," "F," "G," "S," and "T") and one pupa ("a") of the family Lasiocampidae, while two caterpillars ("K" and "Z") and two pupae ("k" and "z") remain unidentified.[106] Six of the caterpillars are shown sitting on their respective food plant and feeding on it. Some bear handwritten notes stating their names, and, in one case, what appears to be an illustration number referring to an edition of the *Geschichte europäischer Schmetterlinge* that is unknown to us.[107]

Schiffermüller captured the movements and details of the animals in astonishing accuracy. Only a few features like the ornamental shape of the grass blades pay tribute to Schiffermüller's artistic background. The fact that the letters were assigned to the images in random order suggests that these sheets were intended to serve as sketch material for a new layout of the *Wiener Verzeichnis* in which the larval stages would be paired with the adult lepidopterans.

In all likelihood, the album of watercolor images formed part of the preparation for the extended illustrated edition of the *Wiener Verzeichnis*, which was thwarted by the abolishment of the *Societas Jesu* in 1773. As discussed above, the editions that were published in 1775/6 contain only very few illustrations.[108] According to his biography, Schiffermüller had, by 1773, "painted after life many rare caterpillars, eventually up to 400, with the greatest possible accuracy" with an eye to using them to illustrate the *Wiener Verzeichnis*.[109] As promised by Denis and Schiffermüller in their introduction to the *Wiener Verzeichnis*, the work was intended to feature plates depicting the various larvae and, on the opposite side, the imagines that emerge from them.[110]

105 Higgins 1959, 50. Sometimes the pencil note is just a question mark.

106 The scientific names of the species depicted in this sheet are discussed in the Appendix.

107 See Higgins 1959, 50. The references provided by Hübner are often confusing, as the numbers of the illustrations neither correspond to the copy of his *Geschichte europäischer Schmetterlinge* held at the Naturhistorisches Museum in Vienna nor to the numbers assigned by Higgins, who appears to have worked with another edition of Hübner's book. Hübner's note to "A" reads "*Catax*"; to "B" reads "*Everia*"; to "F" reads "nicht fertig" (unfinished); to "G" reads "*Neustria*"; to "K" bears two question marks; to "S.2" reads "Quercus?"; and to "T" reads "*Trifolii?*".

108 As stated above, the only illustrations in Denis and Schiffermüller 1775/6 are two plates, a vignette, and the frontispiece.

109 "Schiffermüller hatte damals schon seltener Raupen sehr viele /nach der Zeit bis 400/ mit möglichster Genauigkeit nach dem Leben gemalet geschildert." Anonymous *c.*1806, 13.

110 Denis and Schiffermüller 1776, 8.

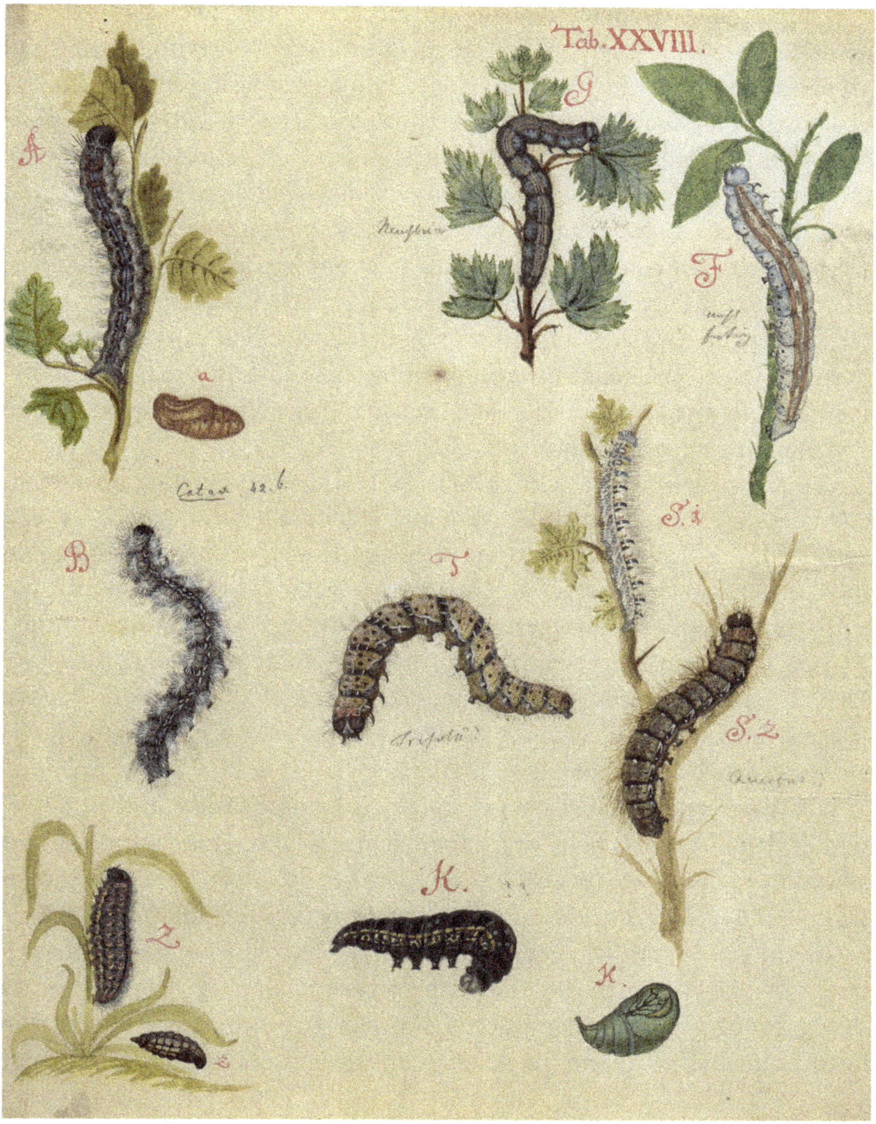

FIGURE 6.1 Ignaz Schiffermüller, larvae and a pupa of the family Lasiocampidae (A, B, F, G, S, T, a) and unidentified larvae and pupae (K, Z, k, z) with handwritten notes, in: *Collection of 53 original watercolor drawings of pupae, caterpillars and plants*, before 1773, Plate 28, watercolor and body color on paper, London, Natural History Museum, Entomology Artwork, SB q S. 3c. Reproduced with the kind permission of the Trustees of the Natural History Museum

Schiffermüller definitely had the artistic background necessary for such an undertaking: In his youth, he had spent his spare time with drawing and miniature painting, which in turn had led him to establish contacts with several artists from Vienna.[111] A rather baroque passage in the *Wiener Verzeichnis* also refers to Schiffermüller's profession as a teacher of architectural drawing: "The ability and desire to paint, and the considerable practice in architectural drawing acquired by One of us, who undertook to draft after life the colored caterpillars that are not well preservable in nature, stood us in good stead."[112] By "not well preservable in nature," the authors likely meant that dried and taxidermized caterpillars tend to lose their colors, which made it necessary to record their appearance through painting.[113] As already mentioned above, there is nothing to suggest that Denis and Schiffermüller's collection of lepidopterans included any dried caterpillars.

The centerpiece of the *Wiener Verzeichnis* is the large section on systematic nomenclature, in which the various lepidopteran species are listed and described.[114] When it comes to the shape and traits of the larvae's bodies, these descriptions are very comprehensive, while the discussion of their colors remains conspicuously superficial. It appears likely that the information on color should have been complemented by the planned illustrations. The authors themselves were certainly convinced of the power of images: "Descriptions, if too brief, remain unclear to some. ... One should rather present the eye with a well-crafted illustration."[115]

Whereas Schiffermüller's images are almost photo-realistic in their depiction of details such as body shape, hair, and legs, the same accuracy is not always in evidence with regard to color. The first example we will discuss to illustrate this point is once again the Viennese emperor, *Saturnia pyri* ([Denis & Schiffermüller], 1775). The London album of watercolors contains a life-size image of one of the species' fully-grown caterpillars (Fig. 6.2a). While its shape, pattern, and characteristic tufts of hair are accurately rendered, the image is unable to capture the striking brilliance of the living animal's color

111 Anonymous *c.*1806, 9.

112 "Die Fähigkeit und Lust zum Malen, und einige Uebung in der architektonischen Zeichnung Jemandes aus unserem Mittel, der es unternahm die in Natur nicht wohl aufbehältlichen Raupen mit Farben nach dem Leben zu entwerfen, kam uns wohl zu statten." Denis and Schiffermüller 1776, 2.

113 See Beth Tobin's chapter in this volume.

114 "Systematische Nomenclatur" Denis and Schiffermüller 1776, 40–187.

115 "Beschreibungen, wenn sie zu kurz gefaßt sind, bleiben manchen undeutlich. ... Man halte dagegen dem Auge ein wohlgeratene Abbildung vor." Denis and Schiffermüller 1776, 7.

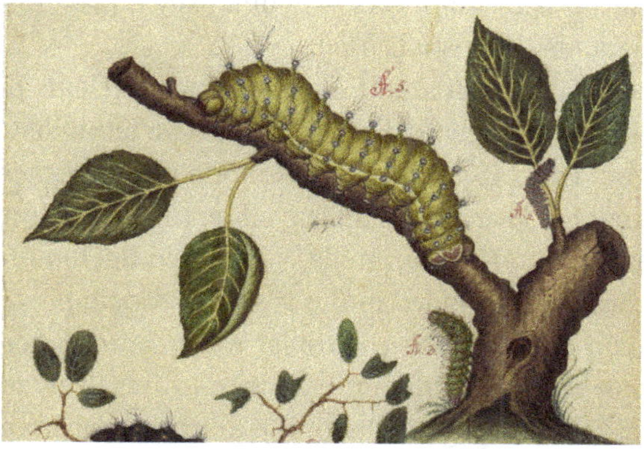

FIGURE 6.2A Ignaz Schiffermüller, larva of a Viennese emperor (*Saturnia pyri*) with two younger stages, in: Ignaz Schiffermüller, *Collection of 53 original watercolor drawings of pupae, caterpillars and plants*, before 1773, detail of Plate 25, Figures A.2, A.3, A.5, watercolor and body color on paper, London, Natural History Museum, Entomology Artwork, SB q S. 3c. Reproduced with the kind permission of the Trustees of the Natural History Museum

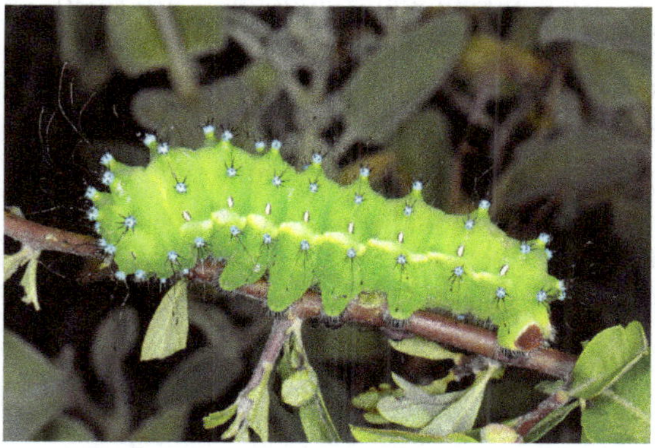

FIGURE 6.2B Fully grown larva of the Viennese emperor (*Saturnia pyri*)
© WOLFGANG WAGNER

(Fig. 6.2b). Since Schiffermüller studied colors so intensely and since the above-quoted description of the caterpillar of *Saturnia pyri* in the *Wiener Verzeichnis* leaves no doubt that he was aware of its vibrant coloration, the most likely explanation for this mismatch is deterioration of the watercolor painting.[116] The color green, in particular, appears very dull, which could be due to discoloration of the pigments or dyestuffs used.

Our second example is the oak hawk-moth, *Marumba quercus* ([Denis & Schiffermüller], 1775). A comparison between the caterpillar in Schiffermüller's watercolor album (Fig. 6.3a) and the corresponding illustration in the *Wiener Verzeichnis* (Fig. 6.3b) shows that the former is again less vibrant in color. As it can be taken for granted that Denis and Schiffermüller supervised the illumination process of the few illustrations included in the *Wiener Verzeichnis*, we must assume that the coloration of the moth's caterpillar was in accordance with Schiffermüller's intentions. While the difference in color is thus likely to have been caused by fading or deteriorating pigments, the exact reasons have yet to be established – in the absence of a detailed scientific analysis, our deliberations must, for the time being, remain speculative.[117]

Schiffermüller's watercolors were first mentioned in Francis Hemming's (1893–1964) monograph *Hübner* (1937). Hemming, who comprehensively studied Hübner's life and entomological works, cites a note dated 1831 by Hübner's editor, Peter Carl Friedrich Geyer (1802–1889), in which the latter stated that Hübner had fashioned the illustrations of his *Geschichte europäischer Schmetterlinge* after the images of Schiffermüller.[118] This pioneering work, self-published in three volumes within an imprecise time frame between 1793 and 1842, was an important milestone for the discipline of lepidopterology.[119] It depicted the development of 734 hand-colored lepidopteran species in the preimaginal stages of egg, caterpillar, and pupa together with their caterpillar host plants, mirroring the holistic perspective of the *Wiener Verzeichnis* and Schiffermüller's watercolor images.[120]

After Hemming, the caterpillar paintings were almost forgotten. It was not until 1959 that the album (then already in London) was brought back to attention by Lionel George Higgins (1891–1985), who was the first to describe the watercolors in depth.[121] Schiffermüller's biography suggests that most of them must have been completed by 1773.[122]

116 Denis and Schiffermüller 1776, 37.

117 On the deterioration of paper, see Daniels 1982, 66; on conservation, Kubik 2010, 2; on the color green, Gröne 2023.

118 Hemming 1937, 68.

119 Hemming 1937, 67; Pfeuffer 2011, 4. We take here as reference the copies held at the Naturhistorisches Museum in Vienna. The first two volumes (see Hübner 1793–1842a; 1793–1842b) have been digitized and are available in https://zenodo.org/record/7936889#.ZGIIy87PouU and https://zenodo.org/record/7936926#.ZGIKxM7PouU.

120 Pfeuffer 2011, 4.

121 Higgins 1959, 49–60; Ackery 1986, 77–78.

122 Anonymous c. 1806, 21.

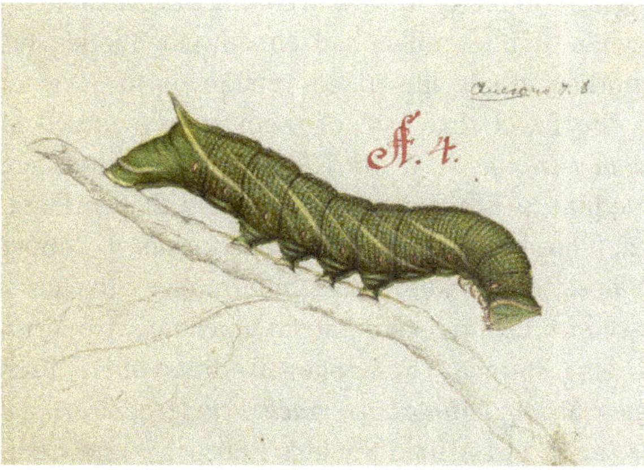

FIGURE 6.3A Ignaz Schiffermüller, larva of the oak hawk-moth (*Marumba quercus*), in: *Collection of 53 original watercolor drawings of pupae, caterpillars and plants*, before 1773, detail of Plate 12, Figure A.4, watercolor and body color on paper, London, Natural History Museum, Entomology Artwork, SB q S. 3c. Reproduced with the kind permission of the Trustees of the Natural History Museum

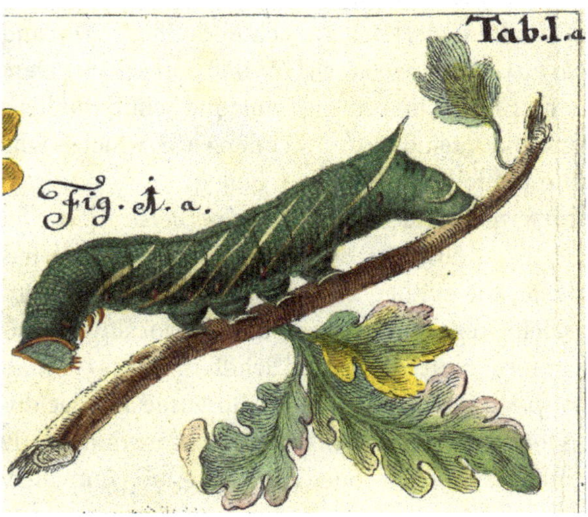

FIGURE 6.3B Ignaz Schiffermüller (illustrator), Ferdinand Landerer (engraver), larva of the oak hawk-moth (*Marumba quercus*), in: Michael Denis and Ignaz Schiffermüller, *Wiener Verzeichnis*, 1775, detail of Plate 1, Figure 1.a, copper engraving and watercolor, Vienna, Naturhistorisches Museum, 85.672

According to Higgins, Schiffermüller's images were made available to Hübner *c.*1797, at a time when Schiffermüller had already abandoned his intention of publishing another, amply illustrated version of the *Wiener Verzeichnis*[123] – eventually, "no fewer than 153" of them were included in *Geschichte europäischer Schmetterlinge*.[124] While Hübner's illustrations are extremely accurate reproductions of Schiffermüller's originals, their orientation on the page is reversed, a not unusual result of their transfer to copper engravings.[125] Higgins also noticed that some of Schiffermüller's images – mediated through Hübner – continued to reverberate in much later works on caterpillars such as Hans Rebel's (1861–1940) new enlarged edition of Friedrich Berge's (1811–1883) *Fr. Berge's Schmetterlingsbuch nach dem gegenwärtigen Stande der Lepidopterologie* (1910) and Arnold Spuler's (1868–1937) new edition of Ernst Hofmann's (1837–1892) *Die Raupen der Schmetterlinge Europas* (1903).[126] While in Berge's first edition of *Schmetterlingsbuch* (1842) contains very different images from those in Rebel's 1910 publication, the images of caterpillars published by Spuler had already appeared in Hofmann's *Die Raupen der Gross-Schmetterlinge Europas* (1893) and were executed by the Stuttgart lithographer Max Seeger (1843–1917).

Two further examples will allow us not only to assess the pictorial accuracy of Schiffermüller's images, but also to trace their reception by Hübner and – the most interesting aspect – their lasting iconographic relevance: His depictions of the sloe emperor moth, *Saturnia spini* ([Denis & Schiffermüller], 1775) and the blue underwing, *Catocala fraxini* (Linnaeus, 1758). While both species were quite frequent in the surroundings of Vienna during Denis and Schiffermüller's lifetime, many once common lepidopterans are now endangered, which is why we will supply some basic information regarding their biology.

Denis and Schiffermüller referred to the sloe emperor moth, *Saturnia spini* ([Denis & Schiffermüller], 1776) as "*Bombyx Spini*."[127] While the genus has since been changed to *Saturnia*, the epithet '*spini*' remains valid today, with Vienna and its surroundings being regarded as the moth's type locality. Its caterpillars live mainly on blackthorn (hence Denis and Schiffermüller's German common name, *Schwarzdornspinner*) and also on other Rosaceae and deciduous trees.[128] The caterpillar is found from late April to June or, more rarely, early July, with maximum from mid-May to mid-June. The larvae are completely

123 Higgins 1959, 50.
124 Higgins 1959, 52.
125 Higgins 1959, 52.
126 Higgins 1959, 50.
127 Denis and Schiffermüller 1776, 49.
128 The old German common name *Schwarzdornspinner* (see Denis and Schiffermüller 1776, 49) has since been supplanted by *Mittleres Nachtpfauenauge*.

black up to and including the penultimate stage; it is only in the last stage that the tubercles become orange or reddish. The species inhabits areas with hot, dry summers and cold winters. The pupa hibernates several times up to seven years. Today, the sloe emperor moth is found in the southeastern parts of Europe, especially in the Balkan Peninsula, but it was once distributed to the northwest at least as far as eastern Austria and Slovakia, according to some sources even as far as Bavaria. The western distribution limit of the species has shifted further and further east in recent decades, and these occurrences are probably all extinct.[129] Its Red List status for Austria is therefore 'regionally extinct' (RE).

As the juxtaposition of Schiffermüller's and Hübner's illustrations shows, the posture of the caterpillar is very similar, with Hübner's being a mirror image of Schiffermüller's (Figs. 6.4a and 6.4b). Both depictions are extremely accurate in their details, as becomes obvious in comparison with a modern photograph (Fig. 6.4c). Characteristic details like the hair-adorned tubercles on each ring feature prominently in both.[130]

What is even more striking, however, is the strong resemblance between Schiffermüller's watercolor and an image from the early twentieth century: Seeger's depiction of the sloe emperor moth in *Die Raupen der Schmetterlinge Europas* (Fig. 6.4d). The similarities to Hübner's version of the motif are even more pronounced, which suggests that the latter served as template for Seeger's lithograph in Hofmann's original *Die Raupen der Gross-Schmetterlinge Europas* (1893) and later in Spuler's new edition thereof – it even quotes the shape of the leaves on the twig the caterpillar is perched upon.

The second example is *Catocala fraxini* (Linnaeus, 1758), which was first described by Linnaeus as "*Phalaena Noctua Fraxini*." Its English common name is blue underwing, after the blue bands on the hind wings that are displayed when the moth extends its fore wings while resting. The specific epithet '*fraxini*' derives from the European ash tree, *Fraxinus excelsior*. In the *Wiener Verzeichnis*, Denis and Schiffermüller referred to this species with the contemporary German common name *Eschenschwärmer*, which is a translation of Linnaeus's Latin name.[131] However, the two Jesuit scholars did not agree with this nomenclature, adding in a footnote that they had found the caterpillar in question only on poplar trees, especially of the black variety. Their observation is confirmed by modern descriptions, which unanimously name poplars as the species' main food plant. The caterpillar is excellently camouflaged by its gray, bark-like markings and thus very difficult to detect. The blue underwing is

129 Wagner 2005–2024.

130 Denis and Schiffermüller 1776, 49.

131 Denis and Schiffermüller 1776, 90.

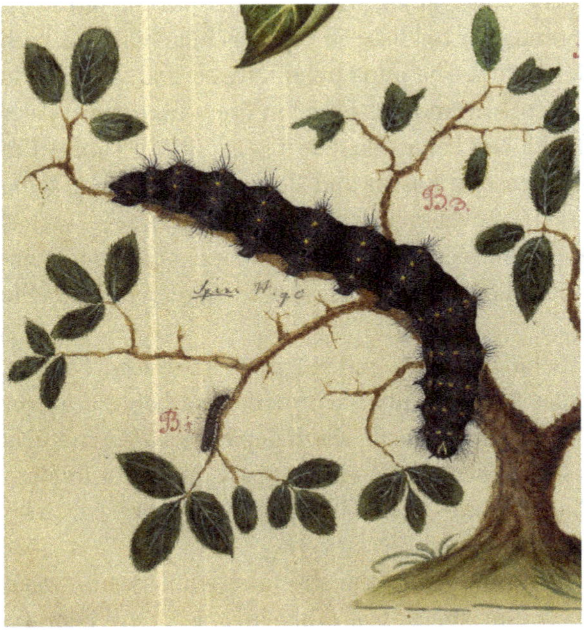

FIGURE 6.4A
Ignaz Schiffermüller, larva of a sloe emperor moth (*Saturnia spini*) with younger stages, in: Ignaz Schiffermüller, *Collection of 53 original watercolor drawings of pupae, caterpillars and plants*, before 1773, detail of Plate 25, Figures B.1, B.2, B.3, watercolor and body color on paper, London, Natural History Museum, Entomology Artwork, SB q S. 3c. Reproduced with the kind permission of the Trustees of the Natural History Museum

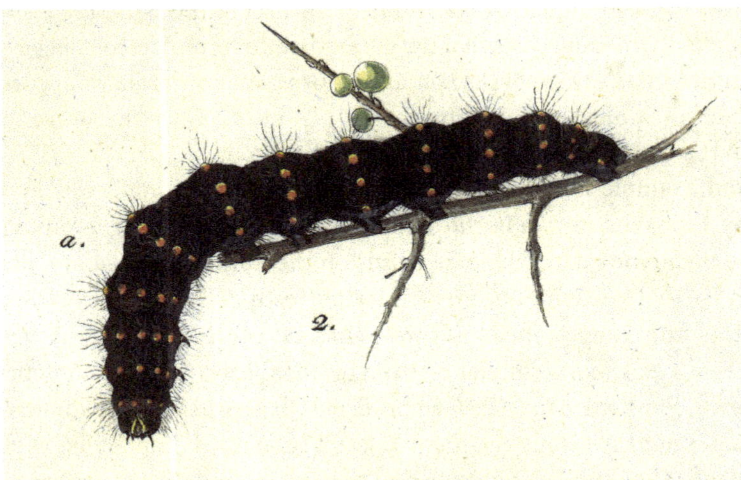

FIGURE 6.4B Jacob Hübner (after Ignaz Schiffermüller), larva of the sloe emperor moth (*Saturnia spini*), in: Jacob Hübner, Geschichte Europäischer Schmetterlinge. Tabulae Vol. I. Larvae lepidodopterorum I.–III. Papiliones, Sphinges, Bombyces, 1793–1842, Plate "Larvae Lepidopt. III, Bombyces II, Veroe 1," Figure 2.a., p. 124 of PDF copper engraving on paper, watercolor, Vienna, Naturhistorisches Museum, Lepidoptera-B © NATURHISTORISCHES MUSEUM WIEN. HTTPS://DOI .ORG/10.5281/ZENODO.7936889

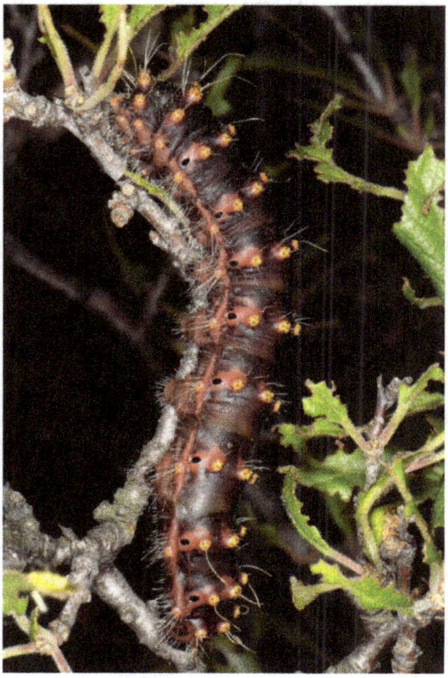

FIGURE 6.4C
Fully grown caterpillar of the sloe emperor moth (*Saturnia spini*)

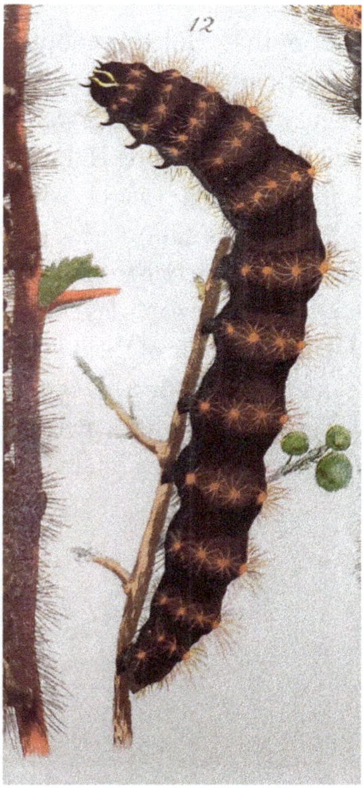

FIGURE 6.4D
Max Seeger (after Jacob Hübner), larva of the sloe emperor moth (*Saturnia spini*), in: Ernst Hofmann and Arnold Spuler, *Die Raupen der Schmetterlinge Europas*, 1903, Plate 18, Figure 12, lithography, Vienna, Naturhistorisches Museum, Lepidoptera-B

the largest European owl moth with a wingspan of up to 10 cm.[132] Its distribution area covers almost all of central and northern Europe, parts of southern Europe, and even Japan. Its Red List status in Austria is 'least concern' (LC).

Evidently based on close observation, Schiffermüller's watercolor painting depicts the caterpillar of the blue underwing in a feeding position, with its head bent upwards in order to reach the leaf of the tree (Fig. 6.5a). Hübner's illustration follows Schiffermüller's template closely. The caterpillar's posture is only slightly different, in that it is depicted stretching towards the leaf's stem (Fig. 6.5b). The grayish blue caterpillar is shown on a gray twig, thus visualizing the amazing ability of the larva to blend in with its surroundings.

Comparing Schiffermüller's and Hübner's images to the corresponding illustration in Ernst Hofmann's and Arnold Spuler's book (Fig. 6.5c) again reveals striking similarities. Here, too, the lithograph hews closely to Schiffermüller's and/or Hübner's model, especially with regard to the caterpillar's posture. But whereas Schiffermüller's and Hübner's caterpillars are feeding on the respective plants, Seeger's illustration gives no logical reason for the raised head of its subject – normally, the caterpillar of *Catocala fraxini* sits pressed tightly against a twig or leaf, turning this into a clear case of mere 'image citation.' A few years later, in 1910, a very similar illustration was published by Friedrich Berge and Hans Rebel (Fig. 6.5d).

Our comparisons confirm Higgins's observation that Hübner adopted Schiffermüller's depictions more or less unchanged in his *Geschichte der europäischen Schmetterlinge*.[133] According to Higgins, Schiffermüller's watercolors were passed on to Geyer, who recorded their presence among Hübner's effects.[134] Mediated through Hübner's versions, they reached a much wider audience. Indeed, as our two examples have shown, it is safe to say that Schiffermüller's lifelike records of caterpillars established an iconographic tradition. Any form of artistic exaggeration was alien to his paintings – fashioned after living specimens, the images are accurate documentations of the respective animals. Hübner remained true to this style, and, by adopting his predecessor's illustrations almost unchanged, ensured that Schiffermüller's pictorial compositions would continue to be used even more than a century later.

132 Forster and Wohlfahrt 1971, 281.

133 Higgins 1959, 50.

134 Higgins 1959, 51.

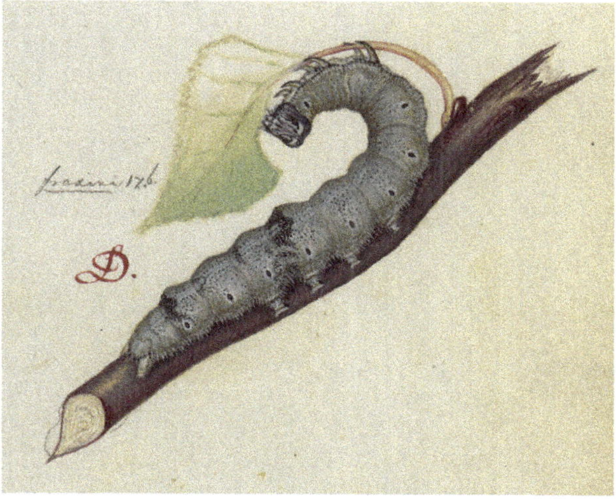

FIGURE 6.5A Ignaz Schiffermüller, larva of the blue underwing (*Catocala fraxini*), in: Ignaz Schiffermüller, *Collection of 53 original watercolor drawings of pupae, caterpillars and plants*, before 1773, detail of Plate 45, Figure D, watercolor and body color on paper, London, Natural History Museum, Entomology Artwork, SB q S. 3c. Reproduced with the kind permission of the Trustees of the Natural History Museum

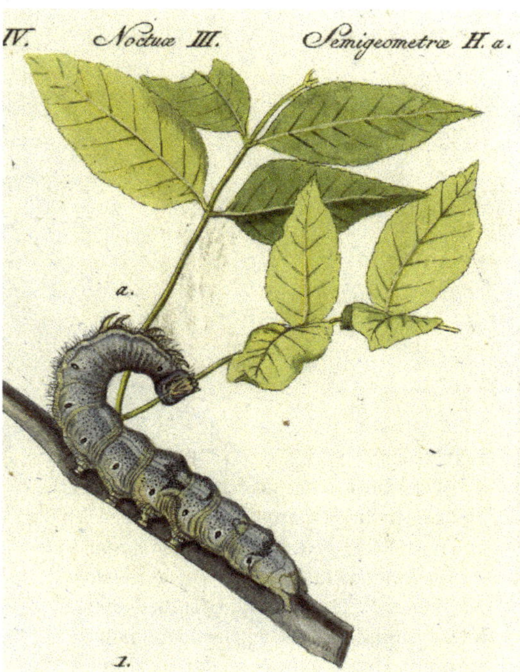

FIGURE 6.5B
Jacob Hübner (after Ignaz Schiffermüller), larva of the blue underwing (*Catocala fraxini*), in: Jacob Hübner, *Geschichte Europäischer Schmetterlinge. Tabulae Vol. 2. Larvae lepidodopterorum, Noctuae*, 1793–1842, detail of Plate "Larva Lepidopt. IV. Noctuae III. Semigeometrae H.a," Figure 1.a, p. 112 of PDF, copper engraving on paper, watercolor. Vienna, Naturhistorisches Museum, Lepidoptera-B
© NATURHISTORISCHES MUSEUM WIEN. HTTPS://DOI.ORG/10.5281/ZENODO.7936889

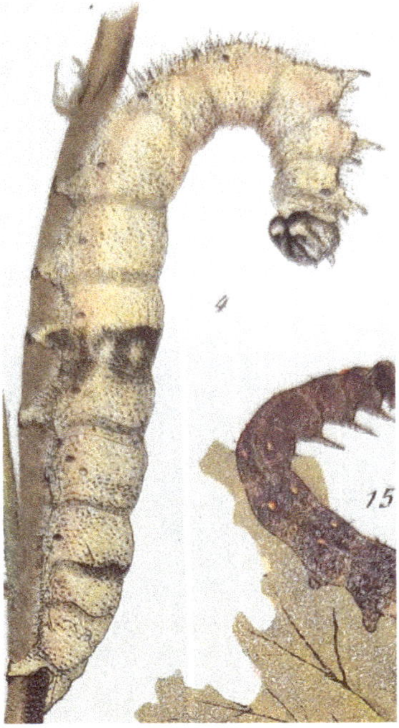

FIGURE 6.5C
Max Seeger (after Jacob Hübner), larva of the blue underwing (*Catocala fraxini*), in: Ernst Hofmann and Arnold Spuler (ed.), *Die Raupen der Schmetterlinge Europas*. 1903, Plate 37, Figure 4, lithography, Vienna, Naturhistorisches Museum, Lepidoptera-B

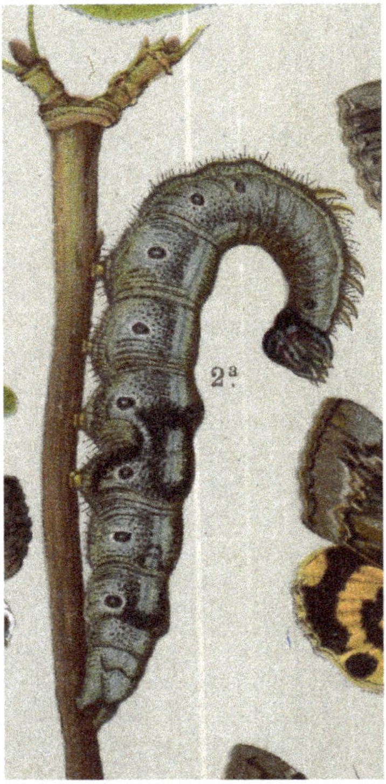

FIGURE 6.5D
Artist unknown (after Jacob Hübner), larva of the blue underwing (*Catocala fraxini*), in Friedrich Berge and Hans Rebel, *Fr. Berges Schmetterlingsbuch nach dem gegenwärtigen Stande der Lepidopterologie*, 1910, Plate 38, Figure 2.a, lithography. Vienna, Naturhistorisches Museum, Lepidoptera-B

7 Schiffermüller's Images – Iconotypes?

In the past, it was quite common to name a new species solely by reference to an image made by another person.[135] Today, however, images are not accepted as types in zoological nomenclature: According to the ICZN, a given species name can only be 'tied' to a single (preserved) specimen, the so-called type specimen. These specimens are, simply put, the "primordial meter" of a species, and as such constitute an immensely valuable resource for the international scientific community – an illustration *per se* cannot replace the specimen.[136] Despite this ruling of the ICZN, images can sometimes function as "proxy name bearers," and they can assist in clarifying taxonomic questions.[137] They therefore remain important historical documents. This is also the reason why Schiffermüller's watercolors carry special importance with regard to the history of zoological nomenclature: As the only remaining evidence of the lost type specimens, they can be interpreted as "iconotypes."[138]

Thanks to the accuracy of Schiffermüller's illustrations, modern entomologists have been able to identify many of the lepidopteran species depicted. However, it must be emphasized that the determination of butterfly species based solely on images of the larvae is quite a challenge even for experts in this field, which is why Denis and Schiffermüller themselves advised to examine both larva and imago.[139]

As mentioned, many of Schiffermüller's watercolors bear later pencil notes, which were attributed to Hübner by Higgins 1959 which provide names for the species depicted.[140] Based on these not always reliable notes, several cursorily described archival sources, and his own observations and comparisons, Higgins compiled a list of the species featured in the paintings.[141] By his own admission, Higgins could not identify 28 figures due to the difficulty of the task.[142] Additionally, many of the species' names have changed since 1959 due to the progress of zoological nomenclature. Based on Higgins's preliminary work and a reassessment of Schiffermüller's watercolor images, we have therefore assembled an updated list of species names (see Appendix).

135 Vane-Wright 2021, 35.
136 Dubois 2011, 51.
137 Vane-Wright 2021, 11–35, esp. 34.
138 Vane-Wright 2021, 11–35, esp. 35.
139 Denis and Schiffermüller 1776, 31 f.
140 Higgins 1959, 50 (see notes 104 and 106 above).
141 Higgins 1959, 53–60.
142 Higgins 1959, 53. According to our research, the actual figure was only 27.

8 Determining the Current Scientific Names – a Complex Challenge

It was clear from the outset that it would be very difficult to match the specimens depicted in Schiffermüller's watercolors with modern scientific names. Given that taxonomic research is in constant flux, such an attempt can only be understood as an approximation and a snapshot in time. Yet despite these limitations, an up-to-date list of names is essential to render Schiffermüller's images accessible to current lepidopterological research. As mentioned before, Denis and Schiffermüller adopted Linnaeus's system with the categories kingdom, class, order, genus, and species, but they also added new categories of their own devising.[143] They divided lepidopterans into three groups: Crepuscular moths or hawkmoths (Sphinges L.), nocturnal moths (Phalaenae L., and diurnal butterflies (Papiliones L.).[144] The Phalaenae were again subdivided into seven "divisions."[145] Some of the terms Denis and Schiffermüller used do not correspond to their modern meaning.[146] In this context, it should be noted that Denis and Schiffermüller were much more interested in the description of new species than in the further development of the hierarchical system, as is made clear by the following remark from the *Wiener Verzeichnis*: "However, we leave it up to everyone to regard the divisions established by us as orders, or whatever else one may wish. Our families may perhaps be considered fitting for what is referred to as genera in the plant kingdom."[147]

The same non-conformity of old and new terms is evident in the species names employed by Hübner: He, too, created a slightly different terminology, which again does not always correspond to modern usage. It was this kind of systematic conundrum that revealed the need for general rules, which ultimately led to the establishment of the above-mentioned *International Code of Zoological Nomenclature*.

Linnaeus described about 4400 animal species, 450 of them lepidopterans. Denis and Schiffermüller increased the latter number – according to the literature and Schiffermüller's biography – to more than 1150.[148] Our own research

143 Sattler and Tremewan 1984, 283 and Sattler 1969, 5–6.

144 'L.' stands for Linnaeus.

145 "Abtheilungen" Denis and Schiffermüller 1776, 48 ff.

146 Mentzer 1984, 63.

147 "Wir lassen es aber einem jeden frey, die Abtheilungen, die wir angesetzt haben, für Ordnungen, oder was man sonst will, anzusehen. Unsere Familien werden vielleicht für das, was im Pflanzenreiche die Gattungen sind, schicklich gelten können." Denis and Schiffermüller 1776, 196 (footnote).

148 On the number of species described by Linnaeus, see Geiser 2008, 28. On the number of species described by Denis and Schiffermüller, see Anonymous *c.*1806, 16 f.; Hoffmann 1952, 59; Sattler and Tremewan 1984, 282.

has shown that a total of 1136 species are described in the *Wiener Verzeichnis*, 740 of which were originally identified by Denis and Schiffermüller (65%), 376 by Linnaeus (33%), 17 by Scopoli (1%), and one each by Eugen Johann Christoph Esper (1742–1810), Johann Reinhold Forster (1729–1798), and Johann Jacob Müller (1743–after 1802). In the meantime, many more species have been described. This process is still ongoing. Yet while modern methods such as DNA sequencing have enabled deeper insights into the relationships of species, the system established by Linnaeus remains important.[149]

To determine the current scientific names of the species featured in Schiffermüller's watercolors, we compared each painting and its annotations with Higgins's data and photographs of living caterpillars. As our main source, we used LepiWiki (Lepiforum),[150] which follows the systematics developed by Ole Karsholt and Józef Razowski.[151] When necessary, we also cross-checked with the online lepidopterological databases Funet and LepIndex.[152] Another source was a two-page list of Geometridae species related to the watercolors, now also held by the NHM London; handwritten by Schiffermüller, it was difficult to decipher and sometimes illegible.[153]

A precise determination of the species depicted was not always possible. In cases of uncertainty, we have provided additional information regarding our assessment. As of now, twelve caterpillar drawings remain unidentified. We estimate that Schiffermüller painted *c.*340 species. Our count does not match the number of images (454), as the same species is sometimes depicted in different life stages.

As our list shows (see Appendix), Denis and Schiffermüller's authorship remains valid for a very large number of cases. The names for more than 100 species newly described in the *Wiener Verzeichnis* and depicted by Schiffermüller are still in current use. This impressive number attests to the precision and accuracy of the methods employed by the two Jesuit scholars. A further 54 names were identified as synonyms, which means that the corresponding species were described as new in the *Wiener Verzeichnis*, but had already been published by earlier authors – these names are now invalid.

149 Fischer 2012, 325.

150 Lepiforum e.V., LepiWiki, https://lepiforum.org/wiki.

151 Karsholt and Razowski 1996. Deviations regarding taxonomy, systematics, and distribution were incorporated into the list based on the relevant literature.

152 Savela, n. d.; Beccaloni et al. 2003.

153 Schiffermüller n.d.

9 Conclusion

As our research has shown, many aspects of Denis and Schiffermüller's *Wiener Verzeichnis* and Schiffermüller's *Versuch eines Farbensystems* are still relevant today. Their efforts to bring order into the highly diverse worlds of butterflies and colors facilitated future standardization processes in the fields of art and the natural sciences. In lepidopterology, their holistic view on the development of lepidopterans with a special focus on the larval stage not only refined Linnaeus's systematics, but sensitized future researchers to the complexity of natural interrelationships, thereby helping to pave the way for today's ecological thinking. Schiffermüller's watercolors are an impressive testimony to his meticulous powers of observation, not only with regard to the animals' appearance, but also their relationship to the respective food plants. As preparatory works for an illustrated reedition of the *Wiener Verzeichnis*, Schiffermüller's paintings were intended to document the appearance of larvae, plants, and pupae as true to nature as possible. This pictorial documentation was especially important because dried specimens lose some of their characteristic traits – above all their colors. Working after nature, he created images which, in their consistent renunciation of aestheticization and their sober reduction to the essential, set the standard for future scientific illustrations. Mediated through Hübner's *Geschichte europäischer Schmetterlinge*, Schiffermüller's almost photo-realistic watercolors continued to have a powerful iconographic impact on caterpillar illustrations more than a century after their creation.

Acknowledgements

Our thanks go to Sabine Gaal-Haszler (head of the Lepidoptera collection), Martin Lödl (former head of the insect department), Mario-Dominik Riedl (archive) and Martin Krenn (head of the archive), all at Naturhistorisches Museum Wien, Gerhard Steininger (Oberösterreichisches Landesarchiv), Magdalena Wieser (Bibliothek, Oberösterreichische Landes-Kultur GmbH), Esther Ockermüller (Entomologie, Biologiezentrum Linz), Grace Touzel and Helen Pethers (Natural History Museum London), Rosemary Pearson (Royal Entomological Society London), Eva Schober (Akademie der Bildenden Künste Wien), Martin Kreuz (Universitätsbibliothek Wien), Dominik Hünniger (Deutsches Hafenmuseum, Hamburg), and André Karliczek (Friedrich-Schiller-Universität Jena) for their support, advice, and expertise as well as to Wolfgang Wagner (Stuttgart) for his generous permission to use his caterpillar photos.

Bibliography

Manuscript Sources

Anonymous. c.1806. *Lebensbeschreibung des hochwürdig-Wohlgeborenen und hochgelehrten Herrn Ignaz Schiffermüller, gewesener Titular = Domherr des bischöflichen Stiftes Linz, dann eben bischöflicher Consistorialrath und würdigsten Dechanten zu Waizenkirchen. Linz,* [undated autobiography handwritten by an unknown and rather clueless scribe. In part it was written after Schiffermüller's death 1806; without pagination]. Oberösterreichischen Landesmuseum Linz, Musealarchiv, Hs. Nr 197.

Schiffermüller, Ignaz. before 1773. Collection of 53 Original Watercolor Drawings of Pupae, Caterpillars and Plants. Natural History Museum London, South Kensington, Entomology Artwork; SB q S.3c.

Schiffermüller, Ignaz. n.d. Two-page List by Schiffermüller on Geometridae Related to the Watercolor Drawings, Natural History Museum London, South Kensington, S 3b ID 000155037.

Printed Sources

Ackery, Phillip R. 1986. "Lionel George Higgins (1891–1985)." *Journal of the Lepidopterists' Society* 40 (1): 77–78.

Beccaloni, George, Scoble, Malcolm, Kitching, Ian, Simonsen, Thomas, Robinson, Gaden, Pitkin, Brian, Adrian Hine and Chris Lyal (ed.). 2003. *The Global Lepidoptera Names Index (LepIndex)*. World Wide Web electronic publication. Accessed June 17, 2023. https://www.nhm.ac.uk/our-science/data/lepindex/lepindex/.

Berge, Friedrich and Rebel, Hans. 1910. *Fr. Berges Schmetterlingsbuch nach dem gegenwärtigen Stande der Lepidopterologie*. Stuttgart: E. Schweizerbart'sche Verlagsbuchhandlung.

Blair, Ann and Von Geyerz, Kaspar. 2020. "Introduction." In *Physico-Theology. Religion and Science in Europe, 1650–1750*, Ann Blair and Kaspar von Geyerz (eds.). Baltimore: Johns Hopkins University Press, 1–20.

Daniels, Vincent. 1982. "Colours Changes of Watercolour Pigments during Deacidification". *Studies in Conservation* 27 Supplement 1: 66–70.

Denis, Michael and Ignaz Schiffermüller, 1775. *Ankündung eines systematischen Werkes von den Schmetterlingen der Wienergegend, herausgegeben von einigen Lehrern am k. k. Theresianum*. Wien: Augustin Bernardi.

Denis, Michael and Ignaz Schiffermüller. 1776. *Systematisches Verzeichniß der Schmetterlinge der Wienergegend, herausgegeben von einigen Lehrern am k.k. Theresianum.* Wien: Augustin Bernardi.

Denis, Michael. 1797a. *Lesefrüchte. 1: A. bis L.* Vienna: Rötzel.

Denis, Michael. 1797b. *Lesefrüchte 2: M. bis Z.* Vienna: Rötzel.

Dubois, Alain. 2011. "The International Code of Zoological Nomenclature Must Be Drastically Improved before It Is Too Late." *Bionomina* 2(1): 1–104. https://doi.org/10.11646/bionomina.2.1.

Etheridge, Kay. 2021. *The Flowering of Ecology: Maria Sibylla Merian's Caterpillar Book.* [Emergence of Natural History 3]. Leiden; Boston: Brill.

Fabricius, Johan Christian. 1775. *Systema entomologiae: sistens insectorvm classes, ordines, genera, species, adiectis synonymis, locis, descriptionibvs, observationibvs,* Flensburg; Leipzig: Kortii.

Fischer, Manfred A. 2012. "Was ist das natürliche System? Überlegungen zum Begriff des biologischen Systems im Zeitalter der Molekularphylogenetik." *Verhandlungen der Zoologisch-Botanischen Gesellschaft in Wien* 148/149: 323–362.

Fitzinger, Leopold Joseph. 1856. *Geschichte des kais. kön. Hof-Naturalien-Cabinetes zu Wien. 1: Älteste Periode bis zum Tode Kaisers Leopold II. 1792*, Wien: kaiserl. königl. Hof- und Staatsdruckerei in Commission bei W. Braumüller.

Fitzinger, Leopold Joseph. 1868. "Geschichte des kais. kön. Hof-Naturalien-Cabinetes zu Wien. 2: Periode unter Franz II. (Franz I. Kaiser von Österreich) bis zu Ende des Jahres 1815: (vorgelegt in der Sitzung am 22. Mai 1868)." *Sitzungsband der k. Akademie der Wissenschaften* 57 (Mai): 1–80.

Forster, Walter and Wohlfahrt, Theodor. 1971. *Die Schmetterlinge Mitteleuropas. Band IV. Eulen (Noctuidae).* Stuttgart: Franckh'sche Verlagshandlung.

Füssli, Johann Caspar. 1781–1786. *Archiv der Insectengeschichte.* Zurich: Winterthur, Füssli und Steiner.

Geiser, Elisabeth. 2008. "Die biologische Systematik vor und nach Linné oder warum Carl von Linné kein Systematiker, sondern ein Informatiker war." *Entomologica Austriaca* 15: 17–32.

Gröne, Matthias. 2023. "Pigment-Geschichte(n)." *Malerblatt.* Accessed May 5, 2023. https://www.malerblatt.de/aus-und-weiterbildung/pigment-geschichten-3/.

Harrison, Peter. 2006. "The Metaphor 'the Book of Nature' and Early Modern Science." In *The Book of Nature in Early Modern and Modern History*, Klaas van Berkel and Arjo Vanderjagt (eds.). Leuven: Peeters, 1–26.

Harvey, Julie M.V., Gilbert, Pamela, and Martin, Kathy. 1996. *A Catalogue of Manuscripts in the Entomology Library of the Natural History Museum*, London: Mansell.

Hemming, Francis. 1937. *Hübner. A Bibliographical and Systematic Account of the Entomological Works of Jacob Hübner and of the Supplements Thereto by Carl Geyer, Gottfried Franz von Fröhlich and Gottlieb August Wilhelm Herrich-Schäffer.* Vol. 1. London: Royal Entomological Society of London.

Hemming, Francis. 1958. "Opinion 516." *Opinions and Declarations Rendered by the International Commission on Zoological Nomenclature* 19 (1): 1–44.

Higgins, Lionel George. 1959. "The Original Larval Figures of Schiffermüller." *Entomologist* 92: 49–60.

Hofmann, Ernst and Spuler, Arnold (ed.). 1903. *Die Schmetterlinge Europas. Band III: Die Raupen der Schmetterlinge.* Stuttgart: E. Schweizerbart'sche Verlagsbuchhandlung.

Hoffmann, Emil. 1952. "Ignaz Schiffermüller." *Zeitschrift der Wiener Entomologischen Gesellschaft* 37: 57–65.

Hope, Frederic William. 1845. "The Autobiography of Johann Christian Fabricius with Additional Notes and Observations." *The Transactions of the Royal Entomological Society of London* 4: i–xvi.

Hübner, Jacob. 1793–1842a. *Geschichte Europäischer Schmetterlinge. Tabulae Vol. 1: Larvae lepidodopterorum I.–III. Papiliones, Sphinges, Bombyces.* Augsburg: Privately published.

Hübner, Jacob. 1793–1842b. *Geschichte Europäischer Schmetterlinge. Tabulae Vol. 2. Larvae lepidodopterorum, Noctuae.* Augsburg: Privately published.

Hübner, Jacob. 1796. *Sammlung europäischer Schmetterlinge.* Band 1: Lepidoptera I, Papiliones I, Nymphales A. Augsburg: Privately published.

Hübner, Jacob. 1836. *Sammlung europäischer Schmetterlinge.* Band 8: Lepidoptera VIII, Tortrices I, Pyralidoides A. Augsburg: Privately published.

Huemer, Peter. 2007. "Rote Liste ausgewählter Nachtfalter Österreichs (Lepidoptera: Hepialoidea. Cossoidea. Zygaenoidea. Thyridoidea. Lasiocampoidea. Bombycoidea. Drepanoidea. Noctuoidea)." In *Rote Liste gefährdeter Tiere Österreichs. Checklisten, Gefährdungsanalysen, Handlungsbedarf. Teil 2: Kriechtiere, Lurche, Fische, Nachtfalter, Weichtiere*, Klaus Peter Zulka (ed.). [Grüne Reihe des Bundesministeriums für Land- und Forstwirtschaft, Umwelt und Wasserwirtschaft (Gesamtherausgeberin Ruth Wallner) Band 14/2]. Vienna: Böhlau, 199–361.

Hünniger, Dominik. 2018. "Nets. Labels and Boards: Materiality and Natural History Practices in Continental European Manuals on Insect Collecting 1688–1776." In *Naturalists in the Field. Collecting, Recording and Preserving the Natural World from the Fifteenth to the Twenty-First Century*, Arthur MacGregor (ed.). [Emergence of Natural History 2]. Leiden: Brill, 686–705. https://doi.org/10.1163/9789004323841_024.

International Commission on Zoological Nomenclature. 1961. *International Code of Zoological Nomenclature, Adopted by the XV International Congress of Zoology / Code international de nomenclature zoologique, adopté par le XVe Congrès international de zoologie.* https://doi.org/10.5962/bhl.title.50303.

Karliczek, André. 2013. "Vom Phänomen zum Merkmal, Farben in der Naturgeschichte um 1800." In *Erkenntniswert Farbe*, Margit Vogt and André Karliczek (eds.). Jena: Ernst-Haeckel-Haus, 83–111.

Karliczek, André. 2016. 'Katalog.' In *Farre. Farbstandards in den frühen Wissenschaften*, André Karliczek and Andréas Schwarz (eds.). Jena: Salana, 283–523.

Karsholt, Ole and Razowski, Józef. 1996. *The Lepidoptera of Europe – A Distributional Checklist.* Stenstrup: Apollo Books.

Kemp, Martin. 1990. *The Science of Art: Optical Themes in Western Art from Brunelleschi to Seurat*. New Haven: Yale University Press.

Kirby, William Forsell. 1781. *A Synonymic Catalogue of Diurnal Lepidoptera*. London; Berlin: John van Voorst & R. Friedlander & Son.

Kubik, Maria E. 2010. "Preserving the Painted Image: The Art and Science of Conservation." *Colour: Design & Creativity* 5: 1–8.

Kudrna, Otakar. 2015. "The Never Ending Story of Schiffermüller's Names – a Long Evaded Nomenclatural Issue of Pressing Urgency and a Special Case for the ICZN (Insecta: Lepidoptera)." *Quadrifina* 12: 17–26.

Kudrna, Otakar and Belicek, Joe. 2005. "On the 'Wiener Verzeichnis' Its Authorship and the Butterflies Named Therein." *Oedippus* 23: 1–32.

Lepiforum e.V. 2022. *Bestimmung von Schmetterlingen und ihren Präimaginalstadien. LepiWiki (Bestimmungshilfe)*. https://lepiforum.org/wiki.

Lersch, Thomas. 1984. "Von der Entomologie zur Kunsttheorie Ignaz Schiffermüllers Versuch eines Farbensystems (1771); Miszellen zur Problemgeschichte der Farbenlehre." In *De Arte et Libris: Festschrift Erasmus*, Abraham Horodisch (ed.). Amsterdam: Erasmus Antiquariat en Boekhandel, 301–316.

Lesser, Friedrich Christian. 1738. *Insecto-theologia, oder: Vernunfft- und Schrifftmäßiger Versuch wie ein Mensch durch aufmercksame Betrachtung derer sonst geachteten Insecten in lebendiger Erkänntniß und Bewunderung der Allmacht; Weißheit, der Güte und Gerechtigkeit des grossen Gottes gelangen könnte*; Frankfurt; Leipzig: Michael Blochberger.

Lesser, Friedrich Christian. 1742. *Théologie des insectes, ou Démonstration des perfections de Dieu dans tout ce qui concerne les insectes. Traduit de l'allemand de M. Lesser, avec des remarques de M. P. Lyonnet*. Vol. 1. The Hague: Jean Swart.

Linnaeus, Carl. 1758. *Systema naturae, per regna tria naturae: secundum classes, ordines, genera, species cum characteribus, differentiis, synonymis, locis*, Tenth edition. Vol. 1. Stockholm: Salvii.

Linnaeus, Carl. 1767. *Systema naturae, per regna tria naturae: secundum classes, ordines, genera, species cum characteribus, differentiis, synonymis, locis*. Thirteenth edition. Vol. 1. Vienna: Thomas.

Melville, Richard Valentine and Smith, J.D.D. 1987. *Official Lists and Indexes of Names and Works in Zoology*. London: ICNZ.

Mentzer, Erik von. 1984. "Die Genera bei Denis & Schiffermüller als Nomenklaturfrage (Lepidoptera)." *Nota lepidopterologica* 7 (1): 59–70.

Ogilvie, Brian W. 2020. "Maxima in minimis animalibus: Insects in Natural Theology and Physico-theology." In *Physico-Theology. Religion and Science in Europe, 1650–1750*, Ann Blair and Kaspar von Geyerz (eds.). Baltimore: Johns Hopkins University Press, 171–182.

Pfeuffer, Renate. 2011. "Jacob Hübner (1761–1826) zum 250. Geburtstag." *Berichte des Naturwissenschaftlichen Vereins für Schwaben* 115: 2–6.

Prange, Friedrich Christian. 1782. *Christian Friedrich Prangens der Weltweisheit und freyen Künste Magister Farbenlexicon: worinn die möglichsten Farben der Natur nicht nur nach ihren Eigenschaften, Benennungen, Verhältnissen und Zusammensetzungen sondern auch durch die wirkliche Ausmahlung enthalten sind. Zum Gebrauch für Naturforscher, Mahler, Fabrikanten, Künstler und übrigen Handwerker, welche mit Farben umgehen. Mit 48 illuminirten Tafeln und einer großen Landschaft.* Halle: Hendel.

Promitzer, Christian and Speta, Franz. 1990. "Naturgeschichte im josephinischen Linz, Ignaz Schiffermüller und der ökonomisch-botanische Garten beim Bergschlößl." *Historisches Jahrbuch der Stadt Linz 1989*. Linz: Archiv der Stadt Linz 1990.

Retzer, Joseph Friedrich Freyherr von. 1801. *Michael's Denis literarischer Nachlaß*, Vol. 1. Vienna: Pichler.

Russell, Peter, J.C., Luca Bartolozzi, Rachel L. Hawkins, W. J. Tennent and T. Léger 2020. "Designation of Lectotypes for Some Spanish and Other Western European *Melitaea* Taxa, Some with Mixed Syntypic Series of *M. phoebe* ([Denis & Schiffermüller],1775) and *M. ornata* Christoph, 1893 (Lepidoptera: Nymphalidae)." *SHILAP Revista de lepidopterología* 48 (191): 449–472. https://doi.org/10.57065/shilap.369.

Russell, Peter, J.C. and Vane-Wright Richard I. 2022. "*Papilio hermione* Linnaeus, Type Species of *Hipparchia* Fabricius (Lepidoptera, Satyrinae): Restoring Stability to the Application of These Names." *Nota Lepidoperologica* 45: 279–294. https://doi.org/10.3897/nl.45.85341.

Sattler, Klaus 1969. "Das 'Wiener Verzeichnis' von 1775." *Zeitschrift der Wiener Entomologischen Gesellschaft* 54 (80): 2–7.

Sattler, Klaus and Tremewan, Walter G. 1984. "The Lepidoptera Names of Denis & Schiffermüller – a Case for stability." *Nota lepidoperologica* 7 (3): 282–285.

Sattler, Klaus and Tremewan, Walter G. 2009. "The Authorship of the So-Called 'Wiener Verzeichnis'." *Nota Lepidopterologica* 32: 3–10.

Savela, Markku. n.d. Lepidoptera and Some Other Life Forms. Accessed May 23, 2023. https://ftp.funet.fi/index/Tree_of_life/insecta/lepidoptera/.

Schiffermüller, Ignaz. 1771. *Versuch eines Farbensystems*. Vienna: Augustin Bernardi.

Schiffermüller, Ignaz. 1772. *Versuch eines Farbensystems*. Vienna: Augustin Bernardi.

Scopoli Giovanni Antonio. 1763. *Entomologia Carniolica exhibens insecta Carnioliae indigena et distributa in ordines, genera, species, varietates. Methodo Linnaeana.* Vienna: Johann Thomas Trattner.

Simonini, Giulia. 2024. *Color Charts in 18th-Century Europe: Natural, Pigmentary, and Trichromatic.* Heidelberg, arthistoricum.net.

Speta, Franz. 2003. "Ignaz Schiffermüller (1727–1806) – Eine Biographie." *Denisia* 8: 11–14.

Steiner, Axel, Ratzel Ulrich, Morten Top-Jensen and Michael Fibiger. 2014. *Die Nachtfalter Deutschlands: ein Feldführer: sämtliche nachtaktiven Großschmetterlinge in Lebendfotos und auf Farbtafeln.* Oestermarie: Bugbook Publishing.

Tóth, Balázs, Katona Gergely and Zsolt Bálint. 2020. "Considerations on the Historic Context Surrounding the Publications Attributed to Denis & Schiffermüller." *Nota Lepidopterologica* 43: 173–179. https://doi.org/10.3897/nl.43.48374.

Vane-Wright, Richard I. 2021. "William Jones & the Birth of the Icones." In *Iconotypes. A compendium of Butterflies & Moths or: Jones's Icones Complete*, Oxford University Museum (ed.). London; Oxford: Thames & Hudson, 10–36.

Wagner, Wolfgang. 2005–2024. *Schmetterlinge und ihre Ökologie*. Accessed June 17. http://www.pyrgus.de/.

Wikström, Bo Peter Huemer, Marko Mutanen, Juha Tyllinen and Lauri Kaila. 2020. "*Pyralis cardinalis*, a Charismatic New Species Related to *P. regalis* [Denis & Schiffermüller], 1775, first recognized in Finland (Lepidoptera, Pyralidae)." *Nota Lepidopterologica* 43: 337–364. https://doi.org/10.3897/nl.43.54916.

Zilli, Alberto. 2021a. "A Flourishing of Lepidopterological Activity." In *Iconotypes. A Compendium of Butterflies & Moths or: Jones's Icones Complete*, Oxford University Museum (ed.). London, Oxford: Thames & Hudson, 265–274.

Zilli, Alberto. 2021b. "The Early Study of Lepidoptera." In *Iconotypes. A Compendium of Butterflies & Moths or: Jones's Icones Complete*, Oxford University Museum (ed.). London, Oxford: Thames & Hudson, 133–139.

Appendix: List of Current Scientific Species Names of the Caterpillars Depicted in Schiffermüller's Watercolors

Usage instructions:

- **Plate:** refers to Schiffermüller's "Tab.", the original Roman numerals were converted to Arabic numerals
- **No:** refers to letter on the plate (the order follows Higgins 1959)
- **Current species name:** current scientific species name
- **Original species name:** species name according to first description
- **Valid (val.):** an 'x' indicates that a species name assigned by Denis and Schiffermüller is still valid
- **Synonym (syn.):** an 'x' indicates that a synonym assigned by Denis and Schiffermüller exists (name not included in the list for spatial reasons)
- **Additional notes:** comments and additional information (The handwritten notes on the plates were attributed to Hübner by Higgins 1959)
- Square brackets mean, that the year, name, or year and name, was unclear or taken from external sources than from the original description. (International Commission on Zoological Nomenclature 1961, 22, recommendation 22A and 51, recommendation 51A)

Plate	no.	Current species name	Original species name	Val	Syn	Additional notes
1	A.	*Parnassius apollo* (Linnaeus, 1758)	*Papilio apollo* Linnaeus, 1758			
1	B.	*Satyrium pruni* (Linnaeus, 1758)	*Papilio pruni* Linnaeus, 1758			
1	C.	*Thecla betulae* (Linnaeus, 1758)	*Papilio betulae* Linnaeus, 1758			
1	D.	*Favonius quercus* (Linnaeus, 1758)	*Papilio quercus* Linnaeus, 1758			
1	E.	*Apoda limacodes* (Hufnagel, 1766)	*Bombyx testudo* [Denis & Schiffermüller, 1775]		x	
1	F.	*Glaucopsyche alexis* (Poda, 1761)	*Papilio alexis* Poda, 1761		x	
1	G.	*Heterogenea asella* ([Denis & Schiffermüller], 1775)	*Bombyx asella* [Denis & Schiffermüller, 1775]	x		
1	H.	*Apotomis lineana* ([Denis & Schiffermüller], 1775)	*Tortrix lineana* [Denis & Schiffermüller, 1775]	x		
1	I.	*Hesperia comma* (Linnaeus, 1758)	*Papilio comma* Linnaeus, 1758			
1	P.	*Pechipogo strigilata* (Linnaeus, 1758)	*Phalaena strigilata* Linnaeus, 1758			
3	A.	*Hipparchia fagi* (Scopoli, 1763)	*Papilio fagi* Scopoli, 1763			Higgins notes "hermione". The question of whether this refers to *Papilio hermione* Linnaeus, 1764 or *Papilio alcyone* [Denis & Schiffermüller], 1775 has been resolved by Russell & Vane-Wright 2022.
3	B.	*Brintesia circe* (Fabricius, 1775)	*Papilio proserpina* [Denis & Schiffermüller, 1775]		x	

(*cont.*)

Plate	no.	Current species name	Original species name	Val	Syn	Additional notes
3	C.	*Aphantopus hyperantus* (Linnaeus, 1758)	*Papilio hyperantus* Linnaeus, 1758			
3	D.	*Lopinga achine* (Scopoli, 1763)	*Papilio achine* Scopoli, 1763			
3	E.	*Coenonympha arcania* (Linnaeus, 1761)	*Papilio arcania* Linnaeus, 1761			
3	F.	*Melanargia galathea* (Linnaeus, 1758)	*Papilio galathea* Linnaeus, 1758			
3	G.	*Colias hyale* (Linnaeus, 1758)	*Papilio hyale* Linnaeus, 1758			
3	H.	*Pontia daplidice* (Linnaeus, 1758)	*Papilio daplidice* Linnaeus, 1758			
3	I.	*Erebia medusa* ([Denis & Schiffermüller], 1775)	*Papilio medusa* [Denis & Schiffermüller, 1775]			
3	K.	*Minois dryas* (Scopoli, 1763)	*Papilio dryas* Scopoli, 1763			
3	F2	*Melanargia galathea* (Linnaeus, 1758)	*Papilio galathea* Linnaeus, 1758			
4	A.	*Fabriciana niobe* (Linnaeus, 1758)	*Papilio niobe* Linnaeus, 1758			
4	B.	*Euphydryas maturna* (Linnaeus, 1758)	*Papilio maturna* Linnaeus, 1758			
4	C.	*Melitaea athalia* (Rottemburg, 1775)	*Papilio athalia* Rottemburg, 1775			
4	D.	*Melitaea cinxia* (Linnaeus, 1758)	*Papilio cinxia* Linnaeus, 1758		x	

Plate	no.	Current species name	Original species name	Val	Syn	Additional notes
4	E.	*Fabriciana adippe* ([Denis & Schiffermüller], 1775)	*Papilio adippe* [Denis & Schiffermüller, 1775]	x		
4	F.	*Melitaea phoebe* ([Denis & Schiffermüller], 1775)	*Papilio phoebe* [Denis & Schiffermüller, 1775]	x		
4	G.	*Fabriciana niob*e (Linnaeus, 1758)	*Papilio niobe* Linnaeus, 1758			
4	H.	*Boloria dia* (Linnaeus, 1767)	*Papilio dia* Linnaeus, 1767			
5	A.	*Callophrys rubi* (Linnaeus, 1758)	*Papilio rubi* Linnaeus, 1758			
5	B.	*Glaucopsyche alexis* (Poda, 1761)	*Papilio alexis* Poda, 1761		x	see also 1F.
5	C.	*Plebejus argus* (Linnaeus, 1758)	*Papilio argus* Linnaeus, 1758		x	
5	D.	*Polyommatus damon* ([Denis & Schiffermüller], 1775)	*Papilio damon* [Denis & Schiffermüller, 1775]	x		
5	E.	*Lycaena virgaureae* (Linnaeus, 1758)	*Papilio virgaureae* Linnaeus, 1758			
5	F.	*Plebejus argus* (Linnaeus, 1758)	*Papilio argus* Linnaeus, 1758			see also 5C.
5	G.	*Satyrium spini* ([Denis & Schiffermüller], 1775)	*Papilio spini* [Denis & Schiffermüller, 1775]	x		G is not in Higgins's list.
6	A.	*Lygephila lusoria* (Linnaeus, 1758)	*Phalaena lusoria* Linnaeus, 1758			
6	B.	*Cucullia lactucae* ([Denis & Schiffermüller], 1775)	*Noctua lactucae* [Denis & Schiffermüller, 1775]	x		
6	C.	*Heliothis viriplaca* (Hufnagel, 1766)	*Phalaena viriplaca* Hufnagel, 1766			
6	D.	*Hadena perplexa* ([Denis & Schiffermüller], 1775)	*Noctua perplexa* [Denis & Schiffermüller, 1775]	x		

(*cont.*)

Plate	no.	Current species name	Original species name	Val	Syn	Additional notes
6	E.	*Abrostola asclepiadis* ([Denis & Schiffermüller], 1775)	*Noctua asclepiadis* [Denis & Schiffermüller, 1775]	x		
6	F.	*Protoschinia scutosa* ([Denis & Schiffermüller], 1775)	*Noctua scutosa* [Denis & Schiffermüller, 1775]	x		
6	G.	*Melanchra persicariae* (Linnaeus, 1761)	*Phalaena persicariae* Linnaeus, 1761			
6	H.	*Periphanes delphinii* (Linnaeus, 1758)	*Phalaena delphinii* Linnaeus, 1758			
7	A.	*Melitaea trivia* ([Denis & Schiffermüller], 1775)	*Papilio trivia* [Denis & Schiffermüller, 1775]	x		
7	B.	*Melitaea cinxia* (Linnaeus, 1758)	*Papilio cinxia* Linnaeus, 1758		x	
7	C.	*Nymphalis antiopa* (Linnaeus, 1758)	*Papilio antiopa* Linnaeus, 1758			
7	D.	*Brenthis daphne* ([Denis & Schiffermüller], 1775)	*Papilio daphne* [Denis & Schiffermüller, 1775]	x		
7	E.	*Speyeria aglaja* (Linnaeus, 1758)	*Papilio aglaja* Linnaeus, 1758			
7	F.	*Aglais io* (Linnaeus, 1758)	*Papilio io* Linnaeus, 1758			
7	G.	*Araschnia levana* (Linnaeus, 1758)	*Papilio levana* Linnaeus, 1758			Hübner notes "Prorsa," which formely stood for the summer generation.
7	H.	*Melitaea phoebe* ([Denis & Schiffermüller], 1775)	*Papilio phoebe* [Denis & Schiffermüller, 1775]	x		
7	K.	*Argynnis paphia* (Linnaeus, 1758)	*Papilio paphia* Linnaeus, 1758			

Plate	no.	Current species name	Original species name	Val	Syn	Additional notes
8	A.1.	*Aglia tau* (Linnaeus, 1758)	*Phalaena tau* Linnaeus, 1758			
8	A.2.	*Aglia tau* (Linnaeus, 1758)	*Phalaena tau* Linnaeus, 1758			
8	B.	*Endromis versicolora* (Linnaeus, 1758)	*Phalaena versicolora* Linnaeus, 1758			
8	C.	*Drymonia querna* ([Denis & Schiffermüller], 1775)	*Bombyx querna* [Denis & Schiffermüller, 1775]	x		Not identified by Higgins; Hübner notes "velitaris" with a question mark; there are several similar-looking *Drymonia* sp., but because of the more or less reticulated painted head and interrupted sideline, this should be *Drymonia querna* ([Denis & Schiffermüller], 1775).
9	A.	*Colias alfacariensis* Ribbe, 1905	*Colias alfacariensis* Ribbe, 1905			This caterpillar is not "Palaeno"/P. palaeno (as noted by Hübner and Higgins), but *Colias alfacariensis*, described as a new species by Ribbe in 1905.
11	A.	*Hyles gallii* (Rottemburg, 1775)	*Sphinx gallii* Rottemburg, 1775		x	
11	B.	*Macroglossum stellatarum* (Linnaeus, 1758)	*Sphinx stellatarum* Linnaeus, 1758			
11	B (2)	*Macroglossum stellatarum* (Linnaeus, 1758)	*Sphinx stellatarum* Linnaeus, 1758			

(*cont.*)

Plate	no.	Current species name	Original species name	Val	Syn	Additional notes
11	C.	*Deilephila porcellus* (Linnaeus, 1758)	*Sphinx porcellus* Linnaeus, 1758			
11	C (2)	*Deilephila porcellus* (Linnaeus, 1758)	*Sphinx porcellus* Linnaeus, 1758			
11	F.	*Agrius convolvuli* (Linnaeus, 1758)	*Sphinx convolvuli* Linnaeus, 1758			
12	A (3)	*Marumba quercus* ([Denis & Schiffermüller], 1775)	*Sphinx quercus* [Denis & Schiffermüller, 1775]	x		
12	A (4)	*Marumba quercus* ([Denis & Schiffermüller], 1775)	*Sphinx quercus* [Denis & Schiffermüller, 1775]	x		
12	B.	*Sphinx ligustri* Linnaeus, 1758	*Sphinx ligustri* Linnaeus, 1758			
13	A (a)	*Laothoe populi* (Linnaeus, 1758)	*Sphinx populi* Linnaeus, 1758			
13	A (b)	*Laothoe populi* (Linnaeus, 1758)	*Sphinx populi* Linnaeus, 1758			
13	B.	*Acherontia atropos* (Linnaeus, 1758)	*Sphinx atropos* Linnaeus, 1758			
13	C.	*Hemaris fuciformis* (Linnaeus, 1758)	*Sphinx fuciformis* Linnaeus, 1758			
13	D.	*Hyles gallii* (Rottemburg, 1775)	*Sphinx gallii* Rottemburg, 1775		x	Younger stage and color variant of 11A.
14	A (1)	*Daphnis nerii* (Linnaeus, 1758)	*Sphinx nerii* Linnaeus, 1758			
14	A (2)	*Daphnis nerii* (Linnaeus, 1758)	*Sphinx nerii* Linnaeus, 1758			
14	A (3)	*Daphnis nerii* (Linnaeus, 1758)	*Sphinx nerii* Linnaeus, 1758			
15	B (1)	*Proserpinus proserpina* (Pallas, 1772)	*Sphinx proserpina* Pallas, 1772		x	
15	B (2)	*Proserpinus proserpina* (Pallas, 1772)	*Sphinx proserpina* Pallas, 1772		x	
15	B (b)	*Proserpinus proserpina* (Pallas, 1772)	*Sphinx proserpina* Pallas, 1772		x	
15	D (2)	*Deilephila elpenor* (Linnaeus, 1758)	*Sphinx elpenor* Linnaeus, 1758			

Plate	no.	Current species name	Original species name	Val	Syn	Additional notes
15	D (3)	*Deilephila elpenor* (Linnaeus, 1758)	*Sphinx elpenor* Linnaeus, 1758			
15	D (b)	*Deilephila elpenor* (Linnaeus, 1758)	*Sphinx elpenor* Linnaeus, 1758			
16	A.	*Tethea or* ([Denis & Schiffermüller], 1775)	*Noctua or* [Denis & Schiffermüller, 1775]	x		
16	B.	*Auchmis detersa* (Esper, [1791])	*Phalaena detersa* Esper, [1791]		x	This could be *Noctua comma* [Denis & Schiffermüller], 1775, a synonym of *Auchmis detersa* (Esper, [1791]), also because of Higgins's reference to the footplant barberry.
16	C.	*Cymatophorina diluta* ([Denis & Schiffermüller], 1775)	*Noctua diluta* [Denis & Schiffermüller, 1775]	x		
17	A (a)	*Acherontia atropos* (Linnaeus, 1758)	*Sphinx atropos* Linnaeus, 1758			
18	A.	*Zygaena filipendulae* (Linnaeus, 1758)	*Sphinx filipendulae* Linnaeus, 1758			
18	A (b)	*Zygaena filipendulae* (Linnaeus, 1758)	*Sphinx filipendulae* Linnaeus, 1758			
18	B.	*Amata phegea* (Linnaeus, 1758)	*Sphinx phegea* Linnaeus, 1758			
18	C.	*Rhagades pruni* ([Denis & Schiffermüller], 1775)	*Sphinx pruni* [Denis & Schiffermüller, 1775]	x		
18	D.	*Spiris striata* (Linnaeus, 1758)	*Phalaena striata* Linnaeus, 1758			

(*cont.*)

Plate	no.	Current species name	Original species name	Val	Syn	Additional notes
18	E.	*Utetheisa pulchella* (Linnaeus, 1758)	*Phalaena pulchella* Linnaeus, 1758		x	There is no note from Hübner here; Higgins published N. pulchra, so this species should be *Noctua pulchra* [Denis & Schiffermüller], 1775, a synonym of *Utetheisa pulchella* (Linnaeus, 1758). Whether this determination is correct could not be verified.
18	F.	*Tyria jacobaeae* (Linnaeus, 1758)	*Phalaena jacobaeae* Linnaeus, 1758			
18	G.	*Eilema complana* (Linnaeus, 1758)	*Phalaena complana* Linnaeus, 1758			
18	H.	*Conistra rubiginea* ([Denis & Schiffermüller], 1775)	*Noctua rubiginea* [Denis & Schiffermüller, 1775]	x		
18	I.	*Lithosia quadra* (Linnaeus, 1758)	*Phalaena quadra* Linnaeus, 1758			
18	K.	*Zygaena ephialtes* (Linnaeus, 1767)	*Sphinx ephialtes* Linnaeus, 1767		x	
18	L.	*Acronicta megacephala* ([Denis & Schiffermüller], 1775)	*Noctua megacephala* [Denis & Schiffermüller, 1775]	x		
18	M.	*Zygaena loti* ([Denis & Schiffermüller], 1775)	*Sphinx loti* [Denis & Schiffermüller, 1775]	x		
18	N.	*Zygaena ephialtes coronillae* ([Denis & Schiffermüller], 1775)	*Sphinx coronillae* [Denis & Schiffermüller, 1775]	x		

Plate	no.	Current species name	Original species name	Val	Syn	Additional notes
18	O.	*Adscita statices* (Linnaeus, 1758)	*Sphinx statices* Linnaeus, 1758			
19	A.	*Zygaena viciae* ([Denis & Schiffermüller], 1775)	*Sphinx viciae* [Denis & Schiffermüller, 1775]	x		
19	B.	*Zygaena carniolica* (Scopoli, 1763)	*Sphinx carniolica* Scopoli, 1763		x	
19	C.	*Adscita statices* (Linnaeus, 1758)	*Sphinx statices* Linnaeus, 1758			
19	D.	*Zygaena minos* ([Denis & Schiffermüller], 1775)	*Sphinx minos* [Denis & Schiffermüller, 1775]	x		
21	A (a)	*Poecilocampa populi* (Linnaeus, 1758)	*Phalaena populi* Linnaeus, 1758			
21	A (b)	*Poecilocampa populi* (Linnaeus, 1758)	*Phalaena populi* Linnaeus, 1758			Younger stage.
21	B (1)	*Gastropacha quercifolia* (Linnaeus, 1758)	*Phalaena quercifolia* Linnaeus, 1758			
21	B (3a)	*Gastropacha quercifolia* (Linnaeus, 1758)	*Phalaena quercifolia* Linnaeus, 1758			
22	A.	*Macrothylacia rubi* (Linnaeus, 1758)	*Phalaena rubi* Linnaeus, 1758			
22	B (1)	*Lasiocampa quercus* (Linnaeus, 1758)	*Phalaena quercus* Linnaeus, 1758			
22	B (2)	*Lasiocampa quercus* (Linnaeus, 1758)	*Phalaena quercus* Linnaeus, 1758			
23	A.	*Dendrolimus pini* (Linnaeus, 1758)	*Phalaena pini* Linnaeus, 1758			
23	B.	*Phyllodesma ilicifolia* (Linnaeus, 1758)	*Phalaena (Bombyx) ilicifolia* Linnaeus, 1758			
23	C.	*Odonestis pruni* (Linnaeus, 1758)	*Phalaena pruni* Linnaeus, 1758			

(*cont.*)

Plate	no.	Current species name	Original species name	Val	Syn	Additional notes
23	D.	*Calliteara abietis* ([Denis & Schiffermüller], 1775)	*Bombyx abietis* [Denis & Schiffermüller, 1775]	x		
24	A.	*Saturnia pyri* ([Denis & Schiffermüller], 1775)	*Bombyx pyri* [Denis & Schiffermüller, 1775]	x		
25	A (2)	*Saturnia pyri* ([Denis & Schiffermüller], 1775)	*Bombyx pyri* [Denis & Schiffermüller, 1775]	x		
25	A (3)	*Saturnia pyri* ([Denis & Schiffermüller], 1775)	*Bombyx pyri* [Denis & Schiffermüller, 1775]	x		
25	A (5)	*Saturnia pyri* ([Denis & Schiffermüller], 1775)	*Bombyx pyri* [Denis & Schiffermüller, 1775]	x		
25	B (1)	*Saturnia spini* ([Denis & Schiffermüller], 1775)	*Bombyx spini* [Denis & Schiffermüller, 1775]	x		
25	B (2)	*Saturnia spini* ([Denis & Schiffermüller], 1775)	*Bombyx spini* [Denis & Schiffermüller, 1775]	x		
25	B (3)	*Saturnia spini* ([Denis & Schiffermüller], 1775)	*Bombyx spini* [Denis & Schiffermüller, 1775]	x		
26	F (2–4)	*Saturnia pavonia* (Linnaeus, 1758)	*Phalaena pavonia* Linnaeus, 1758		x	
27	A.	*Clostera anastomosis* (Linnaeus, 1758)	*Phalaena anastomosis* Linnaeus, 1758			
27	B.	*Dysauxes ancilla* (Linnaeus, 1767)	*Phalaena ancilla* Linnaeus, 1767			

Plate	no.	Current species name	Original species name	Val	Syn	Additional notes
27	C.	*Epatolmis luctifera* ([Denis & Schiffermüller], 1775)	*Bombyx luctifera* [Denis & Schiffermüller, 1775]	x		Hübner notes "Luctifera"; given that the drawing does not seem to have been completed and that there is no reference to Hübner's plate, this determination cannot be conclusively verified.
27	D.	*Spilosoma lubricipeda* (Linnaeus, 1758)	*Phalaena lubricipeda* Linnaeus, 1758		x	Hübner notes "Luctifera, Lubricipeda or Aulica"; none of those species look similar to the painted caterpillar. This species could not be clearly identified.
27	E.	*Eilema palliatella* (Scopoli, 1763)	*Phalaena palliatella* Scopoli, 1763		x	
27	F.	*Thaumetopoea pityocampa* ([Denis & Schiffermüller], 1775)	*Bombyx pityocamps* [Denis & Schiffermüller, 1775]	x		
27	G.	*Eilema depressa* (Esper, [1787])	*Noctua depressa* Esper, [1787]			Hübner notes "Marginea" with a question mark; there is also a note "unita" and Higgins states it is N. complana. In fact, it seems clear that this species is *Eilema depressa* (Esper, [1787]).
27	H.	*Arctia plantaginis* (Linnaeus, 1758)	*Phalaena plantaginis* Linnaeus, 1758			

(*cont.*)

Plate	no.	Current species name	Original species name	Val	Syn	Additional notes
27	L. (= I.)	*Arctia caja* (Linnaeus, 1758)	*Phalaena caja* Linnaeus, 1758			Hübner notes "Unita" (the rest is not well readable). Higgins does not identify the species, but notes "Hübner secut = B. caja juv.", so it could be an early stage of *Arctia caja* (Linnaeus, 1758).
28	A.	*Eriogaster catax* (Linnaeus, 1758)	*Phalaena catax* Linnaeus, 1758			Higgins identifies this species as B. lanestris, but see 28B.
28	B.	*Eriogaster catax* (Linnaeus, 1758)	*Phalaena catax* Linnaeus, 1758			Not identified by Higgins. Hübner notes "Catax" for A. and "Everia" for B. (*Phalaena everia* Knoch, 1781 is a synonym for *Eriogaster catax*). There is a second note in the middle, probably referring to both 28A and B, that reads "Catax 42.b."
28	G.	*Malacosoma neustria* (Linnaeus, 1758)	*Phalaena neustria* Linnaeus, 1758			Hübner notes "Neustria", whereas Higgins identifies this one as B. castrensis (like 28F). Hübner's note seems to be correct.

Plate	no.	Current species name	Original species name	Val	Syn	Additional notes
28	F.	*Malacosoma castrensis* (Linnaeus, 1758)	*Phalaena castrensis* Linnaeus, 1758			There is a note "nicht fertig" [unfinished] by Hübner, but Higgins identifies this drawing as B. castrensis L. (like 28G).
28	T.	*Lasiocampa trifolii* ([Denis & Schiffermüller], 1775)	*Bombyx trifolii* [Denis & Schiffermüller, 1775]	x		Not identified by Higgins. Hübner notes "Trifolii" with a question mark, with which the drawing does not have very much in common. It is possible that the drawing is not completely finished.
28	S (1)	*Lasiocampa quercus* (Linnaeus, 1758)	*Phalaena quercus* Linnaeus, 1758			Not identified by Higgins. Hübner notes "Quercus" with a question mark, which seems to be the correct determination.
28	S (2)	*Lasiocampa quercus* (Linnaeus, 1758)	*Phalaena quercus* Linnaeus, 1758			Like 28S (1).
28	K.	currently not determinable	currently not determinable			
28	Z.	currently not determinable	currently not determinable			
29	A.	*Trichiura crataegi* (Linnaeus, 1758)	*Phalaena crataegi* Linnaeus, 1758			
29	B.	*Trichiura crataegi* (Linnaeus, 1758)	*Phalaena crataegi* Linnaeus, 1758			
29	C.	*Trichiura crataegi* (Linnaeus, 1758)	*Phalaena crataegi* Linnaeus, 1758			
29	P.	*Acronicta auricoma* ([Denis & Schiffermüller], 1775)	*Noctua auricoma* [Denis & Schiffermüller, 1775]	x		
29	T.	*Lemonia dumi* (Linnaeus, 1761)	*Phalaena dumi* Linnaeus, 1761			

(*cont.*)

Plate	no.	Current species name	Original species name	Val	Syn	Additional notes
29	Z.	*Lemonia taraxaci* ([Denis & Schiffermüller], 1775)	*Bombyx taraxaci* [Denis & Schiffermüller, 1775]	x		
30	A (2)	*Arctia caja* (Linnaeus, 1758)	*Phalaena caja* Linnaeus, 1758			Schiffermüller's numbering is not continuous on plate 30; there is no A, D, D1, and D2. Moreover, species with the same numbers exhibit different notes, which makes it difficult to determine the drawings correctly.
30	A (3)	*Arctia caja* (Linnaeus, 1758)	*Phalaena caja* Linnaeus, 1758			
30	B (a)	*Spilosoma lubricipeda* (Linnaeus, 1758)	*Phalaena lubricipeda* Linnaeus, 1758		x	30B (a) notes "fuliginosa," (b) "Menthastri, Lubricep."; Higgins calls B. (a) "menthastri" and does not identify (b). As Schiffermüller provides both drawings with a number, we can assume that those are variations or different stages of one species (but see before), which could be *Bombyx menthastri* [Denis & Schiffermüller], 1775, a synonym of *Phalaena lubricipeda* (Linnaeus, 1758).
30	C.	*Arctia villica* (Linnaeus, 1758)	*Phalaena villica* Linnaeus, 1758			

Plate	no.	Current species name	Original species name	Val	Syn	Additional notes
30	E (a)	*Diacrisia purpurata* (Linnaeus, 1758)	*Phalaena purpurata* Linnaeus, 1758			
30	E (b)	*Diacrisia purpurata* (Linnaeus, 1758)	*Phalaena purpurata* Linnaeus, 1758			
30	B (b)	*Spilosoma lubricipeda* (Linnaeus, 1758)	*Phalaena lubricipeda* Linnaeus, 1758		x	See 30B (a).
30	C (2)	*Arctia caja* (Linnaeus, 1758)	*Phalaena caja* Linnaeus, 1758			Juvenile caterpillar, the fully grown caterpillar is shown under C(3)
30	C (3)	*Arctia caja* (Linnaeus, 1758)	*Phalaena caja* Linnaeus, 1758			Not listed by Higgins.
30	D (3)	*Arctia festiva* (Hufnagel, 1766)	*Phalaena festiva* Hufnagel, 1766			Hübner notes "Hebe"; *Bombyx hebe* (Linnaeus, 1758) is a synonym.
30	D (4)	*Arctia festiva* (Hufnagel, 1766)	*Phalaena festiva* Hufnagel, 1766			We are not sure if this drawing is actually one of the two species "Caja" or "Hebe". It is probable that this drawing is not finished.
31	A.	*Callimorpha dominula* (Linnaeus, 1758)	*Phalaena dominula* Linnaeus, 1758			
31	B (1)	*Euplagia quadripunctaria* (Poda, 1761)	*Phalaena quadripunctaria* Poda, 1761			
31	B (2)	*Euplagia quadripunctaria* (Poda, 1761)	*Phalaena quadripunctaria* Poda, 1761			

(cont.)

Plate	no.	Current species name	Original species name	Val	Syn	Additional notes
31	H.	*Phalera bucephala* (Linnaeus, 1758)	*Phalaena bucephala* Linnaeus, 1758			
31	Z.	*Pieris brassicae* (Linnaeus, 1758)	*Papilio brassicae* Linnaeus, 1758			
31	M.	*Acronicta euphorbiae* ([Denis & Schiffermüller], 1775)	*Noctua euphorbiae* [Denis & Schiffermüller, 1775]	x		
31	R.	*Acronicta rumicis* (Linnaeus, 1758)	*Phalaena rumicis* Linnaeus, 1758			
31	F.	*Leucoma salicis* (Linnaeus, 1758)	*Phalaena salicis* Linnaeus, 1758			
31	G.	*Moma alpium* (Osbeck, 1778)	*Phalaena alpium* Osbeck, 1778		x	
31	X.	*Apaustis rupicola* ([Denis & Schiffermüller], 1775)	*Noctua rupicola* [Denis & Schiffermüller, 1775]			Not identified by Higgins. The meaning of Hübner's note "Ypsilon" is unclear; The species depicted could be *Noctua rupicola* [Denis & Schiffermüller], 1775, now *Apaustis rupicola* ([Denis & Schiffermüller], 1775).
31	K.	*Acronicta tridens* ([Denis & Schiffermüller], 1775)	*Noctua tridens* [Denis & Schiffermüller, 1775]	x		
31	L.	*Acronicta psi* (Linnaeus, 1758)	*Phalaena psi* Linnaeus, 1758			
31	Q.	*Lymantria monacha* (Linnaeus, 1758)	*Phalaena monacha* Linnaeus, 1758			
32	A.	*Ptilodon capucina* (Linnaeus, 1758)	*Phalaena capucina* Linnaeus, 1758			

Plate	no.	Current species name	Original species name	Val	Syn	Additional notes
32	B.	*Craniophora ligustri* ([Denis & Schiffermüller], 1775)	*Noctua ligustri* [Denis & Schiffermüller, 1775]	x		
32	C.	*Brachionycha nubeculosa* (Esper, [1785])	*Bombyx nubeculosa* Esper, [1785]			
32	D.	*Dryobotodes eremita* (Fabricius, 1775)	*Noctua eremita* Fabricius, 1775		x	
32	E.	*Acronicta leporina* (Linnaeus, 1758)	*Phalaena leporina* Linnaeus, 1758			
32	F.	*Pheosia tremula* (Clerck, 1759)	*Bombyx tremula* Clerck, 1759			
32	G.	*Euplexia lucipara* (Linnaeus, 1758)	*Phalaena lucipara* Linnaeus, 1758			
33	A.	*Orgyia recens* (Hübner, [1819])	*Gynaephora recens* Hübner, [1819]			
33	B.	*Orgyia antiqua* (Linnaeus, 1758)	*Phalaena antiqua* Linnaeus, 1758			
33	C.	*Gynaephora fascelina* (Linnaeus, 1758)	*Phalaena fascelina* Linnaeus, 1758			
33	D.	*Calliteara pudibunda* (Linnaeus, 1758)	*Phalaena pudipunda* Linnaeus, 1758			
33	E.	*Colocasia coryli* (Linnaeus, 1758)	*Phalaena coryli* Linnaeus, 1758			
33	G.	*Clostera curtula* (Linnaeus, 1758)	*Phalaena curtula* Linnaeus, 1758			
33	I. (1)	*Clostera pigra* (Hufnagel, 1766)	*Phalaena pigra* Hufnagel, 1766		x	
33	F.	*Clostera anachoreta* ([Denis & Schiffermüller], 1775)	*Bombyx anachoreta* [Denis & Schiffermüller, 1775]	x		
33	K.	*Euthrix potatoria* (Linnaeus, 1758)	*Phalaena potatoria* Linnaeus, 1758			
35	A.	*Cossus cossus* (Linnaeus, 1758)	*Phalaena cossus* Linnaeus, 1758			
35	B.	*Notodonta ziczac* (Linnaeus, 1758)	*Phalaena ziczac* Linnaeus, 1758			

(*cont.*)

Plate	no.	Current species name	Original species name	Val	Syn	Additional notes
35	S.	*Achlya flavicornis* (Linnaeus, 1758)	*Phalaena flavicornis* Linnaeus, 1758			Higgins notes "(olive yellow)"; see also 35R.
35	L.	*Catocala promissa* ([Denis & Schiffermüller], 1775)	Noctua promissa [Denis & Schiffermüller, 1775]	x		
35	Z.	*Tenthredo* sp. Linnaeus, 1758	*Tenthredo* sp. Linnaeus, 1758			Sawfly species.
35	R.	*Achlya flavicornis* (Linnaeus, 1758)	*Phalaena flavicornis* Linnaeus, 1758			Hübner's note "Contigua" with a question mark refers to *Lacanobia contigua* ([Denis & Schiffermüller], 1775), formerly *Noctua contigua* [Denis & Schiffermüller], 1775, but Higgins defines it as a "green" variant of *Achyla flavicornis* (Linnaeus, 1785) – see also 35S. Whether this determination is correct could not be verified.
35	T.	*Utetheisa pulchella* (Linnaeus, 1758)	*Phalaena pulchella* Linnaeus, 1758		x	
35	M.	*Catocala pacta* (Linnaeus, 1758)	*Phalaena pacta* Linnaeus, 1758			Hübner notes "Pacta" with a question mark, Higgins N. pacta. Whether this determination is correct could not be verified. It could also be *Catocala elocata* (Esper, [1787]); however, this species does not occur in the *Verzeichnis*.

Plate	no.	Current species name	Original species name	Val	Syn	Additional notes
35	N.	*Catocala nupta* (Linnaeus, 1767)	*Phalaena nupta* Linnaeus, 1767			
35	E.	*Cerura vinula* (Linnaeus, 1758)	*Phalaena vinula* Linnaeus, 1758			
35	F.	*Furcula bicuspis* (Borkhausen, 1790)	*Bombyx bicuspis* Borkhausen, 1790			Hübner notes "bicuspis", Higgins B. bicuspis. The colors do not match, possibly the drawing is not finished.
35	G.	*Furcula furcula* (Clerck, 1759)	*Phalaena (Bombyx) furcula* Clerck, 1759			
36	A.	*Cilix glaucata* (Scopoli, 1763)	*Phalaena glaucata* Scopoli, 1763		x	
36	B.	*Drepana falcataria* (Linnaeus, 1758)	*Phalaena falcataria* Linnaeus, 1758		x	
36	C.	*Falcaria lacertinaria* (Linnaeus, 1758)	*Phalaena lacertinaria* Linnaeus, 1758		x	
36	D (1)	*Dicranura ulmi* ([Denis & Schiffermüller], 1775)	*Noctua ulmi* [Denis & Schiffermüller, 1775]	x		
36	D (2)	*Dicranura ulmi* ([Denis & Schiffermüller], 1775)	*Noctua ulmi* [Denis & Schiffermüller, 1775]	x		
36	e.	*Notodonta tritophus* ([Denis & Schiffermüller], 1775)	*Bombyx tritophus* [Denis & Schiffermüller, 1775]	x		
36	E.	*Notodonta dromedarius* (Linnaeus, 1767)	*Phalaena dromedarius* Linnaeus, 1767			
36	F.	*Harpyia milhauseri* (Fabricius, 1775)	*Bombyx milhauseri* Fabricius, 1775		x	
36	G.	*Watsonalla binaria* (Hufnagel, 1767)	*Phalaena binaria* Hufnagel, 1767			
36	H.	*Thyatira batis* (Linnaeus, 1758)	*Phalaena batis* Linnaeus, 1758			

(*cont.*)

Plate	no.	Current species name	Original species name	Val	Syn	Additional notes
36	I.	*Euclidia glyphica* (Linnaeus, 1758)	*Phalaena glyphica* Linnaeus, 1758			
37	A.	*Valeria oleagina* ([Denis & Schiffermüller], 1775)	*Bombyx oleagina* [Denis & Schiffermüller, 1775]	x		
37	B.	*Spatalia argentina* ([Denis & Schiffermüller], 1775)	*Bombyx argentina* [Denis & Schiffermüller, 1775]	x		There is a pupa without number (belongs to the caterpillar 37B).
37	C.	*Polia nebulosa* (Hufnagel, 1766)	*Phalaena nebulosa* Hufnagel, 1766			This drawing probably is of *Polia nebulosa* (Hufnagel, 1766), syn. *Noctua plebeja* Hübner, [1803]. Noted by Hübner and Higgins, *Phalaena polyodon* Clerck, 1759, now *Actinotia polyodon* (Clerck, 1759), looks different.
37	D.	*Falcaria lacertinaria* (Linnaeus, 1758)	*Phalaena lacertinaria* Linnaeus, 1758		x	
37	E.	*Stauropus fagi* (Linnaeus, 1758)	*Phalaena fagi* Linnaeus, 1758			
37	F.	*Lamprosticta culta* ([Denis & Schiffermüller], 1775)	*Noctua culta* [Denis & Schiffermüller, 1775]	x		
37	G.	*Meganephria bimaculosa* (Linnaeus, 1767)	*Phalaena bimaculosa* Linnaeus, 1767			
38	A.	*Diloba caeruleocephala* (Linnaeus, 1758)	*Phalaena caeruleocephala* Linnaeus, 1758			

Plate	no.	Current species name	Original species name	Val	Syn	Additional notes
38	A (2)	*Diloba caeruleocephala* (Linnaeus, 1758)	*Phalaena caeruleocephala* Linnaeus, 1758			
38	B.	*Cucullia tanaceti* ([Denis & Schiffermüller], 1775)	*Noctua tanaceti* [Denis & Schiffermüller, 1775]	x		
38	C.	*Cucullia verbasci* (Linnaeus, 1758)	*Phalaena verbasci* Linnaeus, 1758			
38	D.	*Calophasia lunula* (Hufnagel, 1766)	*Phalaena lunula* Hufnagel, 1766		x	
38	F.	*Abrostola triplasia* (Linnaeus, 1758)	*Phalaena triplasia* Linnaeus, 1758			
38	G.	*Cucullia absinthii* (Linnaeus, 1761)	*Phalaena absinthii* Linnaeus, 1761			
38	H.	*Cucullia chamomillae* ([Denis & Schiffermüller], 1775)	*Noctua chamomillae* [Denis & Schiffermüller, 1775]	x		
38	L.	*Cucullia umbratica* (Linnaeus, 1758)	*Phalaena umbratica* Linnaeus, 1758			
38	M.	*Pachetra sagittigera* (Hufnagel, 1766)	*Phalaena sagittigera* Hufnagel, 1766		x	
38	Y.	*Iphiclides podalirius* (Linnaeus, 1758)	*Papilio podalirius* Linnaeus, 1758			
38	Z.	*Papilio machaon* Linnaeus, 1758	*Papilio machaon* Linnaeus, 1758			
39	A.	*Scoliopteryx libatrix* (Linnaeus, 1758)	*Phalaena libatrix* Linnaeus, 1758			
39	B.	*Cosmia diffinis* (Linnaeus, 1767)	*Phalaena diffinis* Linnaeus, 1767			
39	C.	*Peridea anceps* (Goeze, 1781)	*Phalaena (Noctua) anceps* Goeze, 1781			This one does not seem to be "tremula" as noted by Hübner and Higgins (see 32F), but rather *Peridea anceps* (Goeze, 1781).

(*cont.*)

Plate	no.	Current species name	Original species name	Val	Syn	Additional notes
39	D.	*Orthosia gracilis* ([Denis & Schiffermüller], 1775)	*Noctua gracilis* [Denis & Schiffermüller, 1775]	x		
39	E.	*Drymonia dodonaea* ([Denis & Schiffermüller], 1775)	*Bombyx dodonaea* [Denis & Schiffermüller, 1775]	x		
39	F.	*Drymonia ruficornis* (Hufnagel, 1766)	*Bombyx ruficornis* Hufnagel, 1766		x	Hübner notes "querna" and Higgins B. querna, but this one should be *Drymonia chaonia* ([Denis & Schiffermüller], 1775), a synonym of *Drymonia ruficornis* (Hufnagel, 1766).
39	H.	*Conistra vaccinii* (Linnaeus, 1761)	*Phalaena vaccinii* Linnaeus, 1761		x	
39	I.	*Conistra vaccinii* (Linnaeus, 1761)	*Phalaena vaccinii* Linnaeus, 1761		x	
39	K.	*Agrotis segetum* ([Denis & Schiffermüller], 1775)	*Noctua segetum* [Denis & Schiffermüller, 1775]	x		
39	L.	*Noctua fimbriata* (Schreber, 1759)	*Phalaena fimbriata* Schreber, 1759			
40	A.	*Diachrysia chrysitis* (Linnaeus, 1758)	*Phalaena chrysitis* Linnaeus, 1758			This larva does not seem to be *Cornutiplusia circumflexa* (Linnaeus, 1767) as noted by Hübner and Higgins ("Circumflexa??"/N. circumflexa), but probably *Diachrysia chrysitis* (Linnaeus, 1758).
40	B.	*Sideridis rivularis* (Fabricius, 1775)	*Noctua rivularis* Fabricius, 1775		x	

Plate	no.	Current species name	Original species name	Val	Syn	Additional notes
40	C.	*Polia bombycina* (Hufnagel, 1766)	*Phalaena bombycina* Hufnagel, 1766		x	
40	I.	*Chersotis rectangula* ([Denis & Schiffermüller], 1775)	*Noctua rectangula* [Denis & Schiffermüller, 1775]	x		
40	G.	*Pachetra sagittigera* (Hufnagel, 1766)	*Phalaena sagittigera* Hufnagel, 1766		x	
40	H.	*Agrochola humilis* ([Denis & Schiffermüller], 1775)	*Noctua humilis* [Denis & Schiffermüller, 1775]	x		
40	K.	*Dichagyris signifera* ([Denis & Schiffermüller], 1775)	*Noctua signifera* [Denis & Schiffermüller, 1775]	x		
40	L (1–3)	*Noctua pronuba* (Linnaeus, 1758)	*Phalaena pronuba* Linnaeus, 1758			
40	L (4)	*Noctua pronuba* (Linnaeus, 1758)	*Phalaena pronuba* Linnaeus, 1758			
40	M.	*Lacanobia contigua* ([Denis & Schiffermüller], 1775)	*Noctua contigua* [Denis & Schiffermüller, 1775]	x		
40	N.	*Apamea sordens* (Hufnagel, 1766)	*Phalaena sordens* Hufnagel, 1766			
41	A (a)	*Griposia aprilina* (Linnaeus, 1758)	*Phalaena aprilina* Linnaeus, 1758		x	Although there are three drawings numbered with "A," A(a) is another species (Hübner notes "Runica", which is a synonym of *Griposia aprilina* (L.)), then Ab(1) and Ab(2); see notes there.

(*cont.*)

Plate	no.	Current species name	Original species name	Val	Syn	Additional notes
41	Ab (1)	*Anorthoa munda* ([Denis & Schiffermüller], 1775)	*Noctua munda* [Denis & Schiffermüller, 1775]	x		Hübner notes "munda" with a question mark. This species seems to be *Anorthoa munda* ([Denis & Schiffermüller], 1775) at different stages or variations (see also 41Ab(2)).
41	B.	*Allophyes oxyacanthae* (Linnaeus, 1758)	*Phalaena oxyacanthae* Linnaeus, 1758			
41	C.	*Orthosia miniosa* ([Denis & Schiffermüller], 1775)	*Noctua miniosa* [Denis & Schiffermüller, 1775]	x		
41	D.	*Agrochola helvola* (Linnaeus, 1758)	*Phalaena helvola* Linnaeus, 1758			
41	F.	*Eupsilia transversa* (Hufnagel, 1766)	*Phalaena transversa* Hufnagel, 1766			
41	G.	*Dicycla oo* (Linnaeus, 1758)	*Phalaena oo* Linnaeus, 1758			
41	H.	*Orthosia cruda* ([Denis & Schiffermüller], 1775)	*Noctua cruda* [Denis & Schiffermüller, 1775]	x		
41	I.	*Asphalia ruficollis* ([Denis & Schiffermüller], 1775)	*Noctua ruficollis* [Denis & Schiffermüller, 1775]	x		
41	K.	*Apterogenum ypsillon* ([Denis & Schiffermüller], 1775)	*Noctua ypsillon* [Denis & Schiffermüller, 1775]	x		
41	L.	*Agrotis exclamationis* (Linnaeus, 1758)	*Phalaena (Noctua) exclamationis* Linnaeus, 1758			

Plate	no.	Current species name	Original species name	Val	Syn	Additional notes
41	M.	*Apamea monoglypha* (Hufnagel, 1766)	*Phalaena monoglypha* Hufnagel, 1766		x	
41	Ab (2)	*Anorthoa munda* ([Denis & Schiffermüller], 1775)	*Noctua munda* [Denis & Schiffermüller, 1775]	x		See 41Ab(1).
41	N.	*Cryphia algae* (Fabricius, 1775)	*Noctua algae* Fabricius, 1775		x	
42	A.	*Cosmia trapezina* (Linnaeus, 1758)	*Phalaena trapezina* Linnaeus, 1758			
42	B.	*Lithophane ornitopus* (Hufnagel, 1766)	*Phalaena ornitopus* Hufnagel, 1766		x	
42	C.	*Asteroscopus sphinx* (Hufnagel, 1766)	*Phalaena sphinx* Hufnagel, 1766		x	
42	D.	*Agrochola lychnidis* ([Denis & Schiffermüller], 1775)	*Noctua lychnidis* [Denis & Schiffermüller, 1775]	x		See also 44K. Hübner notes "Pistacina 56. a" and for "Lychnidis 56. b" (in the existing edition of his caterpillar book, the plate number is 62 and the figures are 1.a. and 3.a.). It seems that the specimens belong to the same species but to different subspecies or variants. The current name is *Agrochola lychnidis* ([Denis & Schiffermüller], 1775), synonyms are *Noctua lychnidis* [Denis & Schiffermüller], 1775 and *Noctua pistacina* [Denis & Schiffermüller], 1775.

(*cont.*)

Plate	no.	Current species name	Original species name	Val	Syn	Additional notes
42	E.	*Orthosia cerasi* (Fabricius, 1775)	*Noctua cerasi* Fabricius, 1775		x	
42	F.	*Amphipyra pyramidea* (Linnaeus, 1758)	*Phalaena (Noctua) pyramidae* Linnaeus, 1758			
42	G.	*Autographa gamma* (Linnaeus, 1758)	*Phalaena gamma* Linnaeus, 1758			
42	H.	*Scoliopteryx libatrix* (Linnaeus, 1758)	*Phalaena libatrix* Linnaeus, 1758			
42	I.	*Ptilophora plumigera* ([Denis & Schiffermüller], 1775)	*Bombyx plumigera* [Denis & Schiffermüller, 1775]	x		
42	L.	*Opigena polygona* ([Denis & Schiffermüller], 1775)	*Noctua polygona* [Denis & Schiffermüller, 1775]	x		
42	M.	*Phlogophora meticulosa* (Linnaeus, 1758)	*Phalaena meticulosa* Linnaeus, 1758			
42	N.	*Orthosia incerta* (Hufnagel, 1766)	*Phalaena incerta* Hufnagel, 1766		x	
42	K.	*Pterostoma palpina* (Clerck, 1759)	*Phalaena palpina* Clerck, 1759			
42	P.	*Anarta trifolii* (Hufnagel, 1766)	*Phalaena trifolii* Hufnagel, 1766		x	
42	Q.	*Lacanobia oleracea* (Linnaeus, 1758)	*Phalaena oleracea* Linnaeus, 1758			
43	A.	*Ceramica pisi* (Linnaeus, 1758)	*Phalaena pisi* Linnaeus, 1758			
43	B.	*Panolis flammea* ([Denis & Schiffermüller], 1775)	*Noctua flammea* [Denis & Schiffermüller, 1775]	x		
43	D.	*Agrotis exclamationis* (Linnaeus, 1758)	*Phalaena (Noctua) exclamationis* Linnaeus, 1758			

Plate	no.	Current species name	Original species name	Val	Syn	Additional notes
43	G.	*Caradrina clavipalpis* (Scopoli, 1763)	*Phalaena clavipalpis* Scopoli, 1763		x	Hübner notes "Cubicularis", Higgins N. cubicularis. *Noctua cubicularis* [Denis & Schiffermüller], 1775, a synonym of *Caradrina clavipalpis* (Scopoli, 1763), at a younger stage could be the correct identification.
43	H.	*Mythimna albipuncta* ([Denis & Schiffermüller], 1775)	*Noctua albipuncta* [Denis & Schiffermüller, 1775]	x		
43	I.	*Dichagyris forcipula* ([Denis & Schiffermüller], 1775)	*Noctua forcipula* [Denis & Schiffermüller, 1775]	x		
43	K.	*Mythimna conigera* ([Denis & Schiffermüller], 1775)	*Noctua conigera* [Denis & Schiffermüller, 1775]	x		
43	L.	*Agrochola nitida* ([Denis & Schiffermüller], 1775)	*Noctua nitida* [Denis & Schiffermüller, 1775]	x		
43	M.	*Diarsia mendica* (Fabricius, 1775)	*Noctua mendica* Fabricius, 1775		x	
44	A.	*Hadena capsincola* ([Denis & Schiffermüller], 1775)	*Noctua capsincola* [Denis & Schiffermüller, 1775]	x		
44	B.	*Jodia croceago* ([Denis & Schiffermüller], 1775)	*Noctua croceago* [Denis & Schiffermüller, 1775]	x		
44	C.	*Cucullia chamomillae* ([Denis & Schiffermüller], 1775)	*Noctua chamomillae* [Denis & Schiffermüller, 1775]	x		

(*cont.*)

Plate	no.	Current species name	Original species name	Val	Syn	Additional notes
44	D.	*Cucullia artemisiae* (Hufnagel, 1766)	*Phalaena artemisiae* Hufnagel, 1766		x	
44	E.	currently not determinable	currently not determinable			
44	F (a)	*Dichonia convergens* ([Denis & Schiffermüller], 1775)	*Noctua convergens* [Denis & Schiffermüller, 1775]	x		
44	F (b)	*Dichonia convergens* ([Denis & Schiffermüller], 1775)	*Noctua convergens* [Denis & Schiffermüller, 1775]	x		
44	G.	*Tiliacea citrago* (Linnaeus, 1758)	*Phalaena citrago* Linnaeus, 1758			This is more likely to be *Tiliacea citrago* (Linnaeus, 1758) as in Hübner's note than *Tiliacea aurago* ([Denis & Schiffermüller], 1775) as in Higgins's.
44	H.	*Cerastis rubricosa* ([Denis & Schiffermüller], 1775)	*Noctua rubricosa* [Denis & Schiffermüller, 1775]	x		
44	I.	*Enargia paleacea* (Esper, [1788])	*Phalaena paleacea* Esper, [1788]	x		This one does not look like *Tiliacea sulphurago* ([Denis & Schiffermüller], 1775) as noted by Hübner and Higgins, but could be *Noctua paleacea*, now *Enargia paleacea* (Esper, [1788]).
44	K.	*Agrochola lychnidis* ([Denis & Schiffermüller], 1775)	*Noctua lychnidis* [Denis & Schiffermüller, 1775]	x		See also 42D.

Plate	no.	Current species name	Original species name	Val	Syn	Additional notes
45	A.	*Lygephila craccae* ([Denis & Schiffermüller], 1775)	*Noctua craccae* [Denis & Schiffermüller, 1775]	x		
45	B.	*Minucia lunaris* ([Denis & Schiffermüller], 1775)	*Noctua lunaris* [Denis & Schiffermüller, 1775]	x		
45	C.	*Catocala hymenaea* ([Denis & Schiffermüller], 1775)	*Noctua hymenaea* [Denis & Schiffermüller, 1775]	x		
45	D.	*Catocala fraxini* (Linnaeus, 1758)	*Phalaena fraxini* Linnaeus, 1758			
46	A.	*Xestia baja* ([Denis & Schiffermüller], 1775)	*Noctua baja* [Denis & Schiffermüller, 1775]	x		
46	B.	*Diarsia brunnea* ([Denis & Schiffermüller], 1775)	*Noctua brunnea* [Denis & Schiffermüller, 1775]	x		
46	C.	*Euxoa obelisca* ([Denis & Schiffermüller], 1775)	*Noctua obelisca* [Denis & Schiffermüller, 1775]	x		
46	D.	*Euxoa nigricans* (Linnaeus, 1761)	*Phalaena (Noctua) nigricans* Linnaeus, 1761		x	
46	E.	*Mythimna turca* (Linnaeus, 1761)	*Phalaena turca* Linnaeus, 1761			
46	G.	*Xestia ditrapezium* ([Denis & Schiffermüller], 1775)	*Noctua ditrapezium* [Denis & Schiffermüller, 1775]	x		
46	H.	*Anaplectoides prasina* ([Denis & Schiffermüller], 1775)	*Noctua prasina* [Denis & Schiffermüller, 1775]	x		
46	I.	*Epilecta linogrisea* ([Denis & Schiffermüller], 1775)	*Noctua linogrisea* [Denis & Schiffermüller, 1775]	x		

(*cont.*)

Plate	no.	Current species name	Original species name	Val	Syn	Additional notes
47	B.	*Campaea margaritaria* (Linnaeus, 1761)	*Phalaena margaritaria* Linnaeus, 1761			
47	C.	*Acontia trabealis* (Scopoli, 1763)	*Phalaena trabealis* Scopoli, 1763		x	
48	A.	*Crocallis elinguaria* (Linnaeus, 1758)	*Phalaena elinguaria* Linnaeus, 1758			
48	B.	*Angerona prunaria* (Linnaeus, 1758)	*Phalaena prunaria* Linnaeus, 1758			
48	C.	*Gnophos furvata* ([Denis & Schiffermüller], 1775)	*Geometra furvata* [Denis & Schiffermüller, 1775]	x		
48	D.	*Tephronia sepiaria* (Hufnagel, 1767)	*Phalaena sepiaria* Hufnagel, 1767		x	
48	E.	*Scotopteryx luridata* (Hufnagel, 1767)	*Phalaena luridata* Hufnagel, 1767		x	
48	F.	*Cleorodes lichenaria* (Hufnagel, 1767)	*Phalaena lichenaria* Hufnagel, 1767			
48	G.	*Ascotis selenaria* ([Denis & Schiffermüller], 1775)	*Geometra selenaria* [Denis & Schiffermüller, 1775]	x		
48	H.	*Pareulype berberata* ([Denis & Schiffermüller], 1775)	*Geometra berberata* [Denis & Schiffermüller, 1775]	x		
48	I.	*Plagodis dolabraria* (Linnaeus, 1767)	*Phalaena dolabraria* Linnaeus, 1767			

Plate	no.	Current species name	Original species name	Val	Syn	Additional notes
49	A (a)	*Opisthograptis luteolata* (Linnaeus, 1758)	*Phalaena luteolata* Linnaeus, 1758			
49	A (b)	*Opisthograptis luteolata* (Linnaeus, 1758)	*Phalaena luteolata* Linnaeus, 1758			
49	B.	currently not determinable	currently not determinable			Unfinished drawing.
49	C.	*Ourapteryx sambucaria* (Linnaeus, 1758)	*Phalaena sambucaria* Linnaeus, 1758			
49	D.	*Thalera fimbrialis* (Scopoli, 1763)	*Phalaena fimbrialis* Scopoli, 1763		x	
49	E.	currently not determinable	currently not determinable			This one does not seem to be *Phigalia pilosaria* ([Denis & Schiffermüller], 1775), as Hübner notes with a question mark. The species is currently not determinable (Higgins notes "Not identified").
49	F.	*Biston betularia* (Linnaeus, 1758)	*Phalaena betularia* Linnaeus, 1758			Possibly a brown color variant of *Biston betularia* (Linnaeus, 1758), which is very variable. See also 51I in the green form.
49	G.	*Ennomos erosaria* ([Denis & Schiffermüller], 1775)	*Geometra erosaria* [Denis & Schiffermüller, 1775]	x		
49	H.	*Eulithis prunata* (Linnaeus, 1758)	*Phalaena prunata* Linnaeus, 1758			

(*cont.*)

Plate	no.	Current species name	Original species name	Val	Syn	Additional notes
49	I.	*Chlorissa viridata* (Linnaeus, 1758)	*Phalaena* (*Geometra*) *viridata* Linnaeus, 1758			
50	A.	*Eupithecia linariata* ([Denis & Schiffermüller], 1775)	*Geometra linariata* [Denis & Schiffermüller, 1775]	x		
50	B.	*Eupithecia absinthiata* (Clerck, 1759)	*Phalaena absinthiata* Clerck, 1759		x	It is unclear if both depicted specimens belong to *Geometra minutata* [Denis & Schiffermüller], 1775, a synonym of *Eupithecia absinthiata* (Clerck, 1759), as only 50F is listed as such in Schiffermüller's remaining list of Geometridae. 50B does not appear in this list, and it is possible that they belong to two different species.
50	C.	*Eupithecia centaureata* ([Denis & Schiffermüller], 1775)	*Geometra centaureata* [Denis & Schiffermüller, 1775]	x		
50	D.	*Xanthorhoe ferrugata* (Clerck, 1759)	*Phalaena ferrugata* Clerck, 1759		x	
50	E.	*Rhodostrophia vibicaria* (Clerck, 1759)	*Phalaena vibicaria* Clerck, 1759			
50	F.	*Eupithecia absinthiata* (Clerck, 1759)	*Phalaena absinthiata* Clerck, 1759		x	See 50B.

Plate	no.	Current species name	Original species name	Val	Syn	Additional notes
50	G.	*Camptogramma bilineata* (Linnaeus, 1758)	*Phalaena bilineata* Linnaeus, 1758			
50	H.	*Aspitates gilvaria* ([Denis & Schiffermüller], 1775)	*Geometra gilvaria* [Denis & Schiffermüller, 1775]	x		
50	I.	*Lythria purpuraria* (Linnaeus, 1758)	*Phalaena purpuraria* Linnaeus, 1758			
50	K.	*Idaea pallidata* ([Denis & Schiffermüller], 1775)	*Geometra pallidata* [Denis & Schiffermüller, 1775]	x		Hübner notes "glarearia", Higgins G. glarearia. *Heliomata glarearia* ([Denis & Schiffermüller], 1775) does not seem to be the correct determination. Although the caterpillar of this species is likewise brown, this drawing looks more similar to *Idaea pallidata* ([Denis & Schiffermüller], 1775).
50	L.	*Ematurga atomaria* (Linnaeus, 1758)	*Phalaena atomaria* Linnaeus, 1758			
50	N.	*Scopula immutata* (Linnaeus, 1758)	*Phalaena immutata* Linnaeus, 1758			This drawing does not look like *Scopula immutata* (Linnaeus, 1758), but rather like *Eupithecia insigniata* (Hübner, 1790) at an earlier stage. However, Schiffermüller lists it as "Immutata" in his list.

(*cont.*)

Plate	no.	Current species name	Original species name	Val	Syn	Additional notes
50	O.	*Idaea humiliata* (Hufnagel, 1767)	*Phalaena humiliata* Hufnagel, 1767		x	Hübner notes "osseata" with a question mark, Higgins G. osseata. It is difficult to confirm that this drawing really is *Geometra osseata* [Denis & Schiffermüller], 1775, now *Idaea humiliata* (Hufnagel, 1767).
51	A.	*Ennomos alniaria* (Linnaeus, 1758)	*Phalaena alniaria* Linnaeus, 1758			
51	B.	*Ennomos quercinaria* (Hufnagel, 1767)	*Phalaena quercinaria* Hufnagel, 1767		x	
51	C.	currently not determinable	currently not determinable			There is a question mark on the plate. Higgins determines the species as *Geometra badiata* [Denis & Schiffermüller], 1775, now *Earophila badiata* ([Denis & Schiffermüller], 1775). Regarding E. badiata, see also 53C and 53K.
51	D.	*Ennomos alniaria* (Linnaeus, 1758)	*Phalaena alniaria* Linnaeus, 1758			
51	E.	*Phigalia pilosaria* ([Denis & Schiffermüller], 1775)	*Geometra pilosaria* [Denis & Schiffermüller, 1775]	x		

Plate	no.	Current species name	Original species name	Val	Syn	Additional notes
51	F.	*Peribatodes rhomboidaria* ([Denis & Schiffermüller], 1775)	*Geometra rhomboidaria* [Denis & Schiffermüller, 1775]	x		
51	G.	*Selidosema plumaria* ([Denis & Schiffermüller], 1775)	*Geometra plumaria* [Denis & Schiffermüller, 1775]	x		It is uncertain whether Higgins's determination as *Selidosema plumaria* ([Denis & Schiffermüller], 1775) is correct (Hübner, too, notes "plumeria"). This species does not appear in Schiffermüller's Geometridae list.
51	H.	*Odontopera bidentata* (Clerck, 1759)	*Phalaena bidentata* Clerck, 1759			This species is not, as noted by Hübner and Higgins, *Geometra dentaria* Fabricius, 1784, a synonym of *Falcaria lacertinaria* (Linnaeus, 1758), but probably *Odontopera bidentata* (Clerck, 1759).
51	I.	*Biston betularia* (Linnaeus, 1758)	*Phalaena betularia* Linnaeus, 1758			See also 49F.
51	K.	*Selenia lunularia* (Hübner, 1788)	*Phalaena lunularia* Hübner, 1788		x	
52	A.	*Artiora evonymaria* ([Denis & Schiffermüller], 1775)	*Geometra evonymaria* [Denis & Schiffermüller, 1775]	x		
52	B.	*Abraxas grossulariata* (Linnaeus, 1758)	*Phalaena (Geometra) grossulariata* Linnaeus, 1758			
52	C.	*Erannis defoliaria* (Clerck, 1759)	*Phalaena defoliaria* Clerck, 1759			

(*cont.*)

Plate	no.	Current species name	Original species name	Val	Syn	Additional notes
52	D.	*Ligdia adustata* ([Denis & Schiffermüller], 1775)	*Geometra adustata* [Denis & Schiffermüller, 1775]	x		
52	E.	*Philereme vetulata* ([Denis & Schiffermüller], 1775)	*Geometra vetulata* [Denis & Schiffermüller, 1775]	x		
52	F.	*Epirrita dilutata* ([Denis & Schiffermüller], 1775)	*Geometra dilutata* [Denis & Schiffermüller, 1775]	x		
52	G.	*Epirrita dilutata* ([Denis & Schiffermüller], 1775)	*Geometra dilutata* [Denis & Schiffermüller, 1775]	x		
52	H.	*Philereme transversata* (Hufnagel, 1767)	*Phalaena transversata* Hufnagel, 1767		x	
52	I.	*Scotopteryx chenopodiata* (Linnaeus, 1758)	*Phalaena chenopodiata* Linnaeus, 1758			
52	K.	*Triphosa dubitata* (Linnaeus, 1758)	*Phalaena dubitata* Linnaeus, 1758			
52	L.	*Operophtera brumata* (Linnaeus, 1758)	*Phalaena (Geometra) brumata* Linnaeus, 1758			
52	M.	*Eupithecia absinthiata* (Clerck, 1759)	*Phalaena absinthiata* Clerck, 1759		x	

Plate	no.	Current species name	Original species name	Val	Syn	Additional notes
52	N.	*Xanthorhoe quadrifasiata* (Clerck, 1759)	*Phalaena quadrifasiata* Clerck, 1759		x	There is a question mark by Hübner, and Higgins states "ligustrata" with a question mark. Schiffermüller notes in a list of Geometridae for G. Ligustrata the plate 54.l:X, but without giving identifying letters. This could be a mistake, and it is probable that this is *Geometra ligustrata* [Denis & Schiffermüller], 1775, a synonym of *Xanthorhoe quadrifasiata* (Clerck, 1759).
52	O.	*Lycia zonaria* ([Denis & Schiffermüller], 1775)	*Geometra zonaria* [Denis & Schiffermüller, 1775]	x		
52	P.	*Cyclophora punctaria* (Linnaeus, 1758)	*Phalaena* (*Geometra*) *punctaria* Linnaeus, 1758			
52	Q.	*Mesotype verberata* (Scopoli, 1763)	*Phalaena verberata* Scopoli, 1763		x	
52	R.	*Aplocera plagiata* (Linnaeus, 1758)	*Phalaena* (*Geometra*) *plagiata* Linnaeus, 1758			
52	Z.	*Epirrita dilutata* ([Denis & Schiffermüller], 1775)	*Geometra dilutata* [Denis & Schiffermüller, 1775]	x		

(*cont.*)

Plate	no.	Current species name	Original species name	Val	Syn	Additional notes
52	S.	*Epirrita dilutata* ([Denis & Schiffermüller], 1775)	*Geometra dilutata* [Denis & Schiffermüller, 1775]	x		Hübner notes "?"; Higgins does not identify the species. This one is probably also *Geometra dilutata* [Denis & Schiffermüller], 1775, now in the genus *Epirrita* (see also 52F, 52G, and 52Z).
53	A.	*Alsophila aescularia* ([Denis & Schiffermüller], 1775)	*Geometra aescularia* [Denis & Schiffermüller, 1775]	x		
53	B.	*Cyclophora pendularia* (Clerck, 1759)	*Phalaena pendularia* Clerck, 1759			See also 53E.
53	C.	*Earophila badiata* ([Denis & Schiffermüller], 1775)	*Geometra badiata* [Denis & Schiffermüller, 1775]	x		Hübner lists 53C as "Badiata," Higgins as *Geometra centaureata* [Denis & Schiffermüller], 1775, now in the genus *Eupithecia*. 53K could therefore also be another form of E. *badiata*, but does not appear in his list.
53	D.	*Cabera pusaria* (Linnaeus, 1758)	*Phalaena pusaria* Linnaeus, 1758			
53	E.	*Cyclophora pendularia* (Clerck, 1759)	*Phalaena pendularia* Clerck, 1759			See also 53B.
53	F.	*Cabera exanthemata* (Scopoli, 1763)	*Phalaena exanthemata* Scopoli, 1763			
53	G.	currently not determinable	currently not determinable			

Plate	no.	Current species name	Original species name	Val	Syn	Additional notes
53	H.	currently not determinable	currently not determinable			Hübner notes "pendularia"; Higgins does not identify the species. For *Cyclophora pendularia* (Clerck, 1759), see 53B and 53 E., which corresponds with Schiffermüller's Geometridae list. 53H is absent or illegible here.
53	I.	*Ectropis crepuscularia* ([Denis & Schiffermüller], 1775)	*Geometra crepuscularia* [Denis & Schiffermüller, 1775]	x		Hübner notes "Crepusc" with a question mark, Higgins determines this one as *Ascotis selenaria* ([Denis & Schiffermüller], 1775). It seems that *Ectropis crepuscularia* ([Denis & Schiffermüller], 1775) may be correct, especially since A. selenaria can also be found on plate 48G.
53	K.	*Earophila badiata* ([Denis & Schiffermüller], 1775)	*Geometra badiata* [Denis & Schiffermüller, 1775]	x		See 53C.
53	L.	*Cosmorhoe ocellata* (Linnaeus, 1758)	*Phalaena ocellata* Linnaeus, 1758			Schiffermüller notes drawings of *Cosmorhoe ocellata* (Linnaeus, 1758) on the plates 53. L. and 54. N. in his Geometridae list, but on plate 54 the identification letter "N" is missing.

(cont.)

Plate	no.	Current species name	Original species name	Val	Syn	Additional notes
54	A.	*Pungeleria capreolaria* ([Denis & Schiffermüller], 1775)	*Geometra capreolaria* [Denis & Schiffermüller, 1775]	x		
54	B.	*Thera variata* ([Denis & Schiffermüller], 1775)	*Geometra variata* [Denis & Schiffermüller, 1775]	x		
54	C.	*Eupithecia pusillata* ([Denis & Schiffermüller], 1775)	*Geometra pusillata* [Denis & Schiffermüller, 1775]	x		
54	D (a)	*Macaria notata* (Linnaeus, 1758)	*Phalaena notata* Linnaeus, 1758		x	
54	D (b)	*Macaria notata* (Linnaeus, 1758)	*Phalaena notata* Linnaeus, 1758		x	
54	D (c)	*Macaria notata* (Linnaeus, 1758)	*Phalaena notata* Linnaeus, 1758		x	
54	E.	*Agriopis leucophaearia* ([Denis & Schiffermüller], 1775)	*Geometra leucophaearia* [Denis & Schiffermüller, 1775]	x		Hübner notes "?"; Higgins does not identify the species. This one is probably *Agriopis leucophaearia* ([Denis & Schiffermüller], 1775).
54	F.	*Jodis lactearia* (Linnaeus, 1758)	*Phalaena (Geometra) lactearia* Linnaeus, 1758		x	
54	G.	*Thera juniperata* (Linnaeus, 1758)	*Phalaena juniperata* Linnaeus, 1758			
54	H (1)	currently not determinable	currently not determinable			
54	H (2)	currently not determinable	currently not determinable			

Plate	no.	Current species name	Original species name	Val	Syn	Additional notes
54	I.	*Charissa obscurata* ([Denis & Schiffermüller], 1775)	*Geometra obscurata* [Denis & Schiffermüller, 1775]	x		
54	K.	*Thalera fimbrialis* (Scopoli, 1763)	*Phalaena fimbrialis* Scopoli, 1763		x	
54	L.	*Gandaritis pyraliata* ([Denis & Schiffermüller], 1775)	*Geometra pyraliata* [Denis & Schiffermüller, 1775]	x		
54	M.	*Pseudoterpna pruinata* (Hufnagel, 1767)	*Phalaena pruinata* Hufnagel, 1767		x	
55	A	currently not determinable	currently not determinable			This species is currently not determinable (unfinished drawing).
57	A.	*Agriphila inquinatella* ([Denis & Schiffermüller], 1775)	*Tinea inquinatella* [Denis & Schiffermüller, 1775]	x		
57	B.	*Hypena rostralis* (Linnaeus, 1758)	*Phalaena rostralis* Linnaeus, 1758			Hübner notes "Levalina" (or Levatina), Higgins Py. (Pyralis) rostralis which is today *Hypena rostralis* (Linnaeus, 1758).
57	C.	*Anchinia daphnella* ([Denis & Schiffermüller], 1775)	*Tinea daphnella* [Denis & Schiffermüller, 1775]	x		Hübner notes "Daphnella" for 57G, which is *Anchinia daphnella* ([Denis & Schiffermüller], 1775). 57C is probably the same species.

(*cont.*)

Plate	no.	Current species name	Original species name	Val	Syn	Additional notes
57	E.	*Colobochyla salicalis* ([Denis & Schiffermüller], 1775)	*Pyralis salicalis* [Denis & Schiffermüller, 1775]	x		Hübner notes "Salicalis", Higgins T. fagana, which is now *Carcina quercana* (Fabricius, 1775), but the correct determination seems to be *Colobochyla salicalis* ([Denis & Schiffermüller], 1775).
57	F.	currently not determinable	currently not determinable			
57	G.	*Anchinia daphnella* ([Denis & Schiffermüller], 1775)	*Tinea daphnella* [Denis & Schiffermüller, 1775]	x		see 57C.
58	A.	*Stenoptilia pterodactyla* (Linnaeus, 1761)	*Alucita pterodactyla* Linnaeus, 1761			
59	A.	*Synanthedon tipuliformis* (Clerck, 1759)	*Sphinx tipuliformis* Clerck, 1759			

CHAPTER 7

Color, Taxonomy, and Exotic Insect Specimens

Beth Fowkes Tobin

1 Introduction

Sir James Edward Smith (1759–1828), botanist, taxonomist, and President and Founder of London's Linnean Society, criticized Carl Linnaeus (1707–1778), Johan Christian Fabricius (1745–1808), and others for basing their taxonomic descriptions of insect species on damaged specimens. Smith's critical remarks were published in *The Rarer Lepidopterous Insects of Georgia* (1797), a book he co-authored with its illustrator John Abbot (1751–1840), an English naturalist artist. Smith served as this book's editor and taxon author, and Abbot was responsible for providing the drawings and notes on the life cycles, behaviors, and geographical locations of the pictured butterflies and moths. Abbot had left London in the early 1770s to travel to the southeastern United States to collect and draw insects for his British clients, among them insect collectors Dru Drury (1724–1803) and John Francillon (1744–1816). Abbot never returned to England, remaining in Georgia, where he continued to collect and draw insects for clients for the rest of his long life.[1] With *Insects of Georgia*, Smith used Abbot's drawings to point out errors made by taxonomists in their published descriptions of North American Lepidoptera.

Several of Smith's critiques centered on the appearance of cabinet specimens, in particular the quality of their colors. For instance, in his discussion of the pink-striped oakworm moth (*Anisota virginiensis*), Smith asserted: "Mr. Drury's *virginiensis* seems to be the female of our *pellucida*, done from a faded specimen."[2] Smith here was contrasting the dull brown moth drawn, engraved, and colored by the entomologist, author, and natural history illustrator Moses Harris (1731c.–1788) in the second volume of Drury's *Illustrations of Exotic Insects* (1770–1782) with Abbot's version of this female moth in plate LVIII of *Insects of Georgia*, which accurately displays

1 For more on Abbot's life and work, see Wilkinson 1981, 1982, and 1984; Rogers-Price 1983; Gilbert 1998; Neri, Nummedal, and Calhoun 2019; Tobin 2020 and 2021. For a thorough examination of Abbot's butterfly drawings, specimens, and publications, see Calhoun 2006 and 2019.

2 Smith and Abbot 1797, 2: 115.

 | DOI:10.1163/9789004684553_008

the orange-colored abdomen and luscious pink wings characteristic of the female of this species (Figs. 7.1 and 7.2). Moreover, Drury's description of the moth is inaccurate in referring to the upper side of its wings as "a faint fox colour" and the underside as a "yellowish fox-colour."[3] Drury's use of a faded specimen to represent this species, not scientifically named or described before, calls attention to the importance of color when it comes to those particular insects used as type specimens, the standard by which to identify an insect species. Entomologists, then and now, pay attention to color when describing the morphological characteristics of insects as a part of the process of locating them within the web of life.[4]

This chapter takes up Smith's concern about the color of cabinet specimens to flag the risks for illustrators and taxonomists who relied on cabinets such as Drury's for access to exotic specimens. Cabinets, often unbeknownst to their owners, could contain faded, blackened, and damaged insects, a problem endemic to the movement of insect specimens to London from the remote reaches of the British empire and the territories of its trading partners. Artists and illustrators working in Britain who wanted to draw exotic insects were dependent on such collections as Drury's, and as a result, the accuracy of their drawings suffered when they reproduced the colors of faded or blackened specimens. With little access to the living insects or even multiple dead specimens of a rare species, illustrators could not guarantee that their drawings reproduced the colors of the living insects. The risk of error existed as well for taxonomists who based their descriptions on cabinet specimens of exotic insects. When describing the morphology of non-European insects, most eighteenth-century European taxonomists, Linnaeus and Fabricius included, did not have access to the living insects or their larvae and did their work of naming and classifying insects based on dead specimens held in cabinets. Because color is an important feature in their descriptions, especially of Lepidoptera, and color is part of the process by which species are differentiated, faded and blackened specimens could result in errors, causing confusion over the identity of species that closely resembled one another. As we shall see, Smith called into question several of Linnaeus's and Fabricius's descriptions that seem to have been based on damaged specimens. Smith's concerns about the quality of cabinet specimens went well beyond faded or blackened examples, contending that taxonomists whose work was confined to dead specimens failed to reap the rewards that could arise from working with living ones, which included the opportunity to see the insects in their various states

3 Drury 1773, 2: 23.

4 See chapter by Kay Etheridge in this volume.

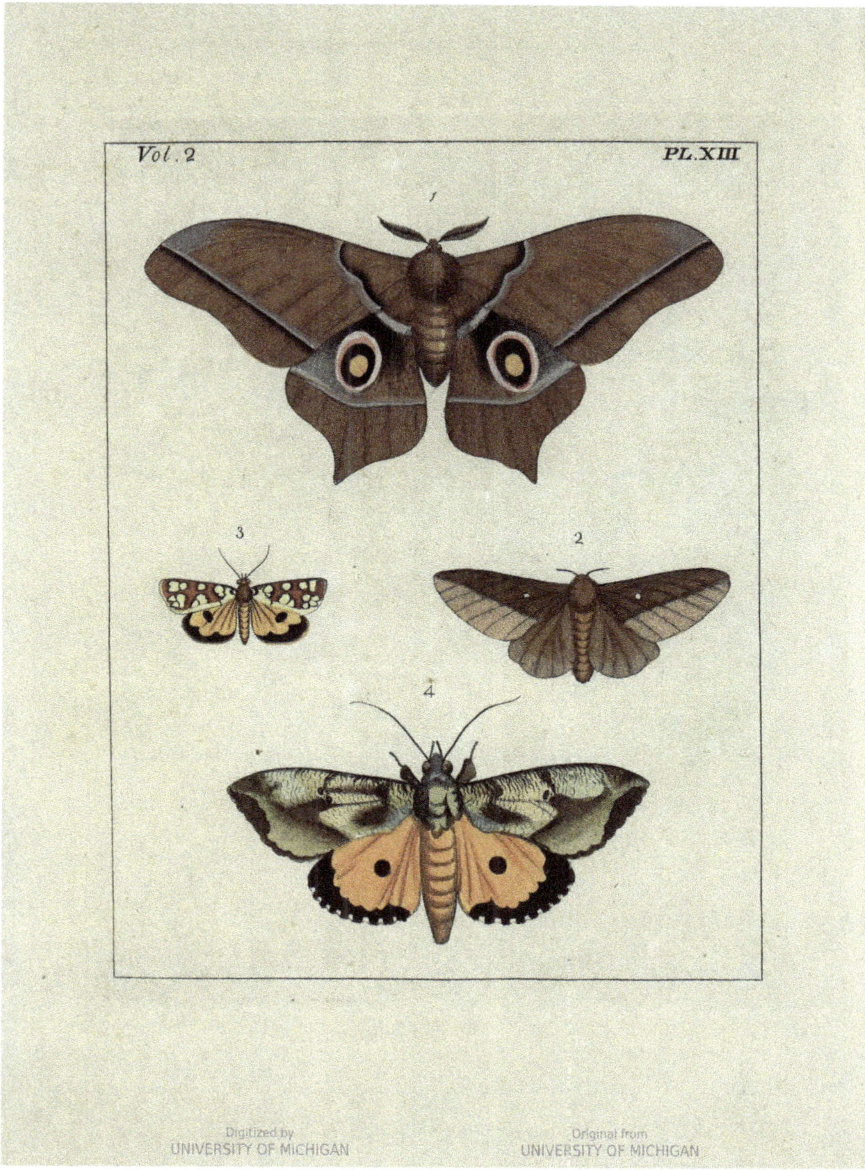

FIGURE 7.1 Moses Harris, imago of a "*Phalaena virginiensis*," the pink-striped oakworm moth (*Anisota virginiensis*), in: Dru Drury, *Illustrations of Natural History; wherein are exhibited upwards of two hundred and forty Figures of Exotic Insects*, 1773, vol. 2, p. 23, Plate 13, Fig. 2, engraving on paper, watercolor and body color, quarto, Ann Arbor, University of Michigan, HathiTrust, Public Domain Mark 1.0
HTTPS://BABEL.HATHITRUST.ORG/CGI/PT?ID=MDP.39015063449964&SEQ=61

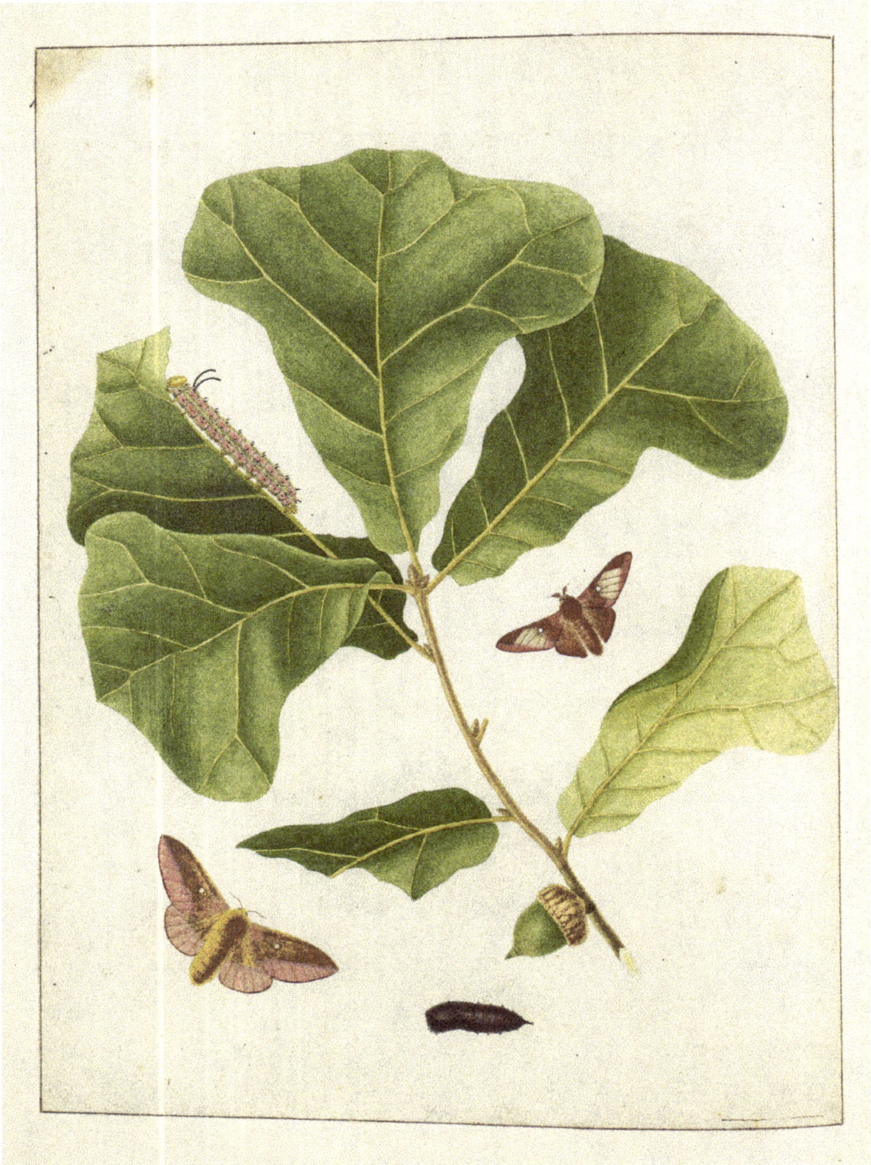

FIGURE 7.2 John Abbot, larva, pupa, male and female imagines of a "Transparent-winged White Spot Moth (*Phalaena Pellucida*)," the pink-striped oakworm moth (*Anisota virginiensis*), on its host plant, in: James Edward Smith and John Abbot, *The Natural History of the Rarer Lepidopterous Insects of Georgia*, 1797, Plate 58, graphite and watercolor drawing on paper, Baltimore, MD, John Work Garrett Library, Sheridan Libraries, The Johns Hopkins University, call no. QL551.G4 S64 1797 FOLIO C. 1, vol. 2
COURTESY OF JOHN WORK GARRETT LIBRARY

of metamorphosis. Smith thought that taxonomic descriptions of butterflies and moths would be enhanced if larvae and their host plants could also be incorporated by taxonomists into their thinking about a species' place in "a natural arrangement."[5]

Abbot's original drawings for *Insects of Georgia* are offered here as an example of how an artist's engagement with living insects, in this case lepidopterans, was productive for the taxonomist. Smith's commentary focuses on systematics and nomenclature, detailing how he went about his taxonomic task for each species he identified and named. It is in these exegetic passages that we find reference to Drury's faded specimen along with other comments on the sorry state of cabinet specimens in general. With these critiques and his insistence on the importance of caterpillars to identification and classification, Smith challenged how exotic specimens had been depicted and described in prestigious publications (the notable exception being Maria Sibylla Merian's (1647–1717) 1705 *Metamorphosis Insectorum Surinamensium*).[6] Following a discussion of Smith's doubt about quality of the color of several cabinet specimens and the illustrations based on them, this chapter turns to Abbot's colorful caterpillars and the importance of drawing from life.[7]

2 Damaged Specimens

Reading Smith's comments on the faded condition of his *Anisota virginiensis* must have been upsetting for Drury because he regarded his *Illustrations of Exotic Insects* as a way to rescue his entomological specimens, especially moths and butterflies, from "oblivion, by thus delineating them on paper." Because moths and butterflies are "of such tender and delicate natures," sunlight destroys "their colours" and air will "totally consume every part of them, leaving nothing behind but a little dust." Drury explained: "however pleasing and agreeable they may be to our sight, they are not easily preserved in all their gay and striking plumage."[8] As the author and publisher of these volumes filled with illustrations of the insects he had collected, Drury saw himself as engaged in a battle with the forces of time and decay. To have offered his readers an image of a faded specimen would have run counter to his goals with

5 Smith and Abbot 1797, 1: iv.

6 For an overview of Merian's career, see Schmidt-Loske 2020 and for Merian's work on metamorphosis, see Etheridge 2021.

7 See introduction to this book.

8 Drury 1770, 1: xiii.

this book. Even more painful for Drury might have been the fact that he had mentored Abbot as a young man in London, inviting him to draw insects in his collection, and had encouraged him to relocate to the American South to collect and draw insect specimens for him. Drury was the first to name and publish a (faulty) description of *Anisota virginiensis*, resulting in the citation "(Drury, 1773)" forever attached to this species. Abbot's illustration and Smith's description undercut Drury's authority in this instance. It is likely that Drury did not know in the 1770s that the specimen was faded, as there probably were no others to compare it with in other London collections, and apparently no other (correct) illustrations of it until Abbot's appeared in 1797.

Drury cared deeply about the accuracy of the illustrations in his multi-volume opus *Illustrations of Exotic Insects*. His desire for accuracy is conveyed with much force in the first volume's preface, with Drury claiming that his "plan of giving *just* and accurate figures" has been carried out in these volumes:

> the utmost care and nicety has been observed, both in the outlines, and engraving. Nothing has been strained, or carried beyond the bounds nature has set; and whoever will compare the engravings with the originals, I flatter myself will allow, that nothing is borrowed from fancy; or any colour given to an insect that does not really exist in the subject intended to be represented.[9]

Drury was vigilant in overseeing the production of his book.[10] He insisted that Harris, who drew, engraved, and colored most of the illustrations in all three volumes, correct the errors he had made in the process of coloring the illustrations. Drury took care to describe the colors of the insects Harris depicted with the idea that readers could compare the verbal description of an insect with the illustration of it. He hoped his descriptions of the insects' colors would eliminate the potential for errors: "the descriptions will be such a guide for colouring the prints, that capital errors will not be able to find admittance; the grossness of colouring a part yellow that should be red, or green that ought to be blue, would immediately be detected."[11] Drury worried that without his superintendence colorists would fail to color the illustrations properly, announcing: "if the reader should chance to meet with any part among them, that does not entirely correspond with the colour given in the print, he will impute it to its

9 Drury 1773, 2: iv.

10 For Drury as publisher, see Noblett 1988, and for Drury's relationship to Harris, see Tobin 2012.

11 Drury 1770, 1: xvi.

proper cause, the painter."[12] He even worried that when he would die, a publisher might get hold of the book's uncolored illustrations, which would then be in danger of being "execrably coloured" without Drury's superintending presence.[13] Despite his vigilance, however, errors did creep into his book that had nothing to do with the quality of pigments and the skills of hand-colorists.

Ultimately, Drury had inadequate control over the quality of his specimens due to the circumstances under which they were obtained and made their way to London from the distant regions of the world, China, South Asia, Africa, and the Americas. Drury paid travelers, including ship captains, colonial bureaucrats, and merchants, for specimens that they had acquired from local people during their time abroad. As Drury's nineteenth-century memorialist Sir William Jardine (1800–1874) noted:

> It was Drury's practice, whenever he became acquainted with any one about to visit a foreign country and willing to take some trouble in collecting insects, to supply him with a series of boxes ... as well as instructions how to proceed. ... Such an individual was generally empowered to engage others, after reaching the place of his destination, on similar terms.[14]

Most travelers would have outsourced specimen hunting to those, such as indigenous, indentured, and enslaved people, who knew much more about local insect life. Drury's agents, uninformed about the specimens they had acquired on his behalf, were unreliable sources of information. As a result, Drury knew very little about his specimens as living insects – their origins, life cycles, and ecological interactions with their environments. "In this respect the work was necessarily very meager," so concluded Jardine, who realized that most of Drury's agents "were contented with transmitting specimens of the perfect insects without furnishing any account of their previous states or general habits." Jardine argued that Drury wished to know more about the insects he had collected but had no means by which to remedy this structural problem. Drury wrote to a correspondent in Quebec about his beloved insects:

> Nothing could be more agreeable to me than to know their history, manner of life, what they feed upon, where they live, what enemies they have, how they secure themselves from the power of those enemies, what time of the year they make their appearance, how often in the year they breed,

12 Drury 1770, 1: xiv.

13 Drury 1770, 1: xvi.

14 Jardine 1842, 32.

> and in short many other circumstances that it is impossible for me to have any idea of, and which only being familiar with the subjects, and seeing them alive, can possibly furnish.[15]

Sadly, these questions could not be answered without possessing knowledge of the living insect.

In addition to his assessment of Drury's faded specimen, Smith spotted other problems with cabinet specimens that had serious consequences for classification. He was not afraid to point out errors in color made by famous taxonomists and entomological illustrators. Smith was able to sit in judgment on descriptions of North American Lepidoptera made by Drury, Linnaeus, Fabricius, Pieter Cramer (1721–1776), Carl Alexander Clerck (1709–1765), and other published authors because he had before him Abbot's original drawings along with the artist's descriptive notes as he worked through the process of identifying and naming the 104 lepidopteran species that Abbot had drawn for their co-authored *Insects of Georgia*.

Smith valued Abbot's skills as a talented natural history illustrator and an excellent observer of insect life, as well as regarding him an expert on North American butterflies and moths. Abbot was also an accomplished taxidermist and a prolific breeder of lepidopterans, having raised over 300 species, including all of the book's butterflies and moths. In addition to drawing the larvae and pupae of each species and the hostplants on which the caterpillars fed for all of the book's plates, Abbot provided notes on the life histories of the pictured lepidopterans, which Smith incorporated into the book's text. *Insects of Georgia* was regarded as one of the most important entomological publications of its era with much praise coming from contemporaries. The author of an anonymous review waxed rhapsodic over Smith and Abbot's book, stating:

> the produce of judicious researches ... the volumes before us are of the highest value. What is given here are not fragments of knowledge, but the result of a series of finished observations. If the work allure and delight by splendour of appearance, and uniform elegance of execution, it still more surprises and instructs by the richness and novelty of its contents, the lucid order with which they are digested, the precision and vivacity of the designs, the modesty of method, and spirit of philosophy, that pervade the whole.[16]

15 Jardine 1842, 45–50.
16 Anonymous 1798, 2.

William Swainson (1789–1855), a prominent naturalist and prolific natural history author, who had hoped to collaborate with Abbot on a sequel to *Insects of Georgia*, remarked that this book "is unquestionably one of the most beautiful and the most valuable that this or any country can boast of .... if the reader possesses this work, and the three volumes of Sepp, his library contains the two best illustrative publications upon Insects that have ever been given to the world."[17]

Insects of Georgia was printed in London by the bookseller James Edwards (1720–1816). Abbot, given his location, did not, of course, oversee the engraving of his drawings or the coloring of the engraved illustrations. Some of the illustrations bear the mark of the engraver John Harris (1767–1832), Moses Harris's son, who may have been responsible for coloring these illustrations in an incredibly beautiful copy of this book on vellum.[18] It is reasonable to assume that the colorists responsible for applying watercolor paint to the illustrations had access to Abbot's drawings so that they could copy the colors properly and attempt to achieve his aquarelle style.[19] They also may have had access to the specimens on which Abbot had based his drawings, since Abbot frequently sent specimens along with his drawings to his British clients. Smith explained in the book's preface: all of "the new insects figured in this work may be found ... in the exquisite collection of Mr. Francillon, transmitted by Mr. Abbot himself."[20] While the exact circumstances under which *Insects of Georgia* was brought into print remain unclear, we know that Smith had in his possession Abbot's notes and drawings while he was working on the manuscript, which involved editing Abbot's prose, naming new species, and disputing established nomenclature. Writing to his friend and Linnean Society Fellow Richard Pulteney (1730–1801), Smith was quite enthusiastic about working on this book:

---

17 Swainson 1834, 66–67. For Swainson's plans to publish a sequel to Smith and Abbot's *Insects of Georgia*, see Neri, Nummedal, and Calhoun 2019. Christiaan Sepp (1710–1775) and his son Jan Christiaan Sepp (1739–1811), published in Amsterdam a series of volumes with the title: *Beschouwing der Wonderen Gods, in de minstgeachte schepzelen of Nederlandsche Insecten* (*Contemplation of the Wonders of God in the least respected creatures, or Dutch insects*). The illustrations and engravings of the first two volumes (1762 and 1763) were by Christiaan Sepp; the third volume's title page indicates that this volume was co-authored and published in 1786 (after the father's death). These are the three volumes to which Swainson was referring.

18 This vellum copy had been the property of Dumbarton Oaks but was culled from the collection and sold. It resides now in Hargrett Library at University of Georgia. For details about this, see Calhoun 2006, 17.

19 For more on Abbot's style, see Tobin 2021.

20 Smith and Abbot 1797, iii.

> I have been much ad[d]icted to butterflies of late – In a few days a splendid work on the Rarer Lepidopterous Insects of Georgia will appear from the shop of Edwards in Pall Mall – indeed it is more splendid than I would have made it – & there are some plants in the plates for w.ch (having often been figured) I w.d have substituted rarer ones, but the plates were engraved before I was applied to – There are however among them some quite new plants, & several of those in Hort. Kew. w.ch have never yet been fig.d – The original drawings were the most *perfect* things I ever saw.[21]

One theory about this collaborative project is that Smith may have been asked by Edwards, the publisher, to oversee the writing and editing of the textual component of these volumes. Edwards, not Smith, according to this theory, had purchased Abbot's drawings and notes probably from Francillon, who in addition to being a client of Abbot's was also his agent.[22] Whatever may have actually happened, Abbot's "perfect" drawings survived this publication process and are now safely housed in the Special Collections of the Sheridan Libraries, Johns Hopkins University.

When Smith challenged received descriptions of butterflies and moths, he did so girded by his confidence in Abbot's accurate and beautiful drawings.[23] These drawings allowed him to see the morphological characteristics necessary to the tasks of identification and classification. He freely acknowledged his debt to Abbot, saying: "we must rely on the accuracy of Mr. Abbot";[24] "we certainly cannot but trust to Mr. Abbot";[25] and "we presume, on the authority of Mr. Abbot."[26] When Abbot's drawings differ from illustrations in Cramer's or Drury's books, Smith does not doubt the accuracy of Abbot's depiction; in fact, as in the case of the "Wild-Honeysuckle Hawk-Moth" on plate XXVII, he ventures to assert that Abbot's drawing represents a new species, which he names "*Sphinx azalea*" (*Darapsa choerilus*).

> We suspect this to be the *Choerilus* of Cramer, though the outline as well as colouring of his are too different from our's to be certainly taken for the same. We have therefore thought it best to give our insect a name

21 Smith 1797.

22 For more on the publication of *Insects of Georgia*, see Calhoun 2019.

23 Smith wrote of Abbot's drawings: "The result of his observations he has delineated in a style of beauty and accuracy which can scarcely be excelled." Smith and Abbot 1797, 1: ii.

24 Smith and Abbot 1797, 1: 3.

25 Smith and Abbot 1797, 1: 31.

26 Smith and Abbot 1797, 1: 41.

> applicable to its food, and quote the above figure [Cramer's] with a mark of doubt.[27]

He has put a question mark after the Cramer citation to indicate doubt about whether Cramer's and Abbot's drawings represent the same species.

Smith used Abbot's drawings and descriptions as the standard against which he compared published illustrations and descriptions of the same species. For instance, when Smith tries to sort out the differences among three similar tiger moths pictured in *Insects of Georgia*, plates LXII "Large American Tiger Moth (*Phalaena Virgo*)" (Fig. 7.3), LXIII "Streaked American Tiger Moth (*Phalaena Dione*)" (Fig. 7.4), and LXIV "Common American Tiger Moth (*Phalaena Phyllira*)" (Fig. 7.5), he suggests that Linnaeus should not have relied on Clerck's illustration of "*Phalaena Virgo*" (Clerck, TAB 42) to describe what is now known as the virgin tiger moth (*Apantesis virgo*). Smith contends:

> Clerck's figure of this insect is very bad, but it can be intended for no other. He represents the abdomen red above and black below; and Linnaeus seems to have described it so from his figure, for that description is not applicable to our species, nor any one of those near allied to it.[28]

In contrast to Clerck's rendition of "*Phalaena Virgo*," Abbot's drawing of this moth displays a red abdomen with black dots that form a vertical black stripe down the top of the abdomen (see Fig. 7.3).[29] Relying on color to distinguish among these three tiger moths, Smith calls attention to the colors of the forewings of the "*Phalaena Phyllira*" (historical name of the phyllira tiger moth, *Apantesis phyllira*) to differentiate it from the two other tiger moths (see Fig. 7.5):

> This is closely related to the two preceding, and would be very like *Virgo* if the nerves of the upper wings were, as in that and *Dione*, of the same light colour with the branching lines which cross or cover them. The abdomen differs much from that of Virgo, having three rows of black spots as in the female *Dione*.[30]

---

27 Smith and Abbot 1797, 1: 53.

28 Smith and Abbot 1797, 2: 123. I am uncertain about the modern scientific and common name of the moth Smith labels "Phalaena Dione." My best guess is *Apantesis doris* (doris tiger moth).

29 Abbot's specimen of the "*Phalaena Phyllira*," TAB LXIV, is in Linnean Society's collection.

30 Smith and Abbot 1797, 2: 127.

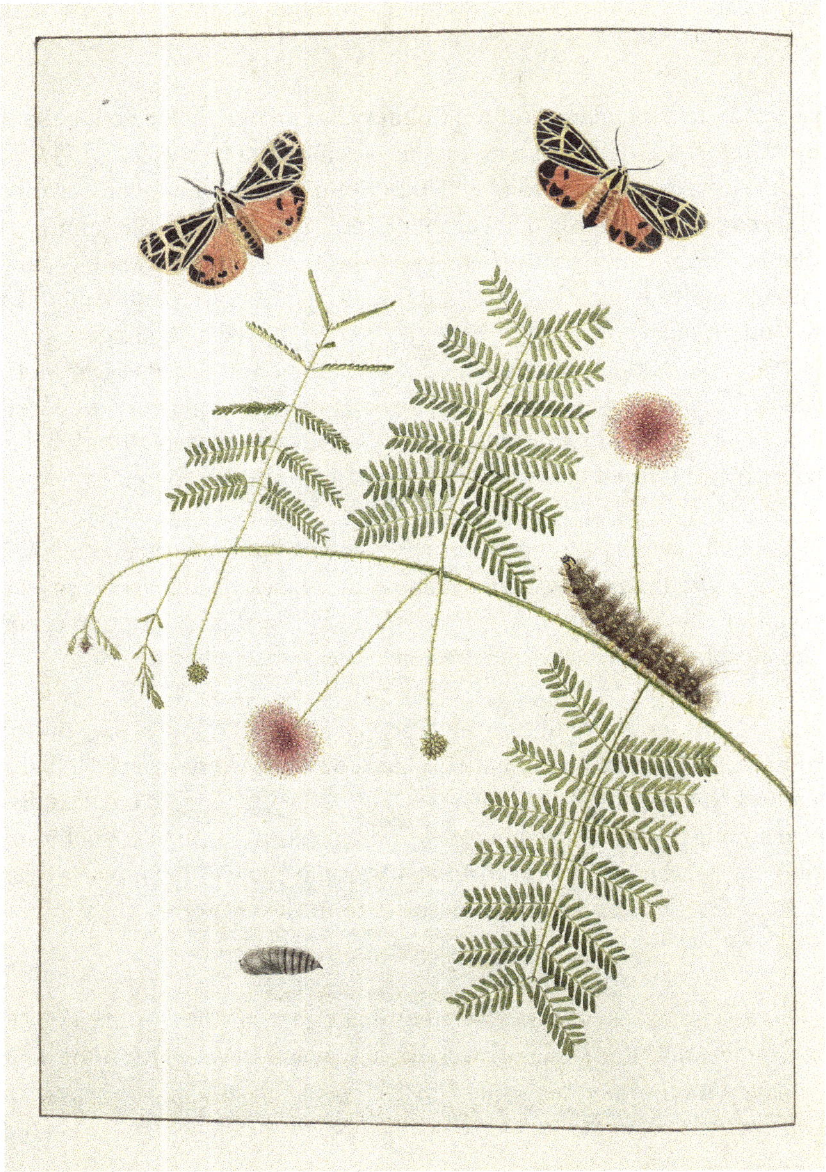

FIGURE 7.3 John Abbot, larva, pupa, male and female imagines of a "Large American Tiger Moth (*Phalaena virgo*)," the virgin tiger moth (*Apantesis virgo*), on its host plant, in: James Edward Smith and John Abbot, *The Natural History of the Rarer Lepidopterous Insects of Georgia*, 1797, Plate 62, graphite and watercolor drawing on paper, Baltimore, MD, John Work Garrett Library, Sheridan Libraries, The Johns Hopkins University, QL551.G4 S64 1797 FOLIO c. 1, vol. 2
COURTESY OF JOHN WORK GARRETT LIBRARY

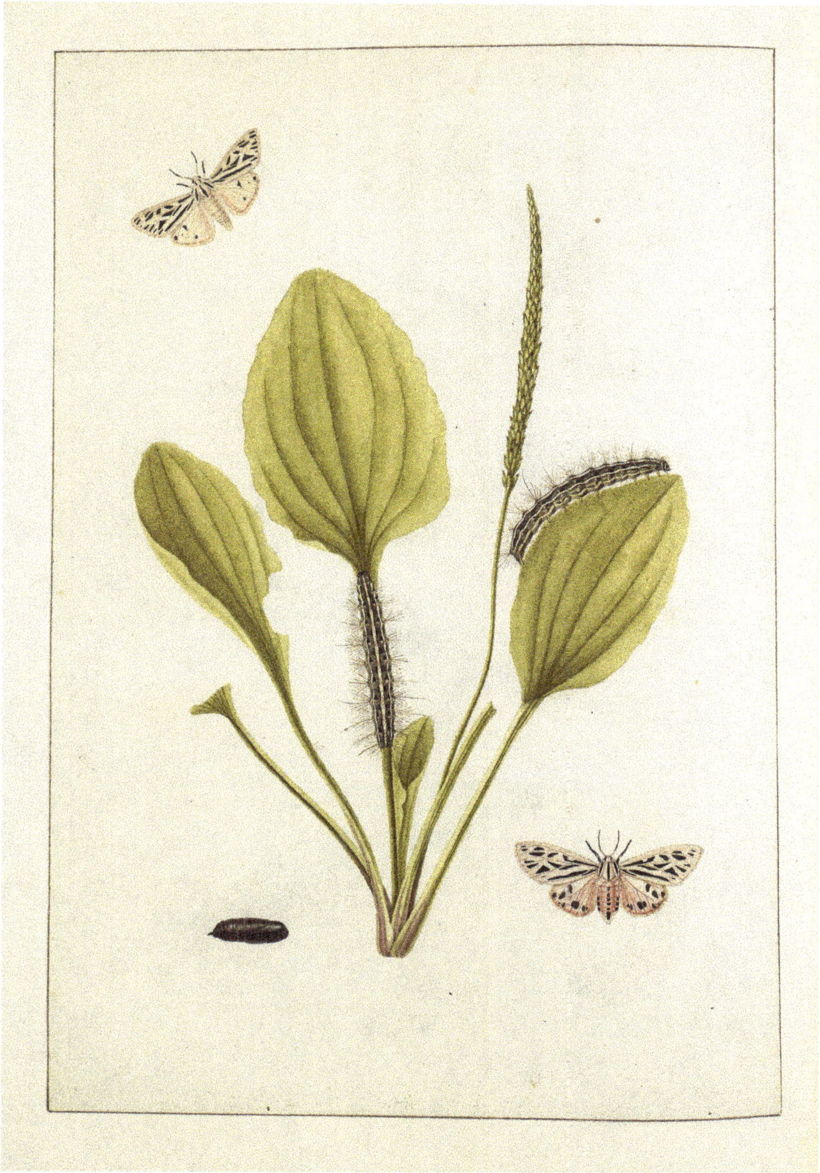

FIGURE 7.4 John Abbot, larvae, pupa, male and female imagines of a "Streaked American Tiger Moth (*Phalaena Dione*)," the doris tiger moth (*Apantesis doris*), on its host plant, in: James Edward Smith and John Abbot, *The Natural History of the Rarer Lepidopterous Insects of Georgia*, 1797, Plate 63, graphite and watercolor drawing on paper, watercolor and body color, Baltimore, MD, John Work Garrett Library, Sheridan Libraries, The Johns Hopkins University, QL551.G4 S64 1797 FOLIO C. 1. vol. 2
COURTESY OF JOHN WORK GARRETT LIBRARY

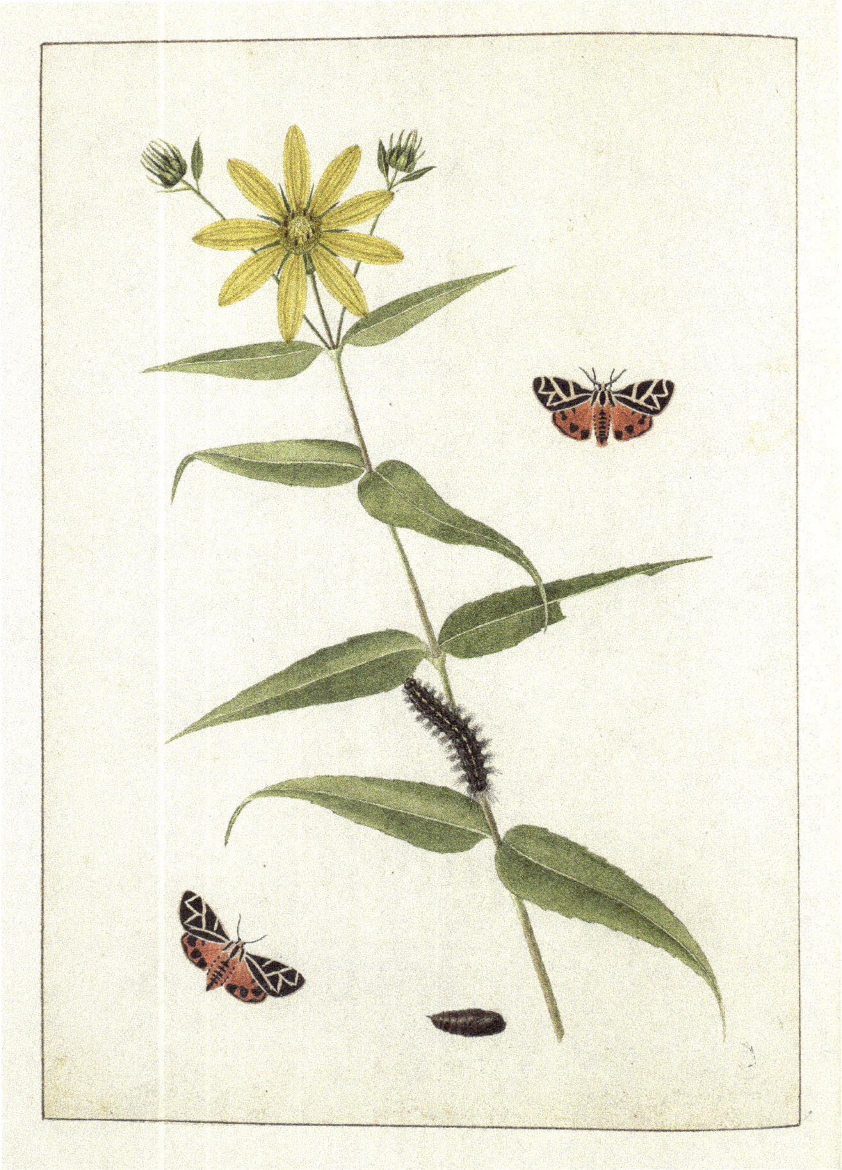

FIGURE 7.5 John Abbot, larvae, pupa, male and female imagines of a "Common American Tiger Moth (*Phalaena Phyllira*)," the phyllira tiger moth (*Apantesis phyllira*), on its host plant, in: James Edward Smith and John Abbot, *The Natural History of the Rarer Lepidopterous Insects of Georgia*, 1797, Plate 64, graphite and watercolor drawing on paper, Baltimore, MD, John Work Garrett Library, Sheridan Libraries, The Johns Hopkins University, QL551.G4 S64 1797 FOLIO c. 1. vol. 2
COURTESY OF JOHN WORK GARRETT LIBRARY

Not satisfied with his passing critique of Linnaeus's dependence on Clerck's flawed illustration of the "*Phalaena Virgo*," Smith continues to criticize Linnaeus but this time with mock politeness:

> The moth here represented [(TAB. LXIII), *Phalaena Dione*] is nearly allied to the last, and was perhaps by Linnaeus confounded with it, but they may be distinguished in every state [of metamorphosis]. Their great affinity, and the analogy of neighbouring species, must form our apology for the seeming absurdity of defining the upper wings as black, when the light colour certainly predominates.[31]

The blackened abdomen of Clerck's "*Virgo*" and blackened forewings of the "*Phalaena Dione*" could have been the result of damaged specimens in either Linnaeus's or Queen of Sweden Lovisa Ulrika's (1712–1786) insect collections, on which Clerck's drawings were based. It is possible that the blackened abdomen in Clerck's illustration could have been caused by the instability of the pigments used in the gouaches (lead white is commonly known to turn black over time). However, Linnaeus was supervising Clerck during the process of making *Icones Insectorum Rariorum* (1759–1764) and therefore his description referencing the black abdomen suggests that the color black can be attributed to the specimen and not the paint.[32] The *Dione*'s blackened wings could have been the natural result of the insect's death since some insects turn black when they die, as we shall see.

Smith's contention that some of Linnaeus's descriptions are faulty when it comes to color is supported by Edward Donovan's assertion that "many of the Linnaean descriptions of Insects appear defective to such as breed them; we not unfrequently read, *body black*, though we know that part of the Insect is white in every specimen."[33] According to Edward Donovan (1768–1837), a gifted artist, natural history collector, prolific natural history author, and friend of Drury's, "Some Moths are very liable to change colour when placed in the cabinet," noting the following examples:

31 Smith and Abbot 1797, 2: 125.

32 "In nearly all cases the original specimens came from the collection of Her Majesty Queen Ulrica or from that of Linnaeus himself, who, with his pupils Forsskal, Alstromer, and Bergman, supervised the proceedings" (Higgins 1970, 334). These were Pehr Forsskål (1732–1763), Claes Alströmer (1736–1794), and Torbern Olof Bergman (1735–1784).

33 Donovan 1794, 44.

> the body of the Satin Moth is perfectly white when fine, but after it has been killed some time, it becomes black in parts; the body of the Burnet Sphinx is of a very brilliant blue colour, with yellow bands on every annulation, when alive, but changes to a velvety black soon after the Insect dies; the same is observed on the body of the Current Sphinx; and every part of the body of the Hornet Sphinx changes to a jet black, after being some time in the cabinet; although when alive it is a very bright yellow, with a band of purple. Hence also it is that some specimens of very common Insects are valuable, by having preserved their proper colours uninjured.[34]

Artists, like Abbot and Donovan, who caught and raised their own insects, could see how death transformed their colors. Abbot notes that some lepidopterous species are hard to preserve and specimens easily decay in cabinets. The luna moth (*Actias luna*) "is difficult to preserve from decay in boxes or cabinet"[35] and the "corn emperor moth" (*Automeris io*) is "a remarkably tender fly to keep in cabinets, the male sooner decaying than any other."[36] Likewise, John William Lewin (1770–1819), an English artist sent to Australia by Drury, noted how beetles, locusts, and dragonflies have "not so much of their Beauty when they are dry for any length of time. It will be best to paint and etch them, as they … are so various and pleasing when living but turn black as soon as dead."[37] Donovan was well aware of this problem with dragonflies, remarking on how difficult it was to preserve their beautiful colors in his booklet *Instructions for Collecting and Preserving various Subjects of Natural History*: "the colours on the body are exceedingly brilliant in some species, but inevitably change black within a few days after death, unless the collector is particularly attentive to their preparation."[38] Donovan's assertion of the defectiveness of Linnaean descriptions due to blackened specimens chimes in with Smith's ongoing critique of Linnaeus and other established authorities whose descriptions were dependent on sometimes damaged cabinet specimens.

34 Donovan 1794, 44.
35 Smith and Abbot 1797, 1: 95.
36 Smith and Abbot 1797, 1: 97.
37 Lewin 1803.
38 Donovan 1794, 49.

3 Larvae and Pupae

Beyond the problem of damaged specimens, Smith confronted the larger issue generated by Europeans collecting decontextualized specimens from afar, specimens about which they knew very little in terms of life histories, insect-plant entanglements, and geographical distribution. Embedded in Smith's critique of Linnaeus's description of "*Phalaena Dione*" is a reference to metamorphosis: he states that the three different kinds of tiger moths "may be distinguished in every state."[39] Smith suggests here that taxonomists need to consult the larvae and pupae as well as the host plants to make informed decisions about affinities and differences within genera. Smith even criticized Fabricius for relying on collectors' specimens of Lepidoptera and failing to take the pupal and larval forms of these species into account in his masterwork on insect taxonomy *Systema entomologiae* (1775): "Fabricius ... was not particularly acquainted with their metamorphoses, knowing them chiefly in cabinets, and therefore wanting the great clue to a natural arrangement."[40] Smith thought Fabricius's lack of familiarity with all the stages of transition that butterflies and moths go through hindered his ability to see relationships among species as clearly as he might have if he had been able to take the appearance of caterpillars, including their color, into account as he worked on classifying these species. Not being able to access this information about exotic specimens in cabinets had put taxonomists such as Fabricius at a disadvantage. To compensate for this handicap, Smith, who never travelled outside Europe, relied on Abbot's drawings that included imago, larva, pupa, and host plants of the North American lepidopterans he raised and drew.

In his deliberations over distinguishing species Abbot had drawn that closely resembled one another, Smith often referred to the caterpillars, focusing on their colorful patterns of stripes and spots as a way to register difference among species. For instance, in his attempt to distinguish the "Black-Barred Swallowtail Butterfly" (*Papilio ajax*, tiger swallowtail) from its European look alike, "*Papilio Podalirius*" (*Iphiclides podalirius*, scarce swallowtail), it being "very difficult to make a specific character which shall distinguish them,"[41] he turned to their larvae to help make the case for them being different species. Using their "caterpillar state" to argue that these two kinds of swallowtails are distinct species, Smith writes: "in the caterpillar state these two insects differ

39 Smith and Abbot 1797, I: ii.

40 Smith and Abbot 1797, I: iii–iv. For more on Fabricius and his connections with British entomology, see Hancock 2015.

41 Smith and Abbot 1797, I: 7.

considerably, that of the *Podalirius* not having either the horns of ours, nor the transverse black, blue, and yellow stripe."[42] The colors and patterns of these caterpillars were key to his decision to distinguish the European from the North American tiger swallowtail. Likewise, in his discussion of the monarch butterfly (*Danaus plexippus*), Smith suggests that the confusion over the "varieties, as they called them, of *P. Plexippus*" in all the authors he had consulted might have been resolved "if their metamorphosis were known," meaning that without the caterpillar, he would not have been completely convinced that "*Papilio Archippus*," as he named it, was "distinct" from these other "varieties."[43] Abbot's illustration of the monarch, which he called the "Large Black and Orange Butterfly," includes its now widely recognized distinctive stripped caterpillar. Its pupae, also drawn by Abbot, provided even more information to help identify the monarch: they had "beautiful gold spots" that preserved "their colour even in spirits."[44] Another example of how the colors of larvae and pupae figured in his decisions about how to differentiate species returns us to Drury's faded *virginiensis*:

> This and the three proceeding species evidently form a natural family of *Phalaena bombyces*, which agree in the horns on the neck of the caterpillar, and double spines along its back: the straight projecting tail of the chrysalis, and minute prickles placed more or less on several parts of it surface; and finally in the whole habit, yellowish colours, and peculiar marking of the perfect insect, as well as the antennae of the male being feathered at the base only.[45]

Not only did Smith rely on Abbot's depiction of lepidopteran larvae and their pupae to identify and name new species, he used them to address larger questions of affinity and difference at the levels of genera and families. Smith contended that Abbot's illustrations and accompanying notes, in providing important information about the stages of lepidopteran metamorphosis, were engaged in "the philosophical study" of lepidopteran "economy," which included questions such as "where are their various habitations, foods, and manners?"[46]

---

42 Smith and Abbot 1797, 1: 7. See Hemming 1951 for his attempt to suppress Linnaeus's name *Papilio ajax*.

43 Smith and Abbot 1797, 1: 11.

44 Smith and Abbot 1797, 1: 11. "Spirits" in this case means alcohol-based liquid used to preserve soft-tissue animals.

45 Smith and Abbot 1797, 2: 115.

46 Smith and Abbot 1797, 1: i–ii.

With so few well preserved exotic caterpillars, entomologists often had to rely on illustrations to tell them what they wanted to know about colors, patterns, and shapes of these larvae. In Britain, only a few entomological illustrators portrayed British caterpillars and their host plants: the aforementioned Moses Harris and Edward Donovan, Benjamin Wilkes (*fl.*1690–1749), and Eleazar Albin (*fl.*1690–1742), who, it seems, copied Merian's drawings.[47] Smith praised "Harris, Wilkes, Esper, Ernst, the admirable Roesel, and the inimitable Sepp," for exhibiting European lepidopterans "through all their changes."[48] Smith complained that when it came to non-European Lepidoptera, "the splendid works of Clerck, Cramer, and Olivier, and the more exquisite one of Drury, do indeed display the complete insect in a degree of perfection that leaves scarcely anything to be desired," but the illustrations in these volumes lacked important "philosophical" information: "where are the metamorphoses through which these finished forms have passed?" Referring to drawings of the stages of metamorphosis, Smith inquired: "who since the celebrated, though not very accurate, Merian, has laboriously scrutinized those of the remote regions of India or America?"[49] In spite of this backhanded compliment delivered to Merian, Smith correctly assesses the dearth of illustrations that displayed the stages of metamorphosis of non-European butterflies and moths after Merian's bold and beautiful *Metamorphosis Insectorum Surinamensium* (1705), which Abbot knew well.[50] Swainson, writing forty years after Smith, agreed how little was known in Europe about non-European lepidopteran metamorphosis: "There are thousands of exotic *Lepidoptera* familiar to our collections, but of whose metamorphosis we are entirely ignorant. We of course allude to the drawing or preservation of the larva and pupa, and of the plants upon which they feed."[51] *Insects of Georgia* was the first publication after Merian's to begin the process of rectifying the problems created by this dearth of information about North American lepidopteran metamorphosis.

As an expert breeder of lepidopterans, Abbot knew caterpillars well, including what they ate, what they looked like in the various stages of metamorphosis, where they could be found, and how they behaved. Swainson praised Abbot's abilities as a butterfly breeder:

47 Valiant 1993, 475.

48 Smith and Abbot 1797, 1: ii.

49 Smith and Abbot 1797, 1: ii. Guillaume-Antoine Olivier (1756–1814) was a French naturalist and entomologist.

50 Drury possessed "Merian's Surinam Insects" (1705) according to Anonymous (1805, 16). Abbot was mentored by Drury during his formative years in London.

51 Swainson 1840, 15.

> The breeding of exotic insects, more especially *Lepidoptera*, has been prosecuted to a great extent, and with important advantages to science, by a few zealous entomologists: among these, Mr. Abbott in Georgia, North America, and Dr. Horsfield in Java, deserve particular praise. Few can hope to do what they have done in this department, without residing for years in one locality; but every exotic collector may have it in his power to breed a few insects, and thus supply some valuable information.[52]

With great accuracy, Abbot visually recorded the characteristics of the last instar of nearly all of the caterpillars depicted in the illustrations for *Insects of Georgia*.[53] Abbot portrayed caterpillars in all their wondrous oddness, their stripes and spots of bright colors, their tufts of hair, and spiky horns and bulbous protrusions (Fig. 7.6). In contrast to his rather static depictions of butterflies and moths (they tend to look as if pinned), his caterpillars, a lively presence in *Insects of Georgia*, move across the page. They munch their way through the leaves of their hostplants with determination, stretching to reach the next leaf, moving on from leaves stripped of half their green. Abbot's accompanying notes describe their behaviors, hostplants, and times of transformation, as well as the locations where he found them. His notes do not, however, describe what they look like; he depends on the drawings to do the work of conveying size, color, pattern, shape, and texture. No doubt, Smith found Abbot's drawings of caterpillars very helpful in solving the more challenging of the several taxonomic conundrums he encountered as he worked on his task of identifying already named species and classifying unnamed ones.[54]

4 Abbot's Caterpillars

Abbot's familiarity with caterpillars began in London when he was in his teens and would spend his free time capturing and raising them. A record of his experiences has fortunately survived in the form of a manuscript dated 1772 that contains descriptions and drawings of a range of British insects. These drawings, now housed in Harvard University's Houghton Library, include

52 Swainson 1840, 14–15.

53 Ewan 1985, 40.

54 Abbot did not attempt to classify unnamed species of Lepidoptera. Lepidopterist John Calhoun (2006, 17) states that although some of Abbot's drawings of caterpillars have been criticized, "the majority of Abbot's Lepidoptera immatures are accurate, probably having been sketched from living individuals that he reared."

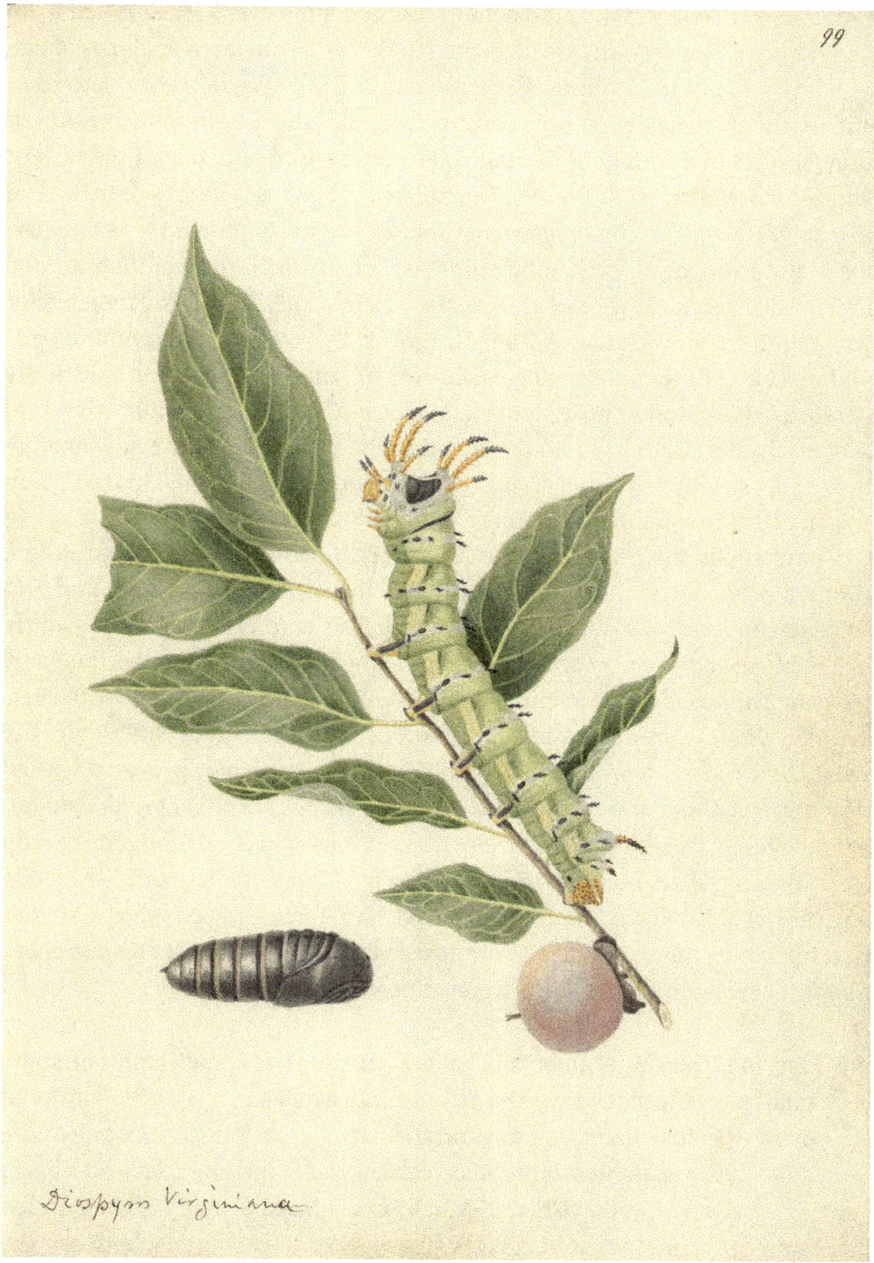

FIGURE 7.6 John Abbot, larvae, pupa, and imago of a "Royal Persimmon Moth (*Phalaena Regia*)," regal moth (*Citheronia regalis*), on its host plant, in John Abbot, "Drawings of Insects and Plants of Savannah," Cambridge, MA, Harvard University, Houghton Library, call no. Ms. Typ. 426.2
COURTESY OF HOUGHTON LIBRARY, HARVARD UNIVERSITY

Coleoptera, Hemiptera, Diptera, and Hymenoptera but focus heavily on Lepidoptera. Two sections are dedicated to the latter, one with drawings of caterpillars, laid out vertically as if in a list, and the other section with butterflies and moths arranged in a simple pattern. Over the course of his long life, Abbot drew insects in a variety of formats, this manuscript being one of his earliest and most geometrical of layouts. Twenty-five years later, Abbot's drawings for *Insects of Georgia* are much more naturalistic, capturing the form and behavior of the living insect, and much more in keeping with the tradition Merian forged with her drawings of lepidopterans, their larvae, pupae, and food plants. In a contemporary review, Adrian H. Haworth (1767–1833), an entomologist and president of the Entomological Society of London, wrote enthusiastically about this inclusion of host plants in *Insects of Georgia*: "this publication is to the full as valuable to the Botanist, as it is to the Entomologist: We never before beheld the sister sciences walk so closely, and so engagingly hand in hand, as in this interesting volume. – It is truly a Flora et Entomologia."[55]

Compared to the graceful images in *Insects of Georgia* of flitting butterflies, pretty flowers, and lively caterpillars, all of which appeals to amateur and professional botanists and horticulturalists as well as entomologists, the drawings of British insects in the 1772 manuscript are not anywhere near as aesthetically pleasing as those in Abbot and Smith's book. However, these early drawings are effective in communicating useful information. It is quite easy to identify what caterpillars Abbot had drawn, in part, because he also provided textual descriptions of the appearance and behavior of the caterpillars in this manuscript volume. Color features in these descriptions, which are designed to help identify caterpillars in much the same way as it does in a modern field guide such as David L. Wagner's *Caterpillars of Eastern North America* (2005). Abbot's description of the caterpillar and pupa of the pale tussock (*Calliteara pudibunda*) focuses on their colorful appearance:

> Fig. 100. This Catterpiller (sic) yellow. Between the annuli, belly and some marks on the sides near the tail black like velvet. It has yellow tufts of hair, with four light pink tufts on the back, and a longer pink one at the tail. ... It spun up in a white silky web [Fig. 101], and changed into a black Chrysalis with yellowish orange hair on the back. A shade of red on each side of the abdomen next the wing, and tip of each wing; between the joints of the belly yellowish orange.[56]

55 Haworth in Calhoun 2006, 17.

56 Abbot 1772, "Description of the Twelfth Drawing," seq. [page] 29.

Another overwintering and very hairy caterpillar, the sycamore moth (*Acronicta aceris*), captured Abbot's attention by incorporating its yellow and scarlet hair into what became a red chrysalis (Fig. 7.7):

> Fig. 110 This Catterpiller [sic] yellowish brown, with tufts of yellow and scarlet hair; and a row of white spots encircled with black down the middle of the back. Head black marked with yellow. It was taken feeding on Sycamore the first of September. I likewise took one on Oak. The 4th it turned orange and spun up mixing its hair with the web, and [Fig. 111] changed into a bright red Chrysalis.[57]

Abbot had raised these caterpillars and clearly was fascinated by how they looked and how they behaved. His art-making was one way that he preserved their colorful appearance; the other way was to taxidermy them and turn them into specimens, which were much desired commodities in the insect trade.

5 Abbot's Specimens and Their Afterlives

Abbot combined the skills of a practical naturalist and specimen hunter with those of an illustrator, producing drawings of the specimens he caught and raised, drawings that he sent to those who purchased specimens from him. He built a business, based on his skills as a taxidermist, selling North American insect (and bird) specimens to British and European natural history collectors.[58] According to Swainson,

> Abbot's specimens were certainly the finest that have ever been transmitted as articles of commerce to this country; they were always sent home expanded, even the most minute. ... His general price for a boxful was sixpence each specimen; which was certainly not too much, considering the beauty and high perfection of all the individuals.[59]

Donovan possessed specimens collected by Abbot, referring to him as "an intelligent person in North America," who collected "very valuable sphinges" [sphinx moths]: "they retain their colours to the utmost degree of perfection,

57 Abbot 1772, "Description of the Thirteenth Drawing," seq. [page] 31.

58 For insect specimen hunting in this period, see Hünniger 2018. For preservation of specimens, see Peck 2018 and Barnard 2018.

59 Swainson 1840, 99.

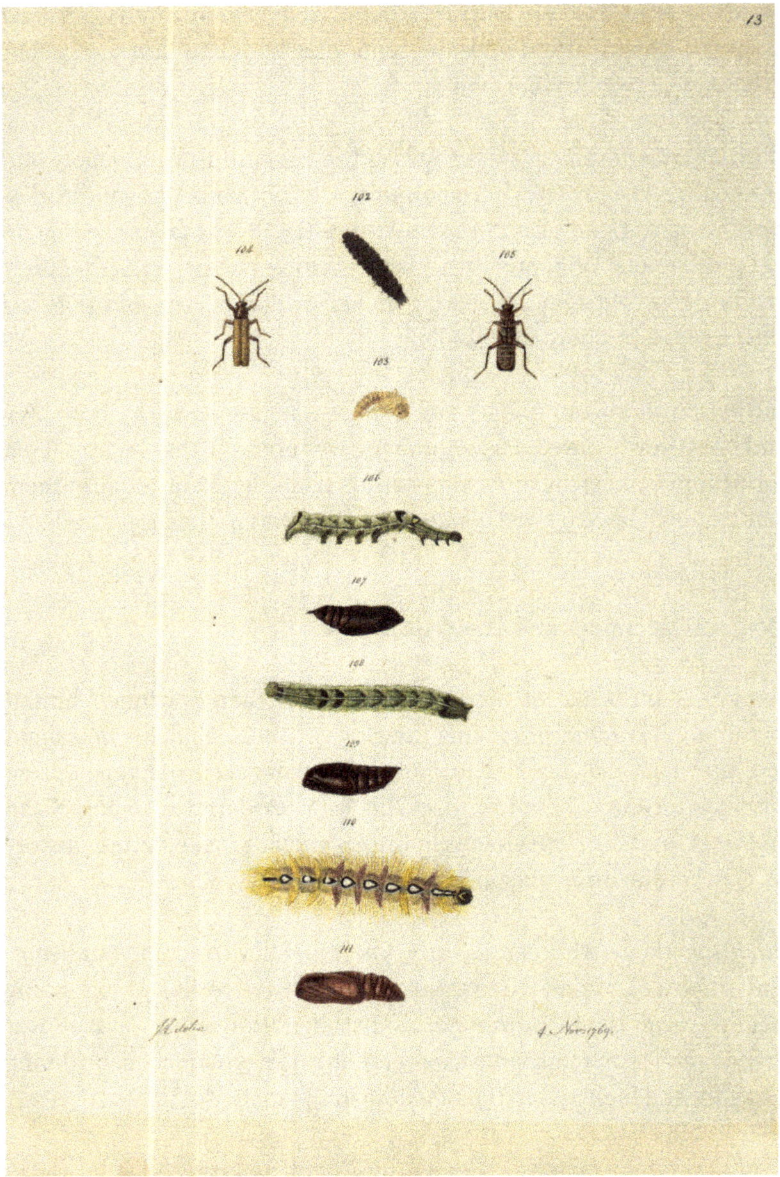

FIGURE 7.7 John Abbot, drawing of the sycamore moth caterpillar (*Acronicta aceris*) (number 110 with the fuzzy hair), in: John Abbot, *A Natural History of Insects. Consisting of Forty two Drawings, Exhibiting Two hundred and thirty five Figures. Drawn and Coloured from Nature. Together with a concise and accurate Description of each*, London, 1772, Plate 13, Cambridge, MA, Harvard University, Houghton Library, Ms. Typ. 426.1
COURTESY OF HOUGHTON LIBRARY, HARVARD UNIVERSITY

and have only been prevented from changing black by this simple preparation," which involved sprinkling powdered clay on the dead insect and putting it in an oven or using a hot poker to dry it out.[60] Abbot's specimens were far superior to those blackened and faded ones that lurked in the cabinets of the great collectors of exotic insects. Unfortunately for Fabricius, his *Systemae Entomologia* (1775) was based in part on Drury's and Sir Joseph Banks's (1743–1820) collections, which at this time did not yet contain Abbot's North American specimens, and, most importantly in terms of providing information about the stages of metamorphosis, before these collections had in their possession Abbot's taxidermized caterpillars, which famously kept the living insects' colors after death.[61] His ability to taxidermize caterpillars was legendary, as indicated by William Kirby (1759–1850) and William Spence (1783–1860), authors of the well-regarded *Introduction to Entomology* (1816–1826), who praised "the ingenious Mr. Abbott of Georgia" for "the admirable manner in which he prepared caterpillars":[62]

> Mr. Abbot of Georgia had an excellent method of preserving caterpillars, so that his specimens retain their colours and other attributes, and look as if they are alive. I am not acquainted with his process, but the following will answer very well. – The animal must first be killed by immersion in spirits of wine; next you must eviscerate it, which is best effected by gradual pressure of the finger and thumb. You must begin at the head, and so proceed till all the fluid contents of the body have passed out at the anus, which you may enlarge with a fine pair of scissors, being careful not to injure the anal prolegs. When you have cleared the skin as much as possible, introduce a fine glass tube, or a piece of hay or slender straw into the anus, round which, as near to the extremity as may be, pass loosely a fine thread; then blowing through the tube, when the skin is fully inflated, withdraw it, at the same time pulling the thread tight and securing it by a knot. The caterpillar will now exhibit its proper shape and colours; to retain which, all that is necessary is to hold it near the flame of

60 Donovan 1794, 45–46.

61 Abbot, whom Drury mentored, was drawing Lepidoptera from Drury's collection in 1767–1772. See Drury 1784, which contains references to items Abbot collected and sent to Drury in 1786 and 1789. This catalogue is in the Hope Library for Entomology at Oxford University Museum of Natural History and a photocopy of it is in the Entomology Library of the Natural History Museum, London.

62 Kirby and Spence 1826, 3: 148–149.

> a lamp until perfectly dry, which will be in a few minutes, when it may be placed in the cabinet along with the image to which it belongs.[63]

Whether any of Abbot's taxidermized caterpillars have survived into the present is unknown, but eleven of his adult lepidopterans, including one of those tricky-to-identify tiger moths, are housed in the collections of the Linnean Society (Fig. 7.8).

The few remnants of the great British insect collections that can be traced into the present day affirm Abbot's skills in preserving the colors of North American insect specimens. In addition to the above-mentioned lepidopterans in the Linnean Society, the Macleay Museum at the University of Sydney, Australia, possesses at least one Abbot specimen, an impressive grasshopper with scarlet wings that seem to glow with vibrancy. This horse lubber grasshopper (*Taeniopoda Stål* 1873) was in the insect collection amassed by Englishman Alexander Macleay (1767–1848), a friend of Smith's and Fellow of the Linnean Society, who took up a government position in Australia as Colonial Secretary of New South Wales.[64] He most likely acquired this grasshopper along with many other insect specimens before leaving England in 1825 from the sale of Drury's massive collection after his death in 1805. Drury had acquired hundreds, if not thousands, of Abbot specimens once the American Revolution ended in 1783 and trade with Britain opened up again. A manuscript catalogue written in his hand lists all the insects he had acquired from 1766 to 1789 and provides the name of the collector if known.[65] In this catalogue, there are long lists of Abbot specimens, which range from showy beetles to stink bugs and mason bees. Drury had asked Abbot for not only rare specimens, but samples of every kind of insect he could find.

Several entomologists and insect collectors acquired Abbot specimens from Drury's auction and from Francillon's auctions in 1817 and 1818. In reference to the latter, Macleay was said to have "purchased little short of half the collection," while Macleay characterized Kirby as "one of the souls of the sale of

---

63 Kirby and Spence 1826, 4: 537–538.

64 For a photograph of this grasshopper, see the lovely book by Stacey and Hay 2007, n.p. The photographs of two trays of sphinx moths look as if they may contain some Abbot specimens.

65 See note 61 for a description of Drury's manuscript catalogue. The annotated sale catalogues of Drury's auction contain information about who purchased what and for how much. Francillon's, Donovan's, Macleay's, and Haworth's name appears frequently in several of these catalogues next to insects from Georgia, which we can assume were collected by Abbot.

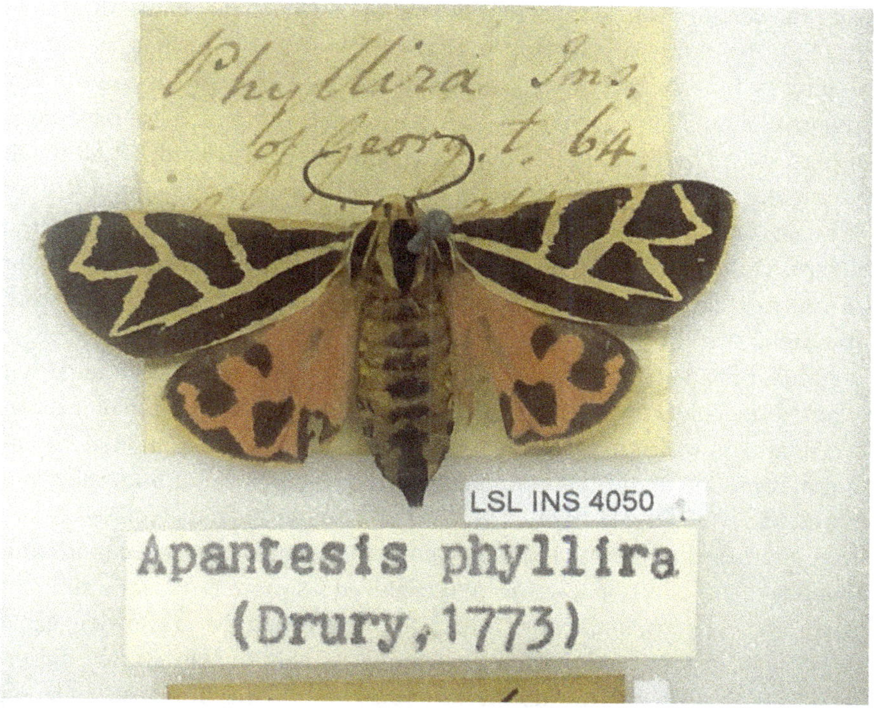

FIGURE 7.8 John Abbot's specimen of the "Common American Tiger Moth (*Phalaena Phyllira*)," phyllira tiger moth (*Apantesis phyllira*), London, Linnean Society, Specimen number: 4050. By Permission of the Linnean Society of London PHOTO © BETH TOBIN

Mr. Francillon's cabinet, giving it life, activity, and above all, value."[66] Haworth also bought insects at Francillon's auction, and Haworth's collection was purchased later by George Milne (?–1839). According to Edward Doubleday (1811–1849), appointed as an assistant responsible for Lepidoptera at the British Museum, his employer purchased "a large number of Abbot's insects"

66 Freeman 1852, 349–350. Kirby (1821, 13) referred to buying Abbot's specimens at Francillon's auction in an article he wrote on "the subfamily of *Fulgorellae*": "Amongst the insects of this family, which I purchased at the sale of the late Mr. Francillon's Museum, I found several that would not well arrange under any of M. Latreille's present genera; but I was particularly struck by a set collected in Georgia, and admirably displayed by that expert collector, accurate delineator and painter, and most intelligent observer of natural objects, Mr. Abbot. Upon examining these attentively, I found that they clearly formed two nondescript and very distinct genera belonging to the subfamily of *Fulgorellae* of the great entomologist before alluded to." Pierre-André Latreille (1762–1833) was an esteemed French entomologist who modernized arthropod systematics and taxonomy.

from the sale of Milne's collection at auction.[67] Doubleday wrote in 1840 to his old Harvard friend Thaddeus W. Harris (1795–1856), with whom he shared an interest in Abbot: "In all old collections are many specimens collected by Abbot; at Francillon's, Donovan's and other sales, some of these have been dispersed, and have crept into collections nominally British only."[68] Although hundreds if not thousands of Abbot specimens are now in the Natural History Museum, they are nearly untraceable as these specimens were folded into the larger entomological collection, where entomologists with little concern for provenance have moved specimens around as they attended to their work on systematics.[69]

Abbot provided his clients in Britain and Europe with specimens that retained the colors of the living creatures. His lepidopteran watercolor drawings complete with caterpillars, chrysalids, and host plants contained important information about the morphology not only of the adult butterflies and moths, but also of the larvae that would aid taxonomists in their classificatory tasks and insect collectors in their attempts to identify their specimens and organize their collections. Swainson recognized how rare it was for a supplier of exotic specimens, in this case from North America, to be so knowledgeable about the insects he sold to clients in Europe, and Smith realized that in having Abbot's notes and drawings in his possession, he was in a position to set the record straight by correcting mistakes about the color of specimens in some of the most famous and prestigious entomological publications of the eighteenth century.

67 Edward Doubleday to Thaddeus W. Harris, Epping May 27, 1840 in Scudder 1869, 147. For Milne's collection, see the auction catalogue (Anonymous 1824). References to Georgia appear on pages 3, 10, 12, 13, and 14, which means that Abbot most probably collected these insects.

68 Edward Doubleday to Thaddeus W. Harris, May 27, 1840 in Scudder 1869, 147. For more on Doubleday and his brother Henry (1808–1875), see Salmon et al. 2000, 150–153.

69 Doubleday reported to Harris that "[t]here are two specimens of this in the British Museum, from Abbot," saying "It is something like *Eudryas unio*, – white, anterior wings with two olivaceous marginal spots and a discoidal one, with a broad lilac border, in which, near the margin, are several sub-ocellated white spots; posterior wings whitish, with a slightly brown margin" (Scudder 1869, 146). I spent a day at the Natural History Museum searching for these specimens to no avail, but that does not mean they are not there.

6 Conclusion

For eighteenth-century taxonomists, color was an important element in the process of identifying and classifying insects. Working from dead specimens, which could fade over time or blacken if poorly preserved, presented taxonomists with problems, especially if those specimens were of rare non-European insects. Exotic lepidopteran specimens arrived in Europe shorn of information about their host plants, life histories, and geographical distribution, often entering into the great insect collections of Britain and Europe as the only representative of an unnamed species. These conditions could jeopardize the accuracy of taxonomists' systematic descriptions of insect morphology. Smith used Abbot's drawings to challenge descriptions of North American Lepidoptera that were based on damaged specimens. Abbot's drawings of butterflies and moths, a product of decades of collecting and raising lepidopterans, emboldened Smith to critique the practice of naming and classifying species based solely on the appearance of exotic cabinet specimens.

Acknowledgements

My work on John Abbot has been supported by a Research Fellowship from the Willson Center for Humanities and Arts at the University of Georgia and by Study in a Second Discipline Fellowship, Provost's Office of Research, University of Georgia. Joseph V. McHugh, Distinguished Professor of Insect Systematics and Evolutionary Biology, very graciously allowed me to sit in on his graduate course "Insect Taxonomy."

Bibliography

Manuscript Sources

Abbot, John. 1772. "A Natural History of Insects. Consisting of Forty Two Drawings, Exhibiting Two Hundred and Thirty Five Figures. Drawn and Coloured from Nature. Together with a Concise and Accurate Description of Each." Cambridge, MA, Houghton Library, Harvard University, Ms. Typ. 426.1. Available online: https://iiif.lib.harvard.edu/manifests/view/drs:444739812$6b.

Abbot, John. 1797. Original Drawings for *The Natural History of Rarer Lepidopterous Insects of Georgia*. Baltimore, Maryland, John Hopkins University, Sheridan Libraries, Garrett Library, QL551.G4 S64 1797 FOLIO C. 1.

Drury, Dru. 1784. "A Catalogue of the Exotic Insects in the Collection of Dru Drury Made 1784." Oxford, Oxford University, Museum of Natural History, Hope Library for Entomology.

Lewin, John William. 1803. Letter by John William Lewin to Dru Drury, March 7, 1803. London, Natural History Museum, Entomology Library, Dru Drury Collection of Miscellaneous Papers. [circa 1758–1803]. Manuscripts, SB f D.6.

Smith, James Edward. Letter by James Edward Smith to Richard Pulteney, May 1, 1797. London, Linnean Society, Smith's Correspondence, GB-110/JES/MS238/17. Available online: https://linnean-online.org/64776/#?s=0&cv=0&z=-0.1233%2C0%2C1.2467%2C1.206.

Printed Sources

Anonymous. 1798. "Review of *Insects of Georgia*." *Analytical Review* 27 (January–June): 1–6.

Anonymous. 1805. *A Catalogue of the Most Capital Assemblage of Insects … Consisting of Upwards of Eleven Thousand Different specimens … collected … by Mr. Dru Drury*. 1805. Auction Catalogue. London: King and Lochee.

Anonymous. 1824. *A Catalogue of an Extensive Collection of Insects … from the Museum of George Milne, F.L.S. which will be sold by auction by Mr. Thomas, 14th May, 1824*. 1824. London: Printed for Richard Taylor.

Barnard, Peter C. 2018. "Bat-Fowlers, Pooters and Cyanide Jars: A Historical Overview of Insect Collecting and Preservation." In *Naturalists in the Field: Collecting, Recording, and Preserving the Natural World from the Fifteenth to the Twenty-First Century*. Ed. Arthur MacGregor. [Emergence of Natural History 2]. Leiden: Brill, 646–685. https://doi.org/10.1163/9789004323841_023.

Calhoun, John V. 2006. "A Glimpse into a 'Flora et Entomologica': *The Natural History of the Rarer Lepidopterous Insects of Georgia* by J. E. Smith and J. Abbot (1797)," *Journal of the Lepidopterists' Society* 60 (1): 1–37.

Calhoun, John V. 2019. "From Oak Woods and Swamps: The Butterflies Recorded in Georgia by John Abbot (1751–c.1840) Based on His Drawings and Specimens." *Journal of the Lepidopterists' Society* 73 (4): 211–256. https://doi.org/10.18473/lepi.73i4.a8.

Donovan, Edward. 1794. *Instructions for Collecting and Preserving Various Subjects of Natural History Together with a Treatise on the Management of Insects in Their Several States*. London: Rivington.

Drury, Dru. 1770, *Illustrations of Natural History; Wherein Are Exhibited Upwards of Two Hundred and Forty Figures of Exotic Insects, According to Their Different Genera*, Vol. 1, London: Printed for the author and sold by B. White.

Drury, Dru. 1773. *Illustrations of Natural History; Wherein Are Exhibited Upwards of two Hundred and Forty Figures of Exotic Insects, According to Their Different Genera*, Vol. 2, London: Printed for the author and sold by B. White.

Etheridge, Kay. 2021. *The Flowering of Ecology: Maria Sibylla Merian's Caterpillar Book*. [Emergence of Natural History 3]. Leiden; Boston: Brill.

Ewan, Joseph. 1985. "The Natural History of John Abbot: Influences and Some Questions," *Bartonia* 51: 37–45.

Freeman, John. 1852. *Life of the Rev. William Kirby, Rector of Barham*. London: Longman.

Gilbert, Pamela. 1998. *John Abbot: Birds, Butterflies, and Other Wonders*. London: Merrell Holberton and the Natural History Museum.

Hancock, E. Geoffrey. 2015. "The Shaping Role of Johann Christian Fabricius (1745–1808): William Hunter's Insect Collection and Entomology in Eighteenth Century London." In *William Hunter's World: The Art and Science of Eighteenth Century Collecting*, Geoffrey Hancock, Nick Pearce, and Mungo Campbell (eds.). [The Histories of Material Culture and Collecting, 1700–1950]. Farnham: Ashgate Publishing, 151–163.

Hemming, Francis. 1951. "On the Proposal that the Trivial Name 'Ajax' Linnaeus, 1758 (as published in the binomial combination 'Papilio Ajax') Should Be Suppressed by the International Commission on Zoological Nomenclature under Its Plenary Powers." *The Bulletin of Zoological Nomenclature* 2 (Part 1): 29–30.

Higgins, L.G. 1970. "Clerck's *Icones Insectorum Rariorum*." *Journal of the Society for the Bibliography of Natural History* 5 (5): 334–339.

Hünniger, Dominik. 2018. "Nets, Labels, and Boards: Materiality and Natural History Practices in Continental European Manuals on Insect Collecting 1688–1776." In *Naturalists in the Field: Collecting, Recording, and Preserving the Natural World from the Fifteenth to the Twenty-First Century*, Arthur MacGregor (ed.). [Emergence of Natural History 2]. Leiden: Brill, 686–705. https://doi.org/10.1163/9789004323841_024.

Jardine, William. 1842. "Memoir of Dru Drury." *The Naturalist's Library*. Vol. 15. Edinburgh: W.H. Lizars.

Kirby, William. 1821."The Characters of *Otiocerus* and *Anotia, Two New Genera of Hemipterous Insects Belonging to the Family* Cicadiadae." Read March 2, 1819. *Transactions of the Linnean Society of London* 13: 12–23.

Neri, Janice, Tara Nummedal, and John V. Calhoun. 2019. *John Abbot and William Swainson: Art, Science, and Commerce in Nineteenth-Century Natural History Illustration*. Tuscaloosa: University of Alabama Press.

Noblett, William. 1988. "Publishing by the Author: A Case Study of Dru Drury's 'Illustrations of Natural History' (1770–1782)." *Publishing History* 23: 67–94.

Peck, Robert McCracken. 2018. "Collecting Abroad, Preserving at Home: Titian Ramsay Peale II, American Entomologist and Collector." In *Naturalists in the Field Collecting, Recording, and Preserving the Natural World from the Fifteenth to the Twenty-First Century*, Arthur MacGregor (ed.). [Emergence of Natural History 2]. Leiden: Brill, 706–729. https://doi.org/10.1163/9789004323841_025.

Kirby, William and William Spence. 1826. *Introduction to Entomology*. Vols. 3–4. London: Printed for Longman, Rees, Orme, Brown, and Green, Paternoster Row.

Rogers-Price, Vivian. 1983. *John Abbot in Georgia: The Vision of a Naturalist Artist, 1751–ca.1840*. Exhibition Catalog. Madison, GA: Madison-Morgan Cultural Center.

Salmon, Michael A. et al. 2000. *The Aurelian Legacy: British Butterflies and Their Collectors*. Berkeley and Los Angeles: University of California Press.

Schmidt-Loske, Katharina. 2020. "Maria Sibylla Merian: A Woman's Pioneering Work in Entomology." In *Women and the Art and Science of Collecting in Eighteenth-Century Europe*. Arlene Leis and Kacie Wells (eds.). [The Histories of Material Culture and Collecting, 1700–1950]. London: Routledge Taylor & Francis Group, 61–77.

Scudder, Samuel H., ed. 1869. *Entomological Correspondence of Thaddeus William Harris, M.D.* Occasional Papers of the Boston Society of Natural History. Boston: Boston Society of Natural History.

Sepp, Christiaan. 1762. *Beschouwing der wonderen Gods, in de minstgeachte schepzelen of Nederlandsche insecten*. Vol. 1. Amsterdam: Gedrukt voor den Auteur.

Sepp, Christiaan. 1763. *Beschouwing der wonderen Gods, in de minstgeachtste schepselen of Nederlandsche insecten*. Vol. 2. Amsterdam: By J.C. Sepp en Zoon.

Sepp, Christiaan and Jan Christiaan Sepp. 1786. *Beschouwing der wonderen Gods, in de minstgeachtste schepzelen of Nederlandsche insecten*. Vol. 3. Amsterdam: By J.C. Sepp en Zoon.

Smith, James Edward and John Abbot. 1797. *The Natural History of the Rarer Lepidopterous Insects of Georgia*. London: Printed by T. Bensley, for J. Edwards.

Stacey, Robyn and Ashley Hay. 2007. *Museum: the Macleays, Their Collections and the Search for Order*. Cambridge: Cambridge University Press.

Swainson, William. 1834. "A Preliminary Discourse on the Study of Natural History." In *The Cabinet Cyclopedia*. Vol. CXXVI. London: Printed for Longman, Rees, Orme, Brown, Green & Longman, and J. Taylor.

Swainson, William. 1840. *Taxidermy; With Biographies of Zoologists and Notices of Their Work*. London: Green and Longman.

Tobin, Beth Fowkes. 2012. "Butterflies, Spiders, and Shells: Coloring Natural History Illustrations in Late Eighteenth-Century Britain." In *The Materiality of Color: The Production, Circulation, and Application of Dyes and Pigments, 1400–1800*. Andrea Feeser, Maureen Daly Goggin, and Beth Fowkes Tobin (eds.). Farnham: Ashgate Publishing, 265–280.

Tobin, Beth Fowkes. 2020. "Collecting John Abbot's Natural History Notes and Drawings." In *After Print: Manuscript Studies and Eighteenth-Century Literature*. Rachael Scarborough King (ed.). Charlottesville (Va): University of Virginia Press.

Tobin, Beth Fowkes. 2021. "Drawing Insects: John Abbot and the Arts of Noticing." *The Eighteenth Century: Theory and Interpretation* 62 (1): 83–105. https://doi.org/10.1353/ecy.2021.0010.

Valiant, Sharon. 1993. "Maria Sibylla Merian: Recovering an Eighteenth-Century Legend." *Eighteenth-Century Studies* 26 (3): 467–479.

Wilkinson, R.S. 1981. "Smith and Abbot, *The Natural History of the Rarer Lepidopterous Insects of Georgia* (1797), and Its Authorship and History." *Entomologists' Record and Journal of Variation* 93: 213–218.

Wilkinson, R.S. 1982. "Smith and Abbot, *The Natural History of the Rarer Lepidopterous Insects of Georgia* (1797), and Its Authorship and History." *Entomologists' Record and Journal of Variation* 94: 122, 159–160.

Wilkinson, R.S. 1984. "John Abbot's London Years, Parts 1–4," *Entomologists' Record and Journal of Variation* 96: 110–123, 165–176, 222–229, 273–285.

Index